Calcium Homeostasis: Hypercalcemia and Hypocalcemia

Calcium Homeostasis:
Hypercalcemia and Hypocalcemia

Second Edition

Gregory R Mundy

MD, FRACP

Professor and Head, Division of
Endocrinology and Metabolism
The University of Texas
Health Science Center
at San Antonio

MARTIN DUNITZ

First published in the United Kingdom in 1989 by Martin Dunitz Ltd, 7–9 Pratt Street, London NW1 0AE

Second edition 1990

A CIP catalogue record for this book is available from the British Library

ISBN 1-85317-036-4

First published in the United States of America and Canada in 1990 by Oxford University Press, 200 Madison Avenue, New York 10016

Library of Congress CIP data applied for

ISBN 0-19-520894-3 (Oxford University Press edition)

Phototypeset by Scribe Design, Gillingham, Kent
Printed and bound by The University Press, Cambridge

C O N T E N T S

ACKNOWLEDGMENTS

I wish to express my gratitude to the following colleagues who helped with the preparation of this book: Nancy Garrett, who was responsible for typing the manuscript and collating the references, and to whom I am grateful for her great patience and expert clerical and secretarial assistance; Nancy Place who was responsible for most of the original artwork; Dr Brendan Boyce, who provided the photomicrographs for Figures 4.4, 5.8 and 8.3; Dr John Bilezikian, who supplied the x-rays for Figures 8.4 and 8.5; Dr Lorraine Fitzpatrick, who provided the photographs for Chapter 10, and Dr Richard Oreffo, who supplied Figure 9.8. I acknowledge the helpful advice and encouragement I received from Mary Banks, Annamaria Ferro and Margaret Brown of Martin Dunitz Ltd. I am also grateful to numerous colleagues in the Division of Endocrinology and Metabolism at the University of Texas Health Science Center in San Antonio for their helpful discussions over the years, in particular, Drs Donald Bertolini, Lynda Bonewald, Brendan Boyce, John Chirgwin, Sharyn D'Souza, James Dunn, Maxine Gowen, Gloria Gutierrez, Kenneth Ibbotson, Michael Katz, Brian MacDonald, James Poser, David Roodman, Massimo Sabatini, Peter Smolens, Katherine Tuttle, Alexandre Valentin, Ashley Yates, and Toshiyuki Yoneda.

G R M 1988

Once more I am grateful to a number of colleagues for their generous help in the preparation of this second edition. Nancy Garrett has again provided invaluable help with the secretarial work. Since the first edition, I have again been fortunate to have a number of colleagues in San Antonio whose perceptive discussions have helped considerably with my thinking in these areas. In the time since the last edition, I would particularly like to acknowledge the contributions of Lynda Bonewald, John Chirgwin, James Dunn, Ross Garrett, Gloria Gutierrez, David Hosking, Charles Reasner, David Roodman, Massimo Sabatini, Mary Samuels, Katherine Tuttle, John Yates and Toshiyuki Yoneda. In addition, I have been fortunate to have a number of postdoctoral fellows whose questioning and enthusiasm have proved remarkably stimulating, including Karen Black, Wolf Gallwitz, Theresa Guise, Chung Ho Lee, Christian Marcelli, Richard Oreffo and Galen Wo.

G R M 1990

PREFACE

The purpose of this book is to provide a simple and straightforward account of the physiological mechanisms which regulate the concentration of calcium in the extracellular fluid, and the diseases which upset the homeostasis of calcium. The author hopes it will be helpful to all who are interested in this area, including not only subspecialists in endocrinology, nephrology and oncology, but also interested internists, housestaff and medical students. The concept arose from discussions with colleagues concerning the lack of a monograph in this area, and the desire of the author to give a personal account of this field. It is not meant to serve as a detailed reference source. The project was begun in the fall of 1986 and completed in early 1988. This coincided with an explosion of interest in calcium homeostasis due to a series of advances, particularly in the field of hypercalcemia. These include the identification of several new factors which cause hypercalcemia (namely, the PTH-related protein, transforming growth factor alpha, lymphotoxin, tumor necrosis factor and interleukin-1) and their potential role in normal calcium homeostasis as well as in other diseases. As a consequence, research in this field will continue to thrive. There has also been a greater understanding of the pathophysiology of disturbances in calcium homeostasis and the way in which extracellular fluid calcium is maintained by fluxes of calcium which occur between the extracellular fluid, bone and the kidney. This is exemplified by the 1988 meeting of the American Society of Bone and Mineral Research, which included numerous abstracts on one factor particularly associated with hypercalcemia, the PTH-related protein. Progress has also occurred in the hypocalcemic disorders but has been less dramatic.

PREFACE TO SECOND EDITION

The second edition has been updated to January 1990 to cover the explosion of data on the mechanisms of hypercalcemia of malignancy in the late 1980s, and to eradicate errors and cover omissions in the first edition. Every effort has been made to include new findings of clinical import, but at the same time to avoid excessive detail so that my major objective of providing a readable account for those who are not directly involved in this area is not lost.

Although the entire book has been revised, obviously some areas have received more attention and have been changed more extensively than others. Chapter 1 has been expanded to include new data on the calcium exchange mechanism between bone and extracellular fluid, and on the evidence of pulsatile release of parathyroid hormone (PTH) which is not directly related to extracellular fluid calcium. Chapter 2 has been altered slightly to include new information on the bone fluid–extracellular fluid exchange mechanism. Extensive additions were required for Chapter 3, particularly on the cytokines which affect bone. More is now known of the effects of interleukin-1, tumor necrosis factor, transforming growth factor β and transforming growth factor α on bone. In addition, several newly described cytokines, namely interleukin-6, leukemia inhibitory factor and differentiation-inducing factor have been shown to affect bone cells, and these factors are also now reviewed. In Chapter 4 a section on the PTH immunoradiometric assay has been included, since this represents a considerable improvement on all previous PTH assays. Chapter 5 now contains extensive new information on the parathyroid hormone-related protein (PTH-rP) and its interactions and synergism with other factors that are often produced in conjunction with it, such as transforming growth factor α and interleukin-1. A new section has also been included in Chapter 5 on the potential role of procathepsin D in the bone destruction associated with breast cancer. In Chapter 6 interleukin-6 has been included as a potential factor in the bone destruction associated with myeloma. In the section on lymphomas, the relative roles of PTH-rP and 1,25-dihydroxyvitamin D have been expanded by new information. Chapter 7 has also been expanded to include more information on the new generation bisphosphonates and gallium nitrate. Bisphosphonates are now used extensively in Europe not only for the treatment of hypercalcemia of malignancy but also for treating metastatic bone disease. There is now a large literature on these agents, which seem likely to be the major treatment for hypercalcemic disorders during the 1990s. Chapter 8 now includes new information on the nature of the cellular abnormality in different types of primary hyperparathyroidism, including familial hyperparathyroidism. There is also new knowledge on the parathyroid gland growth factor which is present in the serum of some patients with primary hyperparathyroidism. The role of newer PTH assays in the differential diagnosis of hypercalcemia, particularly in the diagnosis of primary hyperparathyroidism, has also been included. The controversial role of localization of the parathyroid glands pre-operatively, and the ways in which pre-operative localization may alter the surgical approach, are also now reviewed in more detail. There is new evidence to support the conservative management of patients with primary hyperparathyroidism, and this is also now included in Chapter 8. In Chapter 9 the section on familial hypocalciuric hypercalcemia has been amplified. In Chapter 10 the potential role of altered forms of PTH in the pathophysiology of pseudohypoparathyroidism is covered in more detail.

GRM 1990

General concepts of calcium homeostasis

Introduction

The plasma calcium is maintained at a remarkably constant concentration by a complex homeostatic mechanism which is only partially understood. In this chapter some of the basic principles which underlie our understanding of this mechanism are reviewed, and the following chapters describe in more detail the basic components of this system—namely the fluxes of calcium which occur across the gut, bone, and kidney, and how hormones regulate calcium transport between these organs and the extracellular fluid in order to control the plasma calcium.

A multitude of cell and organ functions are dependent on the maintenance of the extracellular fluid calcium concentration within a very narrow range. A number of important metabolic processes are influenced by small changes in the extracellular ionized calcium concentration. These include: (a) the excitability of nerve function and neural transmission; (b) the secretion by cells of proteins, hormones, and other mediators such as neurotransmitters; (c) the coupling of cell excitation with cell response (for example, contraction in the case of muscle cells and secretion in the case of secretory cells); (d) cell proliferation; (e) blood coagulation, by acting as a cofactor for the essential enzymes involved in the clotting cascade; (f) maintenance of the stability and permeability of cell membranes; (g) modulation of enzyme activity, in particular those

enzymes involved in glycogenolysis, gluconeogenesis, and protein kinases which are calcium dependent; and (h) the mineralization of newly formed bone. Exactly how the concentration of ionized calcium in the extracellular fluid influences many of these processes is not clear, but transport of calcium across the plasma membrane from the extracellular milieu to the cytosol and intracellular structures is likely to be important. This topic is discussed in more detail in other works.[1-3]

When the extracellular fluid calcium concentration varies, there are predictable effects on some essential cell processes, which can be anticipated from knowledge of the effects of calcium on cellular metabolism. For example, increases in extracellular fluid calcium disturb neuronal function. As a consequence hypercalcemia is often accompanied by disturbances in mental state, lethargy, and confusion, and eventually by coma, gastrointestinal disturbances including nausea, vomiting, and constipation, and neuromuscular dysfunction, including muscle weakness and hypotonicity. In contrast, hypocalcemia is associated with nerve-conduction disturbances, and is characterized clinically by paresthesiae, tetany, and convulsions.

Extracellular fluid calcium is controlled closely by fluxes of calcium which occur between the plasma and the gut, bone, and kidney (Figure 1.1). However, although calcium is a tightly regulated variable in extracellular fluid, with variations of

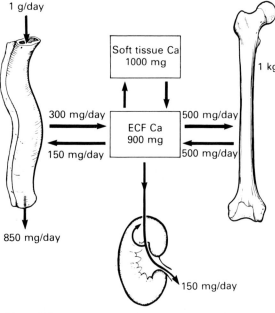

Figure 1.1

Normal adult in zero calcium balance. The numbers are estimates of the daily amount of calcium exchanged between the extracellular fluid and gut, kidney, and bone. The exchange system between bone fluid and the extracellular fluid is not taken into account.

The usual approach to the investigation of the homeostatic mechanism for calcium has been to identify the individual components or loops, in the hope that this knowledge can be used to understand how the homeostatic loops are integrated in order to maintain normal calcium homeostasis. Considerable advances in the understanding of the individual components of the homeostatic mechanism have been made over recent years. The biotechnology revolution has increased our understanding of the individual factors involved in regulating calcium transport, and advances in the in vitro techniques for studying calcium transport have allowed researchers to unravel various segments of the homeostatic system. For example, much is known of 1,25-dihydroxyvitamin D metabolism and production, the synthesis and secretion of parathyroid hormone (PTH) and how it is regulated, and the factors which regulate osteoclast activity. However, this information does not necessarily clarify how these individual components work in concert. How all of these factors and mechanisms cooperate to maintain a constant extracellular fluid calcium concentration, and how the plasma calcium can be corrected when outside influences threaten the homeostatic mechanism remains unclear.

The traditional concept of calcium homeostasis—homeostasis by reactive regulation

Traditionally homeostatic mechanisms are likened to domestic thermostats. Thus, the thermostat (or 'calciostat' in the case of calcium homeostasis) responds to a change in temperature (plasma calcium) and adjusts the temperature (plasma calcium) in the opposite direction. This is simple reactive regulation, and it is basically what happens when a variable such as extracellular fluid calcium is regulated by a long negative-feedback loop such as that mediated by PTH. When the plasma calcium falls, PTH synthesis and secretion are stimulated. Parathyroid hormone acts on bone, kidney and gut, its target organs, to cause an increase in the serum calcium. There is a finely tuned reciprocal relationship between the plasma calcium and plasma

less than 5 per cent occurring about the mean, Parfitt has argued on several occasions that the precision of the calcium homeostatic system is often overestimated.[4-6] The extracellular fluid calcium is controlled no more tightly than the serum sodium, and is possibly less efficiently controlled than the p_{CO_2}, particularly when the magnitude of the disturbances to which the variable is subjected are considered. Although there are clearly overlapping and seemingly redundant homeostatic loops to protect the plasma calcium, the system is neither as rapid nor as precise as the homeostatic mechanisms which protect the p_{CO_2} and the serum sodium, although it is clearly far more efficient than the mechanisms which regulate plasma glucose or plasma phosphate concentrations.

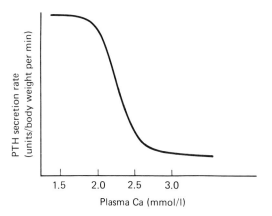

Figure 1.2

The relationship between PTH secretion rate and the plasma calcium. Note that small changes in plasma calcium produce relatively large changes in PTH secretion rate. Also note that PTH secretion is not completely suppressible by extracellular calcium.

interdependent hormonal mechanisms that regulate the influx and efflux of calcium between extracellular fluid and bone, kidney and the gut. The major regulator of calcium homeostasis when the extracellular fluid calcium concentration lies between 7.5 and 11.5 mg/dl (1.9 and 2.9 mmol/l) is almost certainly PTH, but 1,25-dihydroxyvitamin D is also necessary to maintain a normal plasma calcium. When disturbances in extracellular fluid calcium concentration are more profound, the capacity of PTH to regulate calcium homeostasis is relatively limited. This can be seen clearly from the data in Figure 1.2, which shows that PTH release from the parathyroid gland is regulated within this range of serum calcium. Serum concentrations of PTH are not altered further by changes in total serum calcium greater than 11.5 mg/dl (2.9 mmol/l) or less than 7.5 mg/dl (1.9 mmol/l). Outside this range, Parfitt has suggested the major control may be dependent on exchange of calcium between the extracellular fluid and bone (see pages 26–28),[2-4] although renal calcium excretion and gut absorption also play subsidiary roles. Similarly, in severe calcium deficiency, exchange with bone may also be the major protective mechanism. The roles of individual components of the system and the way in which the entire system responds to disturbances are discussed below.

The role of the kidney

The kidney is a major regulator of extracellular fluid calcium concentration. Some years ago, Nordin and Peacock[7] postulated that the kidney was the primary regulator of extracellular fluid calcium homeostasis. They likened the extracellular fluid compartment to a bath, where the kidney was the outlet pipe (see page 12). They argued that decreased renal calcium excretion could clearly increase extracellular fluid calcium, and that the kidney was ideally placed to be the major regulator of calcium homeostasis. However, the capacity of the kidney to clear calcium is severely limited—its maximum capacity is probably not much more than 600 mg per day (15 mmol/day). Thus, under conditions where net calcium input into the extracellular fluid from the gut or bone exceeds this, hypercalcemia will result. Conversely, in hypocalcemic states, the kidney can

PTH, resulting in tight control of the plasma calcium (Figure 1.2). However, this system, like many homeostatic mechanisms, is more finely regulated than most domestic thermostats. Although PTH may be the major regulator of plasma calcium, there are other overlapping long negative-feedback loops (in the case of plasma calcium, mediated not only by PTH, but also by 1,25-dihydroxyvitamin D and possibly, to some limited extent, by calcitonin). The mechanisms by which these individual hormones regulate plasma calcium are discussed in more detail in Chapter 3.

Integration of the components of the homeostatic mechanism

Calcium continuously enters the extracellular fluid, for example, from the gut and bone, but the calcium concentration in the extracellular fluid is maintained relatively constant by the effects of reduce calcium excretion, but usually to not much less than 100 mg calcium per day. Thus, the capacity of the kidney to maintain extracellular fluid calcium homeostasis in hypocalcemic states is also limited. Parfitt has suggested that the kidney may be important in error correction of fluctuations, but that it plays only a supportive role to the bone exchange mechanism in determining the set-point for plasma calcium (see page 10).

The kidney plays an important role in the pathogenesis of hypercalcemia in some situations. For example, in hypercalcemia of malignancy associated with solid tumors, renal tubular calcium reabsorption is enhanced in many cases and this effect potentiates enhanced bone resorption (Chapter 5). Under these circumstances the capacity of the kidney to clear calcium from the extracellular fluid is severely limited, and the result may be profound hypercalcemia.

Abnormalities of renal calcium excretion seem unlikely to be the sole cause of hypercalcemia, except under unusual circumstances. For example, in cases of hypercalcemia associated with thiazide diuretics it may be the major mechanism. In patients with familial hypocalciuric hypercalcemia, impaired renal calcium excretion also seems to be the major mechanism. However, it should be emphasized that even under these conditions it is not clear that this is the only mechanism. Thus, the kidney may be more important for its homeostatic functions to determine whether a steady state can be reached by modulation of calcium excretion, rather than as a primary cause of hypercalcemia or hypocalcemia.

The role of the remodeling system

The remodeling system in bone appears unlikely to be important in calcium homeostasis in most physiological situations as long as rates of bone formation are tightly coupled to rates of bone resorption. Even in many pathological situations the rate of bone turnover may have little influence on the plasma calcium, as long as coupling between bone resorption and bone formation is tight. For example, in Paget's disease and thyrotoxicosis, despite very great increases in rates of bone turnover, the plasma calcium remains normal in most patients. However, when a pathologic disturbance uncouples bone formation from bone resorption, extracellular fluid calcium can be disturbed, particularly if rates of bone resorption are increased. Examples of this phenomenon can be seen in patients with solid tumors complicated by hypercalcemia, in myeloma, and in patients who are immobilized. In such cases bone remodeling is uncoupled. There is less bone formation than is appropriate for the rates of bone resorption. The circumstances may be the following. Where remodeling is increased, resorption and formation remain tightly coupled and hypercalcemia is unlikely. However, if bone formation is impaired for any reason where resorption rates are increased the result is an increase in the plasma calcium. It has been suggested that the remodeling sequence, and particularly osteoclast activation, responds too slowly to play a role in maintaining calcium homeostasis. This argument remains doubtful. Changes in the appearance of the ruffled border of osteoclasts can be detected within minutes of exposure to peptide hormones. Bone resorption in vitro mediated by osteoclasts can also be seen within a very short time (less than 30 min) after stimulation.

The role of the exchange system between bone fluid and extracellular fluid

Exchange of calcium between the bone surface or the specialized fluid of bone and the extracellular fluid is likely to play a role both in the determination of mean plasma calcium and in the correction of fluctuations in plasma calcium. This exchange takes place across the lining cells which cover quiescent bone surfaces. However, absolutely nothing is known about the effects of disease processes on these fluxes or exchanges of calcium. It can be surmised that hormonal effects are important. For example, bone cells retract when exposed to PTH.[8] This may lead to the efflux of calcium from bone to the extracellular fluid. The problem is that until now it has not been possible to quantify calcium in the bone fluid compartment, either in normal or in disease states.

The role of the gut

The gut is certainly not important in the acute control of extracellular fluid calcium, but it may be important in chronic control and is certainly important in the pathophysiology of some causes of hypercalcemia. In states associated with dietary calcium depletion, the net fractional absorption of calcium increases after a period of about 24 h. Calcium absorption from the gut is controlled by 1,25-dihydroxyvitamin D (Chapter 2). In circumstances where serum 1,25-dihydroxyvitamin D is increased, calcium absorption is enhanced. This compensatory mechanism is of course blunted in patients with abnormalities in vitamin D metabolism or those with hypoparathyroidism (in the latter situation due to the absence of the tonic effects of PTH on 1,25-dihydroxyvitamin D production).

In patients who are subjected to a large calcium load—for example, bone destruction by cancer or a marked increase in dietary calcium—net fractional calcium absorption is decreased and 1,25-dihydroxyvitamin D is suppressed. In those with primary hyperparathyroidism 1,25-dihydroxyvitamin D is often in the high normal range or increased, and this leads to enhanced calcium absorption from the gut. However, 1,25-dihydroxyvitamin D concentrations are decreased in most patients with malignant disease and hypercalcemia. There is no satisfactory explanation for this at present. It seems to occur independent of whether there is a PTH-like factor produced by the tumor.

Responses to disturbances in calcium homeostasis (Table 1.1)

Depletion of calcium

When dietary calcium is decreased, the homeostatic response depends on the severity and rapidity of dietary calcium deficiency. If the dietary calcium deficiency is gradual at the onset or relatively minor, then dietary adaptation will occur. This is partly mediated by 1,25-dihydroxyvitamin D, and is relatively slow at onset. A more immediate effect is an increase in

Table 1.1 Defences against hypocalcemia and hypercalcemia—protection is better against hypocalcemia than against hypercalcemia.

Protection against decreased plasma calcium (for example, caused by dietary deficiency or hormonal deficiency)

1 Glomerular filtration—filtered load of calcium decreases

2 Bone and kidney—hypocalcemia stimulates PTH release, which increases plasma calcium by effects on renal tubules and osteoclasts

3 Gut adaptation—increased fractional absorption of dietary calcium, mediated by 1,25-dihydroxyvitamin D

Protection against increases in plasma calcium (caused by bone destruction or large dietary calcium load)

1 Glomerular filtration—filtered load of calcium increases

2 PTH—but no further decrease in secretion if plasma calcium >11.5 mg/dl (2.9 mmol/l)

3 Calcitonin—but no longterm efficacy

4 1,25-dihydroxyvitamin D—but gut effects are slow and limited

5 Possible diuretic effect of chronic hypercalcemia eventually leading to sodium and volume depletion and to decreased calcium excretion[9]

PTH secretion from the parathyroid glands in response to the transient fall in extracellular fluid calcium concentration. The parathyroid glands have the capacity to increase PTH secretion by five- to ten-fold. This will lead to increased renal tubular calcium reabsorption, and to increased transport of calcium from the skeleton into the extracellular fluid. Depletion of calcium will also decrease the filtered load of calcium and reduce renal calcium excretion.

There is a slower additional response to calcium deficiency mediated through 1,25-dihydroxyvitamin D, which takes approximately 24 h. Calcium deficiency, possibly in association with increased PTH secretion, will lead to increased synthesis of 1,25-dihydroxyvitamin D. This will increase the efficiency of gut absorption of calcium, as well as enhancing bone resorption. Under the action of 1,25-dihydroxyvitamin D, fractional absorption of calcium from the gut can

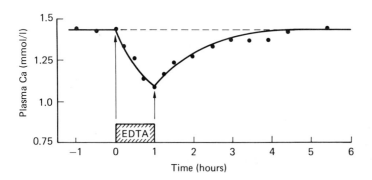

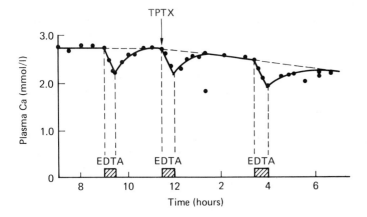

Figure 1.3

Effects of EDTA on plasma calcium in the hypoparathyroid dog. Note the relatively rapid correction of the plasma calcium and the maintenance of the plasma calcium at the predetermined set-point. TPTX = Thyroparathyroidectomy. (Reproduced with permission, copyright 1979 by Pergamon Press.)[5]

increase from 25 per cent (low fractional absorption) to as much as 75 per cent of the ingested calcium.

In situations of severe calcium depletion, such as hypoparathyroidism, there is evidence that other mechanisms must be important in maintaining plasma calcium constant (albeit reduced) or preventing further decreases in plasma calcium. In hypoparathyroid dogs given EDTA to decrease extracellular fluid calcium further, there was a relatively rapid increase in plasma calcium back to the pretreatment concentration (Figure 1.3). Parfitt argues that this response is due to exchanges of calcium between extracellular fluid and the bone fluid mediated by osteocytes, based

on studies with labeled calcium which show translocation of the label across lining cells on bone surfaces.[5]

Calcium excess

A large calcium load can be given experimentally by the intravenous infusion of calcium or by increasing acutely oral intake of calcium. Pathological destruction of bone is the most common clinical situation in which there is an abrupt increase in the extracellular fluid calcium concentration. Large calcium loads will have the immediate effect of maximally suppressing PTH secretion. However, it is not possible to suppress PTH secretion completely if the parathyroid glands are normal (Fig 1.2). The response to increased extracellular fluid calcium will be to increase the renal calcium excretion. This will be mediated by an increase in the filtered load of calcium together with decreased renal tubular calcium reabsorption, which occurs as a consequence of PTH suppression. As noted above, suppression of PTH is maximal (but incomplete) at extracellular fluid calcium concentrations greater than 11.5 mg/dl (2.9 mmol/l).

There is also a more gradual response to the increased influx of calcium into the extracellular fluid. This slower effect is to decrease 1,25-dihydroxyvitamin D production, which will decrease fractional calcium absorption from the gut. However, at extracellular fluid calcium concentrations greater than 11.5 mg/dl (2.9 mmol/l), the major protective mechanism once gut absorption is maximally suppressed and the renal excretion is maximally efficient depends on exchange of calcium between bone and plasma.

Other mechanisms for regulating calcium homeostasis

Although the long negative-feedback loops mediated by systemic hormones are most probably the major mechanisms for maintaining calcium homeostasis, this classical system of reactive regulation cannot account for (a) the constancy of plasma calcium in hypoparathyroid states where there are deficiencies in the systemic hormones (for example, as shown in Figure 1.3), (b) the overall relative constancy of extracellular fluid calcium at times of increased entry of calcium into the extracellular fluid (for example after feeding), and (c) the observation that there are circadian changes in extracellular fluid calcium. These concepts have been expanded recently in publications of Parfitt[10] and Staub et al.[11] Staub and colleagues[11,12] have emphasized that maintenance of calcium homeostasis is much more complicated than the relatively simple negative-feedback model generally accepted. The starting point for their concepts was the observation of the circadian changes which occur in extracellular fluid calcium in the rat. They noted that not only were there apparently 'spontaneous oscillations' or circadian changes in the extracellular fluid calcium but that changes also occurred in the classical hormones. These circadian changes (or 'spontaneous oscillations' as they refer to them) cannot be accounted for by the known negative-feedback relationships between the systemic hormones and extracellular fluid calcium. Thus, circadian variations in plasma calcium (and phosphate) could not be accounted for by changes in the systemic hormones. Examination of temporal changes suggest that in part these circadian changes in plasma calcium may occur in anticipation of feeding (referred to as 'anticipatory regulation'). This would provide a more efficient mechanism than the negative-feedback control or reactive regulation for maintenance of extracellular fluid calcium. For example, the entry of calcium into the extracellular fluid following feeding could be simultaneously counteracted by removal of calcium from the extracellular fluid into bone, a relationship referred to by Staub as 'complementarity'. These events would occur consecutively if reactive regulation alone was responsible for maintenance of extracellular fluid calcium homeostasis.

The concept of 'anticipatory' or 'predictive' regulation (which refers to homeostatic mechanisms which come into operation before a change in the regulated variable) has been discussed in some detail recently by Moore-Ede.[13] Human beings and higher animals have apparently evolved mechanisms to anticipate a change in the internal environment, and they start adapting to the anticipated change before it occurs. For example, they learn to anticipate the changes in the external environment such as the day–night cycle, the sleep–wake cycles and the upright–prone cycles (all of which are probably interdependent), and homeostatic mechanisms may be marshalled before the disturbance is experienced. This may in part explain the circadian variations which have been found in the plasma calcium. Thus, if we have evolved mechanisms to anticipate influxes of calcium in the diet (and subsequently into the gut and extracellular fluid) at regular intervals while we are awake, and therefore usually upright and mobile, we may evoke hormonal secretory mechanisms to anticipate these changes and limit them.

It is now apparent that the relationship between the systemic hormones and extracellular fluid calcium is even more complex. PTH, like most other hormones, has recently been shown to demonstrate episodic secretory behaviour.[14] There is a characteristic secretory pattern in healthy people, indicating that PTH, consistent with the results found with other glandular peptide hormones, is secreted in a pulse amplitude- and frequency-modulated manner. PTH secretion occurs in pulsatile spikes which do not appear to be directly related to the extracellular fluid calcium concentration. There have not been any systematic studies reported as yet on minute-to-minute PTH concentrations in patients with disorders of calcium homeostasis or metabolic bone disease. Harms et al[14] examined three patients with osteoporosis and in these patients there appeared to be both decreased amplitude and frequency of pulsatile PTH secretion. This low pulsatile secretion of PTH may be related to low turnover osteoporosis, but clearly more studies will be needed with appropriate controls. Nothing is yet known as to whether there are disorders in pulsatile PTH secretion in disturbances of calcium homeostasis such as primary hyperparathyroidism, hypercalcemia of malignancy or even in those cases of hypoparathyroidism where parathyroid gland disease or ablation is incomplete. However, it is clear that much of the relationship between extracellular fluid calcium and the systemic hormones is still not understood. The relationship between PTH and extracellular fluid calcium is much more complex than a simple negative-feedback relationship between the systemic hormones and extracellular fluid calcium.

Before a discussion on the individual limbs of the homeostatic mechanism and the way in which they are regulated by calciotropic hormones, several other general aspects of the homeostatic mechanism, and how the homeostatic mechanism responds to deviations, will be considered.

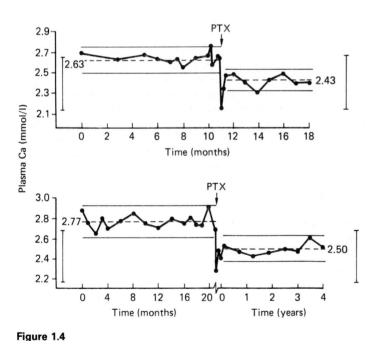

Figure 1.4

Plasma calcium measured serially in two patients with primary hyperparathyroidism. Plasma calcium in patients with primary hyperparathyroidism tends to remain relatively constant, albeit at a higher set-point than normal. PTX = Parathyroidectomy. (Reproduced with permission, copyright 1979 by Pergamon Press.)[5]

The concepts of set-point and error correction

Parfitt has emphasized that in considering mechanisms responsible for calcium homeostasis, it is essential to take into account the separate concepts of set-point and error correction.[4-6] The mechanisms responsible for set-point and error correction are not necessarily the same. Set-point is the level of plasma calcium which the homeostatic mechanism achieves at steady state. It can be estimated in an individual from the mean of serial measurements of the plasma calcium. It appears that one of the objectives of the homeostatic mechanism is to achieve a steady state, regardless of the prevailing concentrations of the hormones which regulate calcium transport or the rates of calcium flux occurring across the limbs of the homeostatic system. The set-point must be under genetic influence, to some extent, since mean plasma calcium varies less between members of the same family.[15] A set-point may be achieved in patients with chronic disorders of the hormonal mechanisms which regulate the homeostatic process although this set-point will not be in the normal range for plasma calcium. The set-point may be at an increased or decreased plasma calcium concentration (for example, in primary hyperparathyroidism or hypoparathyroidism), or at a low plasma calcium concentration (for example, in hypoparathyroidism). In either of these disorders, the plasma calcium is abnormal, but it is usually constant and may remain so for many years with little variation, despite either increased or absent circulating concentrations of PTH (Figures 1.3 and 1.4). Thus, the homeostatic mechanism appears to be designed to keep the plasma calcium constant by mechanisms which aim to maintain a set-point for calcium. The mechanisms by which the set-point is determined, or how the 'calciostat' is set, are still debatable, and are discussed in more detail below.

The other concept to keep in mind is that of error correction. In any individual the homeostatic mechanism will correct the plasma calcium concentration whenever it deviates from the set-point. The extent of this error correction is reflected by the oscillations in plasma calcium about the mean. In the normal individual, error correction can be estimated by the standard deviation of oscillations of the plasma calcium about the set-point.

The set-point and what maintains it (Table 1.2)

As indicated above, the set-point is the absolute plasma calcium concentration which the homeostatic mechanism attempts to maintain. Whatever the rates of entry of calcium into the extracellular fluid are, a mechanism has evolved for adjusting rates of efflux to obtain a steady-state level of plasma calcium. Although the set-point is a theoretical value, and cannot be measured absolutely, in a normal individual it can be estimated from the mean of serial measurements of the plasma calcium. When the 'normal' range for plasma calcium for any individual is considered, this normal range reflects the set-point, and the variations about it which are caused by deviations in calcium intake or calcium loss from the extracellular fluid.

It appears that there are both hormonal and nonhormonal influences on set-point determination. The mechanisms which determine set-point may be slowly acting and take hours to days to achieve a steady-state concentration of plasma calcium. The best evidence that hormonal influences are important comes from observations in patients with primary hyperparathyroidism and in patients with hypoparathyroidism. In primary hyperparathyroidism, although the plasma calcium is increased, it may be maintained constant and relatively stable for many years (Figure 1.4).

Table 1.2 Why is there a set-point for plasma calcium?

1 Exquisite relationship between plasma calcium and PTH secretion in range 7.5–11.5 mg/dl (1.9–2.9 mmol/l)

2 Glomerular filtration (increased plasma calcium leads to increased filtered load and increased renal calcium excretion)

3 Exchange of calcium between plasma and bone fluid

In other words, there appears to be a new, higher set-point in the individual with increased PTH and possibly 1,25-dihydroxyvitamin D in the circulation. Once the source of excess PTH secretion is removed (usually by the resection of a parathyroid adenoma), then the set-point returns to that found in the normal individual. The plasma calcium is also maintained fairly constant at approximately 5.5 mg/dl (1.4 mmol/l) in most individuals when the parathyroid glands are ablated. Even if the extracellular fluid calcium concentration in patients with hypoparathyroidism is lowered further by other means, such as by an infusion of ethylenediamenetetracetic acid (EDTA) which complexes and chelates calcium, the plasma calcium will rapidly return to a set-point of approximately 5.5 mg/dl (1.4 mmol/l) (Figure 1.3).[16] Thus, there must be mechanisms other than those mediated by PTH and 1,25-dihydroxyvitamin D for maintaining the set-point. The exchange of calcium between the bone fluid and the extracellular fluid across the lining cells which comprise the bone membrane is probably important[17] (Chapter 3), and Parfitt suggests that set-point is largely determined by this exchange.[4-6]

The relative importance of the kidneys and bone in set-point determination remains unclear. A primary role for the kidneys is hard to exclude. Certainly, in many patients with primary hyperparathyroidism with stable hypercalcemia and normal renal calcium clearance, the renal tubules must be responsible for the determination of the set-point. Parfitt[6] argues that the special case of pseudohypoparathyroidism makes it unlikely that the kidney plays a major role in the maintenance of the set-point, since the hypocalcemia which occurs in pseudohypoparathyroidism occurs in the presence of no major impairment of renal tubular calcium reabsorption. He points out that there is strong evidence that calcium is exchanged between blood and bone across the 'quiescent' cells that line bone surfaces, and that this is a more plausible mechanism for determining the steady-state level of plasma calcium. However, essentially nothing is known of how fluxes of calcium across bone-lining cells are regulated. Even in the absence of PTH, the amount of calcium which is filtered across the glomerulus will be dependent on the plasma calcium, and renal calcium clearance will be greater when plasma calcium is increased. It is an attractive notion that the set-point could be determined in part both by renal calcium excretion and by the exchange of calcium between the extracellular fluid and bone. Nothing is known of any direct relationship between these two processes, or whether they are mutually independent. However, it appears likely that PTH and other factors that affect bone lining cells and renal tubular cells may regulate calcium fluxes at both sites.

That the remodeling sequence is unlikely to have a major influence on the set-point is clear from the observations that the set-point for calcium appears to be independent of rates of bone remodeling.

Error correction

Whenever there is a deviation in plasma calcium from the set-point, whatever its cause, the homeostatic mechanism attempts to correct this deviation. The term given to the mechanism by which plasma calcium is returned to the set-point is 'error correction', and error correction can be determined by calculation of the standard deviation of oscillations of plasma calcium about the set-point. Error correction is the homeostatic reaction triggered by a disturbance in the set-point for the plasma calcium, and responses to these disturbances are very rapid, usually occurring within minutes. Parfitt[4-6] argues that error correction does not depend on calciotropic hormones, since error correction can occur in the absence of PTH and vitamin D. For example, in hypoparathyroidism error correction can occur with normal efficiency,[4-6] as indicated above in the example of the hypoparathyroid dog made even more hypocalcemic by EDTA (Figure 1.3). However, this does not mean that hormones do not influence error correction. It appears more likely that both hormonal and nonhormonal influences are important. The major hormonal mechanism for error correction is likely to be mediated by the effects of PTH on the renal tubules. For example, an increase in bone resorption with subsequent increased entry of calcium into the extracellular fluid compartment suppresses PTH, which leads to increased fractional excretion of calcium. Later changes such as the effects of decreased circulating PTH on the efflux of calcium from bone and the absorption of calcium from the gut (mediated by 1,25-dihydroxyvitamin D) are slower at the onset and probably less important in this regard.

Steady-state levels of the plasma calcium

The calcium homeostatic mechanism works efficiently to maintain plasma calcium stable at steady state. The maintenance of a steady state occurs not only in normal individuals, but also in chronic disease states such as primary hyperparathyroidism and hypoparathyroidism. In disease states, and even in normals, steady-state plasma calcium concentration is an approximation rather than an absolute reality. In normals the major deviation is caused by calcium intake from the gut following meals. Nevertheless, the mechanism for maintaining a steady state is efficient, and the concept of the careful maintenance of a steady state in both health and disease is important. Of course, in some disease states such as hypercalcemia associated with malignancy, the acute effects of tumor factors on fluxes of calcium across bone and kidney may overwhelm the capacity of the homeostatic mechanism to maintain a constant plasma calcium concentration. This leads to an unstable form of hypercalcemia (see below).

If plasma calcium is at a steady state—whether the absolute concentration is in the normal range, increased, or decreased—the net input of calcium into the extracellular fluid must be equal to the net output. In practical terms this means that the net input from bone and gut equals the net output from kidney. This point has been emphasized by several clinical investigators in this field (including Nordin, Peacock and Kanis). In primary hyperparathyroidism calcium excretion in the urine is often in the normal range. This indicates that the net input of calcium into the plasma from gut absorption and bone resorption cannot be substantially increased (otherwise urine calcium excretion would be increased to reflect this). Thus, the primary mechanism responsible for hypercalcemia in these cases must be increased calcium reabsorption in the renal tubules.

Transient disturbances— 'disequilibrium'

When hypercalcemia is stable and the plasma calcium is maintained at a relatively constant concentration, as often happens in mild hyperparathyroidism, the net inflow of calcium into the extracellular fluid is balanced by outflow, and Parfitt refers to this state as 'equilibrium hypercalcemia'.[5]

However, in some circumstances the plasma calcium concentration rises progressively at a rate which overwhelms the mechanisms of error correction, and a steady state cannot be achieved. This is known also as 'disequilibrium hypercalcemia' and is due to the net input exceeding the capacity of the system to clear this increased input without the plasma calcium increasing steadily. Thus, the mechanisms may be due primarily to decreased losses (impairment of glomerular filtration rate or increased renal tubular calcium reabsorption), to markedly increased entry of calcium (usually enhanced bone resorption), or to a combination of both of these mechanisms.

The tank model of calcium homeostasis

Twenty years ago, Nordin and Peacock[7] likened the homeostatic mechanism for calcium to a tank partially filled with water (Figure 1.5). Others have since drawn on this analogy.[4] This simple but useful model helps to understand the control of calcium homeostasis, and it allows a number of important points regarding the homeostatic mechanism to be made. Using this analogy, the level of fluid in the tank is equivalent to the plasma calcium, the inflow into the tank is equivalent to the net bone resorption and net gut absorption of calcium, and the outflow from the tank is equivalent to renal calcium excretion. The outflow from the tank will be determined by the size of the outflow aperture (the glomerular filtration of calcium) and the height of the outflow aperture (which represents the net tubular reabsorption of calcium).

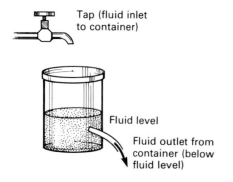

Tap (fluid inlet to container)

Fluid level

Fluid outlet from container (below fluid level)

Figure 1.5

The simple tank model for calcium homeostasis. The size of the aperture corresponds to the glomerular filtration rate and the height of the aperture above the base corresponds to the net tubular calcium reabsorption.

The level of fluid in the tank (the plasma calcium) is determined by the inflow to and the outflow from the tank. There are several points which are immediately apparent when considering this helpful model. If the system works to preserve a relatively constant level of fluid over wide ranges of inflow and outflow and the level of fluid in the tank is stable, then inflow must equal outflow (regardless of whether inflow or outflow is increased or decreased). If the inflow into the tank is increased without a change in outflow, then the level of fluid in the tank will rise. Thus, the level of fluid in the tank will remain constant within the capacity of the outflow to change in response to the inflow.

The relationship between outflow (which at steady state is equivalent to inflow) and the fluid level can be plotted simply. If the aperture for

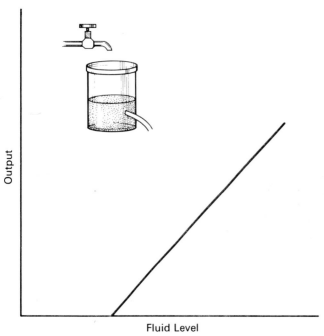

Output

Fluid Level

Figure 1.6

The relationship between the outflow of fluid from the tank and level of the fluid in the tank in the tank model shown in Figure 1.5. Note that the fluid level is dependent on the height of the outlet aperture as well as its size. The size of the outflow aperture represents glomerular filtration and the height net tubular reabsorption of calcium.

outflow is not at the base of the tank, then the outflow will increase with the fluid level in a simple linear fashion, as depicted in Figure 1.6. A similar relationship has been determined between the plasma calcium and urinary calcium excretion by Peacock and Nordin (Figure 1.7). Using intravenous calcium infusions to increase the plasma calcium, they found that there was a curvilinear relationship between calcium excretion and the plasma calcium.

Peacock and Nordin have used this model to emphasize the importance of outflow (the kidney) in controlling the level of fluid (plasma calcium) at steady state.[18,19] As indicated above, at steady state where the plasma calcium is stable, output must equal inflow. They have reasoned therefore that the outflow mechanism is the important site of control, a point which Parfitt[5] believes they have overemphasized. To restate the principles underlying the calcium homeostatic system using this analogy:

(a) As inflow (net gut absorption and net bone resorption) increases, changes in outflow (renal calcium excretion) will determine the eventual level (plasma calcium) reached. If the outflow (renal calcium excretion) does not increase or is impaired, then the plasma calcium or fluid level will increase. Thus, the fluid level will be maintained within the limits of the capacity of the outflow system to increase in response to an increase in inflow. For the kidney, approximately a threefold increase in calcium excretion is possible when renal function is normal.

(b) If the outflow is decreased (because of impaired glomerular filtration or increased renal tubular calcium reabsorption), the fluid level (plasma calcium) will rise unless the inflow (net gut absorption and bone resorption) is decreased.

Although Peacock and Nordin have reasoned that the set-point for calcium will be determined by the threshold for renal tubular reabsorption of calcium, it is extremely difficult to place relative importance on any single organ in the control of plasma calcium, because it is difficult to create an experimental situation where a perturbation in a single limb or component of the homeostatic mechanism can be manipulated independently of all other limbs.

Effects of a calcium infusion on plasma calcium and calcium excretion

As indicated in Figure 1.7, the relationship between plasma calcium and renal calcium excretion can be determined using data obtained from studies utilizing a constant intravenous infusion of calcium in normal individuals. Such an infusion will serve to clarify the relationship between plasma calcium and the homeostatic mechanism which protects it when calcium input into the extracellular fluid is increased. This relationship has frequently been used to assess the contribution of changes in renal tubular calcium reabsorption to any alteration in the

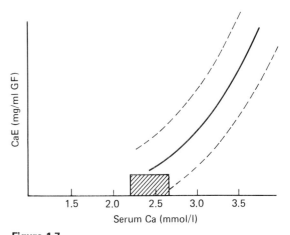

Figure 1.7

The relationship between urinary calcium excretion (CaE) and serum calcium in normal subjects during an acute calcium infusion.[19] The shaded area represents the normal basal range. The solid line and the broken lines represent the mean values ± 2 SD, respectively.

plasma calcium.[7] When calcium is infused acutely into normal subjects at a constant rate and the total plasma calcium or ionized plasma calcium and calcium excretion are evaluated, there is a curvilinear relationship between calcium excretion and plasma calcium (Figure 1.7). The increase in plasma calcium is limited, to some extent, by enhanced renal calcium excretion. In patients with impaired renal function, there is a shift of this curvilinear line to the left, so that any level of plasma ionized-calcium calcium excretion will be relatively greater per nephron or per milliliter of glomerular filtrate.[20] Of course, total calcium excretion will be reduced. When the output of calcium from the extracellular fluid is equal to the input, the plasma calcium level reaches a plateau and there is no further increase.

Clinical assessments of calcium 'input' and 'output'

The absolute rates of calcium transport between bone, gut, and kidney are extremely difficult to assess with any precision. The most difficult of these processes to evaluate is the transport of calcium between bone and the extracellular fluid. Calcium kinetic studies using isotope tracers have been used to give individual assessments, but the methods are cumbersome and the answers imprecise. Many investigators have chosen to use simpler assessments, such as measurements of the net gut absorption of calcium, net bone resorption, glomerular filtration and renal tubular calcium reabsorption. Indirect estimates of net bone resorption and net renal tubular calcium reabsorption can be obtained from careful clinical measurements of the urine calcium. Glomerular filtration rate can be calculated from studies of the creatinine clearance (although these can be misleading, particularly in patients with hypercalcemia of malignancy). Net gut absorption of calcium can be calculated if necessary by isotope tracer studies or by examining calcium excretion following a standard calcium containing meal, as described by Pak et al.[21] The majority of investi-

gators interested in calcium homeostasis have estimated rates of calcium transport due to net bone resorption and net renal tubular calcium reabsorption from careful clinical measurements of the serum and urine calcium. Net bone resorption has been estimated from measurements of renal calcium excretion during periods of fasting. This measurement assumes that the input of calcium from the gut at these times is negligible, which is probably correct after fasting for 12 h. Under these circumstances the fasting urine calcium excretion probably equals the net skeletal loss of calcium.

Renal tubular calcium reabsorption can be estimated from a knowledge of the filtered load of calcium and the amount excreted per filtered load. The filtered load equals the glomerular filtration rate multiplied by the ultrafiltrable calcium, which is approximately 60 per cent of the plasma calcium. The normal curvilinear relationship between the renal calcium excretion and the plasma calcium has been calculated from calcium infusions in normals and was discussed above (Figure 1.7).[19] For any given level of plasma calcium, if renal tubular calcium reabsorption lies to the left of this line there is a decrease in renal tubular calcium reabsorption. If calcium excretion lies to the right of the curvilinear line, then there is increased renal tubular calcium reabsorption. However, this relationship is dependent on normal glomerular function, normal plasma proteins, and the absence of factors which alter renal sodium excretion (for example, dietary or drugs). The relationship between sodium excretion and renal tubular calcium handling is particularly critical. If renal sodium excretion is greater than 4 mmol/l of glomerular filtrate, then renal calcium excretion will be enhanced and the increased renal tubular calcium reabsorption may be obscured. The relationship is also somewhat limited because of the necessity in most patients to measure plasma calcium rather than ultrafilterable calcium. The ultrafilterable calcium may be different from the plasma calcium if changes in pH or abnormal proteins occur. Interpretation of these measurements may be hazardous in patients who have recently received saline infusions, or in patients with renal failure, which may cause differences from the relationship between renal calcium excretion and the plasma calcium observed during an acute calcium infusion.

The most extensive application of this approach to assessing the contribution of renal tubular calcium reabsorption to hypercalcemia has been utilized by the group in Sheffield headed by John Kanis.[22] They have used the relationship between renal calcium excretion and plasma calcium established by the studies of Peacock and Nordin to estimate the relative contributions to the rise in plasma calcium in patients with hypercalcemia from rates of bone resorption, renal tubular calcium reabsorption and impaired glomerular filtration, and have developed an algorithm to assist in estimating the relative contributions to the increase in plasma calcium. There are assumptions made in this method, some of which have already been mentioned. One of the most important is that the normal relationship between renal calcium excretion and plasma calcium has been determined with a short-term calcium infusion, and this does not necessarily mimic the situation in the patient with chronic hypercalcemia. Moreover, rates of bone resorption are assessed by the measurement of fasting urine calcium. Although net bone resorption is certainly a major contribution to the urine calcium in the fasting patient, it may not be the only component. For example, net calcium absorption from the gut may also contribute a part, notwithstanding the fasting state, and it is not known whether renal tubular calcium reabsorption remains constant or varies with the rates of net bone resorption or net gut calcium absorption in these hypercalcemic patients. However, possibly the major limitation is that the role of glomerular function in renal calcium clearance is estimated from measurements of serum creatinine, and this may be a very misleading index of glomerular filtration in, for example, cachectic cancer patients. Despite these unknowns, the concept seems a reasonable approach in the current state of knowledge for the clinical investigator.

Summary of control of calcium homeostasis

How all of these complicated mechanisms work together in concert to maintain a normal plasma calcium, is unknown. However, the author will propose a model which is consistent with the available data. The two major hormones which maintain calcium homeostasis are PTH and 1,25-dihydroxyvitamin D. Both of these hormones are clearly necessary for maintenance of normal plasma calcium. PTH is more important for its rapid effects and 1,25-dihydroxyvitamin D for its longer term effects. However, complete absence of PTH or 1,25-dihydroxyvitamin D does not lead to disappearance of calcium from the extracellular fluid. There is a basal level of plasma calcium which is maintained by fluxes of calcium between the bone fluid/bone surface and the extracellular fluid and is independent of hormonal activity. This level of plasma calcium is approximately 5–5.5 mg/dl (1.25–1.40 mmol/l). The presence of PTH and 1,25-dihydroxyvitamin D is necessary to raise the plasma calcium from 5.5 mg/dl (1.40 mmol/l) to the normal range of approximately 10 mg/dl (2.5 mmol/l). Within the normal range, oscillations in the plasma calcium are dampened both by the exchange mechanism between the bone fluid/bone surface and the extracellular fluid (this dampening effect is observed even in the absence of PTH or 1,25-dihydroxyvitamin D), but also by changes in the PTH and 1,25-dihydroxyvitamin D which are regulated by long negative feedback loops. However, PTH secretion is not further enhanced or suppressed by changes in plasma calcium concentrations outside the range 7.5–11.5 mg/dl, and at these extremes the bone fluid exchange mechanism may be the most important mechanism for correction of deviations in plasma calcium concentration.

Whenever there is a relative increase in the inflow of calcium into the extracellular fluid, the plasma calcium will tend to increase. However, the outflow of calcium from the kidney adjusts in an attempt to maintain the extracellular fluid calcium at a normal level. Whether the increase in plasma calcium is progressive or stable at a new steady state depends on the capacity of the kidney to excrete this calcium load. There is also some capacity of the bone–fluid exchange mechanism to keep the plasma calcium concentration within the normal range. If the capacity of the kidneys to clear calcium is impaired, then plasma calcium will increase and if it is markedly impaired, or if inflow is increased above the capacity of the kidneys to excrete the calcium load, then the increase in plasma calcium will be unstable and a steady state will not be obtained.

However, if the outflow capacity is not exceeded, the kidneys, in combination with the bone exchange mechanism, will maintain the plasma calcium in the normal range. Thus, the concentration of calcium in the extracellular fluid is determined by the capacity of the outflow system (both kidney and bone) to match changes in inflow.

CHAPTER 2

Calcium homeostasis—role of the gut, kidney and bone

Current concepts of calcium influx and efflux between the extracellular fluid and the major organs of calcium homeostasis (gut, bone, and kidney—shown diagrammatically in Figure 1.1) and the hormonal mechanisms which regulate these fluxes are now reviewed. This chapter is lightly referenced, since it summarizes very briefly many years of work by numerous investigators. For extensive citations the reader is referred to a series of review articles in the references for this chapter.

Calcium transport across the gastrointestinal tract

The net absorption of calcium from the gut to the extracellular fluid is approximately 150–200 mg/day. This varies considerably, even in normals, depending on calcium intake in the diet and the circulating concentrations of 1,25-dihydroxyvitamin D and other dietary components which may serve to increase calcium absorption, such as lactose and fatty acids, or to decrease calcium absorption, such as phosphate, oxylate, or phytate. The portion of dietary calcium which is absorbed can vary widely from 20 to 70 per cent, although with a normal diet the average healthy adult absorbs about 40 per cent of dietary calcium. The average daily North American diet contains 800–1000 mg elemental calcium,

although elderly people often ingest less. The major determinants of calcium absorption are dietary calcium, other dietary constituents which determine the availability of calcium in the gut for absorption, and 1,25-dihydroxyvitamin D. When dietary calcium is deficient, the percentage absorption increases, probably due to an increase in serum 1,25-dihydroxyvitamin D. Calcium transport is most efficient in the first part of the small intestine, and particularly in the duodenum. However, the majority of calcium is absorbed in the lower parts of the small intestine, the ileum (approximately 65 per cent) and the jejunum (approximately 17 per cent), since these portions of the small intestine are much longer than the proximal parts. Although the latter are more efficient at calcium absorption, the jejunum and ileum are longer and have a larger surface area, so the duration of exposure of calcium to the epithelium is longer and most of the calcium is absorbed in these segments. Calcium is also secreted back into the lumen of the gut, predominantly in the bile and pancreatic juice, above the sites where the majority of calcium is absorbed. The secretion of calcium into the gut lumen via the intestinal juices is presumably unregulated. It accounts for about 10 per cent of all the calcium within the gut lumen.

Calcium is absorbed from the gut lumen by two mechanisms: an active transport process and a process of facilitated diffusion. The evidence for active transport of calcium is that it occurs against an electrochemical gradient and that the process is saturable. Moreover, calcium transport is inhibited in anaerobic conditions or in the presence of metabolic inhibitors. The evidence that there is a mechanism by which calcium is pumped actively against a concentration gradient and against an electrochemical gradient from mucosal to serosal surface of the gut is now overwhelming. This active transport mechanism has been linked to the synthesis in intestinal epithelium of a calcium-binding protein, which could theoretically be responsible for the translocation of calcium across the intestinal epithelium, and whose synthesis is regulated by 1,25-dihydroxyvitamin D. This calcium-binding protein has been fully characterized in chick intestine, and is a 28 000 dalton protein which has four active binding sites for calcium. Similar calcium-binding proteins have been found in other organs, including the brain, bone, pancreas, kidney, parathyroid glands, and placenta. The calcium-binding protein in mammalian intestine is of 10 000 daltons and contains two high-affinity binding sites per cell. The 1,25-dihydroxyvitamin D-induced calcium-binding proteins share homology with calmodulin, myosin light chain, and troponin C. However, the precise relationship between the intestinal calcium-binding proteins and the effects of 1,25-dihydroxyvitamin D to promote the absorption of calcium from the gut are not entirely clear. There is a poor correlation between the appearance of this protein in response to 1,25-dihydroxyvitamin D and increased intestinal transport. Increases in calcium transport in response to 1,25-dihydroxyvitamin D precede the appearance of the calcium-binding protein, and the calcium-binding protein persists long after the increase in intestinal calcium transport in response to 1,25-dihydroxyvitamin D has subsided. Thus, investigators have looked for other 1,25-dihydroxyvitamin D-regulated proteins which could be involved in calcium transport. Two such proteins have been identified in the brush border of the intestinal mucosa. One of these is an alkaline phosphatase and the other a calcium-dependent adenosine triphosphatase (ATP) activity. However, these two proteins show even less correlation with intestinal calcium transport than the calcium-binding protein, and their relationships to calcium transport and to the calcium-binding protein are not yet clear.

The active transport of calcium is not the only means by which calcium is absorbed from the gut lumen. There is also convincing evidence that facilitated diffusion of calcium which is modulated by 1,25-dihydroxyvitamin D also occurs, since increasing the calcium content in the gut lumen will increase calcium absorption without evidence of saturation. This process is also facilitated by 1,25-dihydroxyvitamin D. Thus, the absorption of calcium from the gut is a complicated process which involves an active transport process and facilitated diffusion, both of which are regulated by 1,25-dihydroxyvitamin D. Fractional calcium absorption is not only modulated by 1,25-dihydroxyvitamin D, it is also influenced by dietary calcium. It has long been recognized that net calcium absorption is dependent on the dietary calcium intake. When dietary calcium falls, the efficiency of absorption of calcium in the intestine increases. Although 1,25-dihydroxyvitamin D is required for this adaptation to occur, it is still not clear that 1,25-dihydroxyvitamin D is the sole mediator of this adaptive response.

For an individual in normal calcium balance the daily net absorption of calcium from the gut is equivalent to the excretion of calcium in the urine, provided the rates of bone resorption and bone formation are balanced. If not, then the net absorption of calcium from the gut is equivalent to the excretion of calcium in the urine and the net bone resorption.

Variations in calcium absorption from the gut have little to do with the acute minute-to-minute regulation of extracellular fluid calcium homeostasis, but are probably of more significance in the chronic maintenance of the serum calcium. As indicated above, the adaptive mechanism whereby the net absorption of calcium is related to the dietary calcium intake is almost mediated predominantly by 1,25-dihydroxyvitamin D. Although dietary calcium excess or decreased calcium intake in man will not lead to an abnormality in extracellular fluid calcium homeostasis as long as the homeostatic mechanisms are intact, there are a number of disease states where calcium absorption is either excessively enhanced or diminished. This is usually dependent on the circulating concentrations of 1,25-dihydroxyvitamin D. In some types of hypercalcemia 1,25-dihydroxyvitamin D concentrations may be increased. These conditions include primary hyperparathyroidism, sarcoidosis and other granulomatous diseases, and T-cell or B-cell lymphomas. Vitamin D intoxication is associated with increased serum 25-hydroxyvitamin D. There are some types of hypercalcemia where serum 1,25-dihydroxyvitamin D is suppressed. This is certainly true in patients with the humoral hypercalcemia of malignancy and in patients with metastatic bone disease or myeloma. In patients with vitamin D deficiency or chronic renal failure hypocalcemia occurs and serum 1,25-dihydroxyvitamin D is suppressed. Calcium absorption from the gut is increased in patients with the absorptive form of idiopathic hypercalciuria. It may be inhibited by drugs such as phosphate and corticosteroids, and by dietary constituents such as phosphate, lipids, and phytates, which limit its availability for absorption.[1-3]

Calcium handling by the kidney

The kidney is a uniquely placed organ to modulate calcium homeostasis. In an average adult approximately 10 000 mg calcium crosses the glomerulus each day, and 98 per cent of this is resorbed in the renal tubules (Figure 2.1). In fact, the actual capacity of the kidneys to regulate extracellular fluid calcium homeostasis is much more limited, since hormonal capacity to re-absorb calcium is limited to less than 1 g/day. Renal calcium handling occurs in two stages, glomerular filtration and tubular handling, and these are discussed individually. This topic has been reviewed thoroughly in recent years.[4,5]

Glomerular filtration

The glomeruli of the kidneys filter about 1 per cent of the total body calcium each day, equivalent to the total calcium in the body which is outside of the skeleton. This process is probably not under any significant hormonal regulation, although it may be altered to a minor degree by renal blood flow, and there is controversial evidence that renal blood flow may be influenced by PTH. The calcium which crosses the glomerulus to reach the tubule lumen is ultrafilterable calcium. In patients with hypercalcemia the calcium which crosses the glomerulus may be decreased due to complexing phosphate or binding to proteins. This has been confirmed in careful studies with calcium infusions. This clearly represents an adverse homeostatic response in a patient with hypercalcemia, since it leads to a reduction in calcium clearance. However, it is probably not a significant factor in most patients.

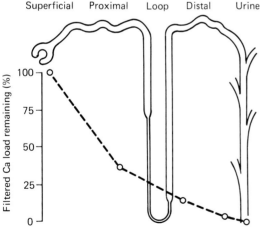

Figure 2.1

Diagrammatic representation of renal handling of calcium. Notice the enormous amount filtered across the glomerulus compared with that reabsorbed in the renal tubules.

Renal tubules

Approximately 98 per cent of the calcium filtered by the glomerulus is reabsorbed in the renal tubules. Of this total, 65 per cent is reabsorbed in the proximal tubules, 20–25 per cent in the ascending limb of Henle's loop, and 10 per cent in the distal convoluted tubules. Calcium which is reabsorbed in the proximal tubule is not under hormonal regulation. The calcium in the tubule lumen is all either ionized (the majority) or complexed. Transport through the lumen depends on the anionic environment and therefore on the transport of other ions, such as bicarbonate, citrate, sulfate, phosphate, and other organic anions, across the tubule lumen.

Calcium is reabsorbed by an active transport mechanism in the proximal tubule. Most of the information which is available about the reabsorption of calcium at this site comes from micropuncture studies. The reabsorption of calcium in the proximal tubules has the following characteristics: it is closely linked with sodium reabsorption; it is unaffected by PTH; the reabsorption of calcium depends on intraluminal calcium concentration and is not limited by the maximal rate (T_m); and the reabsorption is voltage dependent. Several possible mechanisms have been suggested, including a calcium–magnesium dependent ATPase and a sodium–calcium antiport. Neither of these mechanisms is universally accepted. The mechanism by which calcium reabsorption is linked to sodium reabsorption is not clear, but it is critically important. When patients are dehydrated, renal blood flow is decreased and sodium reabsorption is enhanced. Calcium reabsorption accompanies sodium reabsorption, so dehydration can potentiate hypercalcemia through this mechanism.

In the ascending limb of Henle's loop approximately 25 per cent of the total calcium filtered by the glomerulus is reabsorbed. The mechanism responsible for calcium reabsorption at this site is controversial. Thiazide diuretics work at this site to promote calcium reabsorption.

The site at which calcium reabsorption in the renal tubules is finely regulated is in the distal convoluted tubule. Here about 1000 mg is reabsorbed daily. Calcium reabsorption in the distal tubule is under the control of PTH and possibly of 1,25-dihydroxyvitamin D, and these two hormones may work in concert to promote calcium reabsorption. It is still not known if the effects of PTH on calcium reabsorption are mediated by 3',5' cyclic adenosine monophosphate (cAMP). In states of PTH lack, diminished calcium reabsorption occurs and urinary calcium falls to less than 150–200 mg/day. In patients with primary hyperparathyroidism calcium reabsorption is increased, but urinary calcium is relatively high because of the increase in filtered load. The other factors which regulate calcium reabsorption in the renal tubule are obscure. Increased renal tubular calcium reabsorption clearly occurs in some patients with cancer and hypercalcemia, but the mechanisms are not clear. Clearly, neither PTH nor 1,25-dihydroxyvitamin D is involved. It is not known at which site in the kidneys excess calcium reabsorption occurs in malignant disease. It is possible that it may be related to the PTH-like factors which are produced by some cancers associated with hypercalcemia which could interact with PTH receptors in the distal tubule of the kidney.

The relationship between serum calcium and urinary calcium excretion is a curvilinear but continuous line. This relationship is the mechanism by which clinicians have determined that some patients have decreased or increased renal tubular calcium reabsorption (Chapter 1). Studies of this type cannot determine at which site in the renal tubules the alteration in calcium reabsorption occurs. In normal individuals the fractional excretion of calcium increases as the serum (and the filtered load) increases. The curvilinear relationship between the serum calcium and urinary calcium excretion was constructed from studies performed using acute infusions of calcium, so this relationship may not be relevant to the more chronic forms of hypercalcemia and its effects on renal calcium handling.

Since most of the regulation of calcium excretion is performed in the distal tubule, and here reabsorption is less than 1000 mg/day (25 mmol/day), the capacity of the normal kidney to excrete a calcium load is limited, and in fact is usually less than 600 mg/day (15 mmol/day).

Calcium transport across bone

Exactly what the contribution of bone is to the extracellular fluid calcium concentration has been argued for many years. Some people have suggested that it does not provide for rapid changes in mineral homeostasis, but that it may be important for more chronic changes. Others have suggested that there are two pools of calcium in bone, one which is available for rapid exchange and one which is available only for the chronic maintenance of extracellular fluid calcium homeostasis. Still others have suggested that bone plays no important part in normal calcium homeostasis, but rather that fluxes across the kidney are what determine the minute-to-minute control of the serum calcium concentration. These are extremely difficult issues to evaluate. It is clear that bone is a major storehouse of calcium. More than 99 per cent of the total body calcium is present in bone. It is also clear that only small amounts of the skeletal calcium are exchanged daily with the extracellular fluid. It is also apparent that, under most circumstances, the influx and efflux of calcium between the bone store of calcium and extracellular fluid by the process of bone resorption and bone formation are in relative balance. Nevertheless, it seems certain that in some disease states large increases occur in the entry of calcium into the extracellular fluid from bone. This is particularly likely to happen in patients with advanced malignancy associated with bone destruction. However, it must be appreciated that many patients with severe bone destruction associated with malignancy do not have detectable abnormalities in their serum calcium concentration.

There are two mechanisms for the transport of calcium from bone to the extracellular fluid. The first is via the bone remodeling system— osteoclast activation leading to degradation of the bone matrix and release of bone mineral. The second mechanism, which may be quantitatively more important but which is much less well understood, is the exchange of calcium between the extracellular fluid and the crystal surface of bone, via the bone fluid, across the bone-lining cell layer (the bone membrane). These two mechanisms are now discussed in more detail.

The bone remodeling system

The bone volume is regulated by the co-ordinated actions of bone-resorbing cells and bone-forming cells. An extensive review[6] has recently been published on this topic by our group. Bone resorption and bone formation are tightly coupled, and bone is continually being remodeled by this cellular activity. The skeleton consists of discrete packets of cellular activity known as bone-remodeling units. This activity is duplicated at multiple sites throughout the skeleton. The cellular sequence of events involved in remodeling is always the same (Figure 2.2). The initial

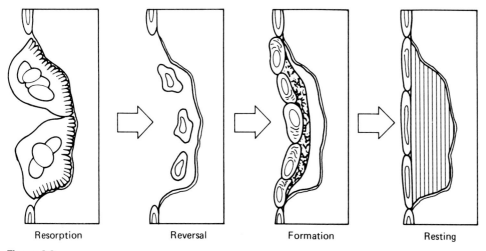

| Resorption | Reversal | Formation | Resting |

Figure 2.2

Sequence of events involved in normal bone remodeling. Bone remodeling is initiated by local events which lead to an increase in osteoclast activity. This is followed eventually by an increased attraction and proliferation of osteoblast precursors, followed by the differentiation of these cells into mature osteoblasts, which lay down new bone and repair the resorption bays formed by the resorbing osteoclasts.[6]

event is an increase in osteoclast activity. The factors or mechanisms which initiate the activation of individual osteoclasts throughout the skeleton are unknown. However, the complete process of osteoclast activation involves the formation of the mature multinucleated osteoclast from its precursor, differentiation of the committed progenitors, and then activation of the preformed multinucleated cell. The osteoclast resorbs bone by a process of extracellular degradation and proteolysis, which occurs across an area of the cell membrane known as the ruffled border. The resorption phase of the remodeling sequence lasts for about 10 days. This is followed by a reversal phase in which osteoclastic resorption ceases and the resorption bay is occupied by mononuclear cells. The bone-formation phase then follows, where a team of osteoblasts accumulates at the site of the defect, replicates, and then produces new bone to repair the defect. The bone-formation phase usually lasts for several months. The result is a new packet of bone. The bone-remodeling sequence is now discussed in more detail.

The first event is formation and activation of the osteoclast to resorb bone. The osteoclast is a large multinucleated cell which ranges in size up to 100 μm in diameter and may contain up to 100 nuclei in pathological conditions such as Paget's disease. It resides on the endosteal bone surface, within Haversian canals and on the periosteal surface of bone beneath the periosteum. Osteoclasts are not frequently seen in normal adult bone, unless they are exposed to a hormonal stimulus. The characteristic feature of the osteoclast is the ruffled border, the specialized area of the cell membrane which consists of a series of fine finger-like processes which invaginate the bone surface. The ruffled border is the site at which the resorption of bone occurs, including the release of mineral and the degradation of bone matrix. These are both extracellular processes. The ruffled border of the osteoclast increases in size and volume when the osteoclast is exposed to factors which stimulate bone resorption, such as PTH. The osteoclast contracts its cell membrane when it is incubated with calcitonin, and eventually disintegrates into mononuclear cells.

The osteoclast arises from a precursor which resides in the marrow cavity. However, there is a phase during its lifespan where it circulates in the blood. The precursor for the osteoclast has not been identified definitively, but the evidence suggests that it is a mononuclear precursor which is a stem cell common to the monocyte–macrophage family. However, not all workers agree with this conclusion. The multinucleated cell is formed by fusion of committed progenitors, and is probably under the regulation of 1,25-dihydroxyvitamin D. Hormones which act as growth factors for osteoclast progenitors include the bone-resorbing cytokines interleukin-1, lymphotoxin, and tumor necrosis factor, as well as transforming growth factor α (TGF α) (Figure 2.3). Parathyroid hormone acts on the more differentiated and committed cell.

Osteoclasts resorb bone by initially causing the release of mineral, then by degrading the demineralized bone matrix. Osteoclasts cause the release of mineral from bone, probably by lowering the pH under the ruffled border. An increase in hydrogen ions underneath the ruffled border causes dissolution of the bone mineral. In some senses the area underneath the ruffled border can be likened to a secondary lysosome. Recently, the presence of a protein on the ruffled border of chick osteoclasts, which cross-reacts with specific antisera to a 100 kdalton lysosomal membrane protein, has been described.[7] The mechanism of hydrogen-ion generation within osteoclasts is unknown, but it may be related to carbonic anhydrase activity. Carbonic anhydrase is present within osteoclasts, and activation of carbonic anhydrase could lead to hydrogen-ion generation. In individuals with a deficiency of one of the isoenzymes of carbonic anhydrase (Type II), osteopetrosis with incompetent osteoclastic bone resorption occurs.

The degradation of the bone matrix is probably due to the release of a family of proteolytic enzymes from the osteoclast or related cells. These enzymes include neutral collagenase as well as lysosomal enzymes. The precise relationship between collagenase and lysosomal enzymes remains unclear. Although collagenolytic activity is produced by resorbing bone cultures, collagenase released by bone cultures can be dissociated from bone resorption, and the only cell type in bone in which collagenase has been clearly shown is the osteoblast. In contrast, there is a close correlation between lysosomal enzyme activity in organ culture media and the presence of bone resorption. Lysosomal enzymes

Model for Osteoclast Formation

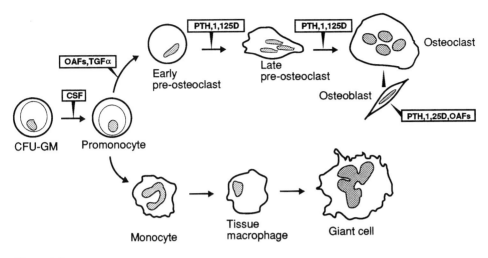

Figure 2.3

Factors probably involved in the formation of the osteoclasts. Osteoclast activating factors (OAFs) and TGFα act as growth factors on osteoclast progenitors. 1,25-dihydroxyvitamin D (1,25D) and PTH act as differentiation agents on committed progenitors. These factors also act indirectly on mature cells (see text for details).

such as cathepsin B-1, β-glucuronidase, β-galactosidase, cathepsin D, and N-acetyl β-glucoaminadase are associated with increased bone resorption.

It is possible that there are other mechanisms for enhancing matrix degradation. One of these is the generation of oxygen radicals. Recently it has been found that bone cultures stimulated to resorb with PTH or interleukin-1 produce oxygen radicals, and that inhibition of oxygen radical generation is associated with decreased bone resorption.[8]

Mechanisms of bone formation

The formation of bone is clearly mediated in different ways in different situations. The mechanism by which bone is formed in the adult during the normal remodeling sequence is probably different from the formation of bone during growth and development. It is also probably different from the formation of bone which occurs during the healing of fractures. Intramembranous bone formation is the process by which the bones of the skull and face form. This type of bone forms as a result of condensation of mesenchymal cells, replacement by bone cells, and the formation of bone independent of cartilage. Intramembranous bone formation also occurs along the periosteal border of long bones. In contrast, endochondral bone formation occurs on a cartilage anlagen. The appearance of cartilage cells precedes the appearance of bone. Cartilage cells hypertrophy, the matrix is invaded by blood vessels, and then the matrix is eventually replaced by marrow and bone. Caplan and Pechak[9] have suggested that endochondral bone formation is related to the blood supply via the nutrient artery in a similar way to muscle cell development. Endochondral bone formation is responsible for the elongation of long bones during growth, and it may be the mechanism involved in the repair of fractures.

During adult life bone formation may occur as appositional growth, which is probably analogous to intramembranous bone formation, and occurs in the periosteum and can occur in pathological situations on trabecular bone surfaces such as at the site of an osteoblastic metastasis, fluoride treatment, or following electrical stimulation. However, bone formation during adult life occurs predominantly as part of the bone-remodeling sequence.

Coupling—the relationship between osteoclastic bone resorption and bone formation

The integrity of the skeleton is maintained by the continued cellular remodeling of bone, which occurs throughout life and which is characterized by an orderly sequence of events, beginning with osteoclastic activation and followed by osteoblastic new bone formation to repair the defect caused by the osteoclasts. Osteoclastic bone resorption and osteoblastic new bone formation are closely linked or coupled, both chronologically and geographically. This process of bone remodeling occurs in discrete units, or packets, throughout the skeleton at any one time, and the weight of evidence strongly suggests that it is primarily under the control of local factors. Until recently very little has been understood about the nature or the source of the locally produced factors which regulate osteoclast and osteoblast activity to carry out this remodeling process. Immune cells (lymphocytes and monocytes) appear to be a likely source of local factors which regulate osteoclast activity.

Each of the focal and discrete packets of bone in the skeleton which are undergoing remodeling are referred to as bone-remodeling units in the terminology of Frost, who first described this sequence more than 20 years ago.[10] These units are geographically and chronologically separated from the other packets of remodeling. Since this has been a repeated observation of all histomorphometrists, this observation has led to the concept that activation of the remodeling sequence is under local control mechanisms—that is, it is caused by autocrine or paracrine factors generated in the bone micro-environment adjacent to the sequence. Clearly, this remodeling sequence is the key phenomenon in bone cell biology for the understanding of regulation of

trabecular bone volume, disorders of which are so important in the bone loss associated with aging and in the pathophysiology of osteoporosis.

Although the mechanism responsible for activation of the osteoclast, which is the initial event in the remodeling sequence, is still unknown, many of the local signals generated in the bone marrow micro-environment, and which could regulate osteoclast activity, are now known. These are discussed in more detail in Chapter 3. The phase of osteoclastic resorption which results in the Howship's lacunae or resorption bay lasts for about 10 days. Following the resorption phase the lacunae is occupied by mononuclear cells of uncertain etiology, while the resorption continues. Following the resorption phase, osteoblast precursors are attracted towards the site of the defect, replicate, and then begin to repair the defect. The formation phase of the remodeling sequence probably takes 3–5 months. It is possible to dissect out the cellular events involved in this phase. These should include chemotaxis of the osteoblast precursors towards the resorption defect, followed by their proliferation. In turn, activation of the osteoblasts must occur so that new bone is formed and mineralized. These events are described in detail by Baron et al, using the alveolar bone of the rat as a model.[7] The cellular events that were observed in this model are similar to those which occur in adult human bone.

This cellular sequence of events in bone remodeling occurs not only in physiological conditions, but also in most pathological bone diseases. Examples are primary hyperparathyroidism, hyperthyroidism, and Paget's disease, where an increase in osteoclast activity is followed by a related increase in new bone formation. In fact, in the normal individual new bone formation in the adult only occurs at sites where bone has recently been resorbed. In some pathological conditions, such as fluorosis or osteoblastic metastases, new bone may be laid down on surfaces which had not been previously resorbed, but these situations are rare.

The mechanisms responsible for coupling bone formation to previous bone resorption remain unclear. Several theories have been suggested. An early theory was that osteoclasts should disaggregate following the conclusion of the resorption phase, and should split into cells which are precursors of the osteoblasts.[11] However, it is now generally accepted that osteoclasts and osteoblasts have different origins. Osteoclasts are hematopoietic cells arising from hematopoietic stem cells in the bone marrow. In contrast, osteoblasts arise from the stromal mesenchymal cell system. Most workers in this area now believe that humoral mechanisms released during the resorption phase or at the conclusion of the resorption phase are responsible for triggering bone formation. This theory was first proposed by Howard et al[12] who identified a skeletal coupling factor present in demineralized bone matrix. Other factors which have been linked with bone formation and which are released during the resorption phase are discussed in more detail in Chapter 3. These include transforming growth factor β (TGF β) and bone-derived growth factor. Another concept is that the factor which stimulates bone resorption also has a delayed effect on osteoblasts to stimulate new bone formation. In contrast with these ideas, Jones and Boyde[13]

have suggested that the bone-forming cells which normally line bone surfaces will reline the bone surfaces at the conclusion of the resorption phase, and will be stimulated by the presence of the resorption defect to repair it, without the necessity of a specific humoral mediator or stimulus being involved.

The concept that bone formation is coupled to previous bone resorption by generation during the resorption phase of specific local humoral osteoblast stimulators (Figure 2.4) is favored by the author. There is now good evidence that such stimulators are produced, and that they act on osteoblasts. This is not to deny that the formation of bone may also be provoked by other factors, such as immune cell products or even the bone surface itself. As mentioned above, this is clearly one of the most important areas of research in the field of bone biology, since it may lead to insights into the disorders of bone volume which are responsible for the common metabolic bone diseases such as osteoporosis.

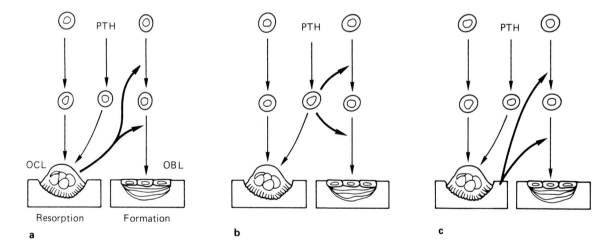

Figure 2.4

Hypothesis for events involved in bone remodeling and the generation of coupling factors during the remodeling process. There are three possible sources for coupling factors. They could be generated by osteoclasts to link bone formation to previous resorption (a), they could be generated by non-osteoclasts but other cells in the bone micro-environment (b), or they could be derived from the bone matrix itself (c). It is possible that each of these mechanisms is involved.[6]

Calcium exchange between the bone surface and the extracellular fluid

There is another mechanism of calcium transfer from bone to extracellular fluid, which is probably very important in the regulation of calcium homeostasis, but is much less well understood than the remodeling sequence. There is a disequilibrium between the state of calcium on the crystal surface of bone and calcium in the extracellular fluid, which is supersaturated with respect to calcium on the bone surface. As a consequence, there is a concentration gradient for calcium between the extracellular fluid and the mineral surface. This has been confirmed by autoradiography experiments, which show calcium fluxes across the cells which line the bone surface that is entirely separate from the remodeling sequence. It has been suspected for many years that the exchange of calcium which occurs between the mineral on the bone surface, the bone fluid, and the extracellular fluid is a major determinant of the extracellular fluid calcium (Figures 1.3 and 2.5). This diffuse transfer of calcium may be responsible in part both for the regulation of the set-point for plasma calcium and for the correction of deviations from the set-point. The general concept is that within the skeleton there is a special bone fluid which should be likened to the cerebrospinal fluid. This bone fluid is surrounded and enclosed by the lining cells which cover all bone surfaces, and which are continuous with the osteocyte canalicular system in the osteocyte lacunae. Ultimately, via this bone fluid, there is exchange between the mineral on the bone surface and the calcium in the extracellular fluid, and this exchange is an important determinant of the extracellular fluid calcium concentration. That exchange of calcium between bone and extracellular fluid across the lining cells occurs has been shown by studies which clearly show fluxes of calcium far greater than those produced by the remodeling sequence. Several theories have been proposed to account for the transport of calcium between bone and the extracellular fluid via the bone fluid, but it has not been possible to test these directly because it has not been feasible to sample the bone fluid and measure the concentration of the ions in the bone fluid directly.

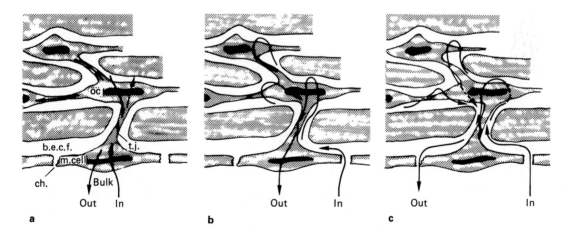

Figure 2.5

Possible ways in which fluid circulates between the bone cells (osteocytes, oc), the bone extracellular fluid (b.e.c.f.), and the bulk extracellular fluids. Tight junctions (t.j.) connect the osteocytes one with another, and with the mesenchymal cells (m.cell) lining the bone surface. (a) The situation in which both fluid and solute transfer into and out of bone through the mesenchymal cells. (b) The case in which fluid and solute enter bone by extracellular channels (ch.) between mesenchymal cells and return via transport within the cell syncytium. (c) The case in which transfer in and out of bone is primarily by extracellular flow.[11]

The bone membrane

The bone fluid is enclosed by cells which form an effective membrane covering the bone surface and across which calcium is transferred. The cells which comprise this bone membrane are separated only by tight or narrow gap junctions. On quiescent surfaces, which in the adult may account for 80–95 per cent of all bone surfaces, the cells which comprise the bone membrane are referred to as bone-lining cells or surface osteocytes. These cells are flattened, spindle-shaped, elongated mesenchymal cells lying in a sheet. Their long axes are parallel. These cells do not seem to be capable of large amounts of protein synthesis, since they contain very little rough endoplasmic reticulum. However, they almost certainly have receptors for PTH, since Jones and Boyde[13] have shown that they change shape (retract) when exposed to PTH. Morphologically, they resemble the osteocytes very deep within bone which are housed in separate osteocytic lacunae, and they communicate freely with the osteocytes within bone through the canalicular network of channels. The lining cells have not (so far as is known) been studied directly in vitro, although it is possible that the human bone cells grown in culture from bone explants are enriched in these cells which line the bone surfaces.[14] However, the responses and functions of these cells are likely to be vitally important to the understanding of the exchange of calcium between bone and extracellular fluid. They are probably cells in the osteoblast lineage, and they may represent osteoblasts which, like bone osteocytes, have undergone further differentiation.

At active bone surfaces the cells lining the bone are osteoblasts which are of uniform size and have extensive rough endoplasmic reticulum similar to that of osteoblasts in other sites. It is not entirely clear that the osteoblasts simply replace the bone-lining cells in the bone membrane, because the junctions between adjacent osteoclasts are not particularly tight, and the lining cells may still be seen covering osteoblasts at the bone surface overlaying the osteoblasts (Figure 2.4).[11]

The existence of a specific bone fluid was suggested for many years by Neuman and his associates, who were able to show by measurement of isotope exchange that there was a bone fluid that is rich in potassium and poor in sodium and calcium. This bone fluid circulates between the bone membrane and the bone surface throughout the canalicular network. It is likely that this fluid compartment between the cells and the bone surface involves a true minicirculation within the canalicular–lacunar system in which the extracellular fluid both enters and leaves the system between the gap junctions in the bone membrane, as suggested by Rasmussen and Bordier.[11] The exact relationship between the bone membrane, the lining cells, and the cells involved in the remodeling process remains unclear.

How, then, is calcium transferred between the bone surface and the extracellular fluid? Several workers have suggested that it could be transported by active transport. Talmage suggested that calcium was pumped across the lining cells by an active excretion process away from the bone surface to the extracellular fluid.[15] Neuman and Scarpace suggested that calcium moved across the lining cell into the extracellular fluid because the lining cell actively pumped potassium into bone from the extracellular fluid, and that calcium movement was passive and along an electrochemical gradient in the opposite direction.[16] However, there may be other mechanisms by which calcium moves from the bone surface to the extracellular fluid independent of these forms of ion transport. Calcium at the bone surface may be rendered more soluble and therefore more readily exchangeable with the calcium in the extracellular fluid. If this was so, the critical transport process would be between the mineral, the bone surface and the bone fluid rather than the bone fluid and the extracellular fluid. A change in solubility of the normal mineral phase at the bone surface could be accomplished by hydrogen ion production by the lining cells. Acidification increases the solubility of bone mineral, and there is evidence that cells in the osteoblast lineage are effective producers of acid, particularly under the direction of PTH. It is possible that the bone surface has variable mineral phases and mineral composition. The increase in the solubility mediated by the lining cells may also be regulated by the production of other proteins by the lining cells.

Recently, using a high-resolution scanning ion microprobe, Bushinsky et al[17] have examined the ionic composition of bone at the mineral surface and deeper within the bone itself in organ cultures of mouse calvariae, and compared living and dead bone. They found that there are differences in ionic composition, and particularly in calcium, between the mineral at the bone surface and the mineral deeper within the bone. The surface of bone is enriched in sodium and potassium relative to calcium, but there is progressively more calcium deeper within bone. This raises the possibility of regulatory processes controlling the mineral phase at the bone surface. When these organ cultures were devitalized, these differences between the mineral deeper within bone and the mineral at the bone surface disappeared. These results suggest that there is a living membrane between the bone fluid and the bone surface, possibly represented by the bone lining cells described above. This is reinforced by the fact that the extracellular fluid is one million-fold supersaturated with respect to the bone surface, so that there must be a barrier which prevents egress of calcium from the plasma to the bone surface.[18] These studies show definitive evidence that a 'bone membrane' in fact exists.

The interested reader is referred to reviews by Parfitt for more details and original citations.[19]

If the process of exchange between the bone surface and the extracellular fluid is as important as it seems to be, then it is possible that it is to some extent hormone regulated and to some extent hormone independent. It is a potential mechanism for the regulation of the set-point when the parathyroid glands are absent, as well as error correction which occurs in the absence of PTH (Chapter 1).

The hormones which regulate transport of calcium and how they act

Parathyroid hormone, calcitonin, 1,25-dihydroxyvitamin D, and other factors which influence calcium transport are discussed in Chapter 3.

Hormonal factors which influence calcium homeostasis

Calcium transport across bone, kidney, and the gut is modulated by local and systemic factors. The concentrations of these factors in their target-cell environment are regulated by the extracellular fluid calcium concentration in the case of the systemic hormones PTH and calcitonin, and the vitamin D metabolites. There are no other hormonal systems which are known to be directly influenced by the extracellular fluid calcium, although some of the local factors are now being discovered, and the control mechanisms which regulate their secretion are still largely uninvestigated. Recent information indicates that some of the local factors produced in the bone microenvironment and which stimulate bone resorption also increase the serum calcium. The best example is interleukin-1. The aim in this chapter is to discuss first the systemic hormones which are known to influence calcium homeostasis, and then to review other factors which are known to affect calcium fluxes, since many of these are clearly involved in disorders of calcium homeostasis.

Parathyroid hormone (PTH)

Parathyroid hormone is an 84 amino acid single-chain polypeptide which is secreted by the chief cells of the parathyroid glands. The PTH gene is a simple gene on chromosome 11, and it contains one intron which separates the 5' from the 3' portion of the gene and the 3' non-coding end. Cloning of the bovine, rat, and human complementary-desoxyribonucleic acid (cDNA) for the parathyroid gene has been achieved in recent years,[1] and has allowed definitive determination of the amino acid sequence, a point of some controversy 10 years ago. The hormone is synthesized on the large endoplasmic reticulum as a precursor of 115 amino acids and 13 000 daltons.[2,3] Its precursor form is known as pre-pro-PTH. The gene encodes a signal peptide which consists of the leader sequence essential for translocating the peptide to the endoplasmic reticulum and through the cisternal space. There it is cleaved by specific proteases to enter the Golgi apparatus. In the rough endoplasmic reticulum it is processed to pro-PTH which consists of 90 amino acids and has a molecular weight of 10 000 daltons.[2] It is cleaved again in the secretory granules to the mature 1–84 form of the hormone which is the major form of the molecule secreted by the parathyroid glands. The biosynthetic pathway is shown diagrammatically in Figure 3.1, and Figure 3.2 shows the amino acid sequence of the hormone.

The primary secretagogue for PTH is the extracellular fluid calcium concentration.[4] Parathyroid hormone is secreted in response to a fall in the ionized calcium concentration bathing the parathyroid gland cells. The mechanism by which calcium controls the synthesis and secretion of PTH remains unclear. Of course, stimulation of hormone release in response to a decrease in

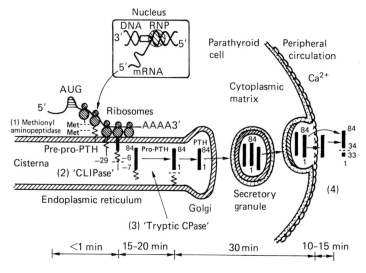

Figure 3.1

Scheme representing the biosynthetic pathway for PTH synthesis and secretion. (Reproduced from Habener et al, Biosynthesis of parathyroid hormone, *Recent Prog Horm Res* (1977), **35**: 249–308)

calcium concentration, and conversely a decrease in PTH release in response to an increase in the extracellular ionized calcium concentration, is in contrast with the effects of calcium on most hormone secretory systems. In vitro and in vivo studies suggest that there are many other secretagogues for PTH, including adrenergic agonists, prostaglandins of the E-series, magnesium, and vitamin D metabolites.[4] Parathyroid hormone secretion is inhibited by chronic severe magnesium depletion and by 1,25-dihydroxyvitamin D. However, the physiological significance of these factors for either the stimulation or the suppression of PTH release from the parathyroid glands (other than calcium) is not known.

The 1–84 PTH peptide secreted by the parathyroids (frequently referred to as the 'intact' molecule) is metabolized extensively in the peripheral tissues, particularly in the kidneys, liver, and bone.[2] The biological activity is contained within the first 34 amino acids of the N-terminal end of the molecule. This complete sequence (1–34) is required for all biological

effects which are currently known. Circulating fragments representing the C-terminal end of the molecule are biologically inert. However, the C-terminal fragments depend on the kidneys for clearance, and these fragments are relatively immunoreactive compared with fragments of the N-terminal region. This has provided numerous difficulties with radioimmunoassays for PTH over the past 20 years. By and large, measurement of N-terminal fragments reflects minute-to-minute secretion by the gland and the assays which measure the C-terminal end give an index of chronic PTH secretion, provided renal function is normal (Chapters 4 and 8).

Biological effects of PTH

Parathyroid hormone regulates the transport of calcium between the extracellular fluid and bone, kidney, and the gut.

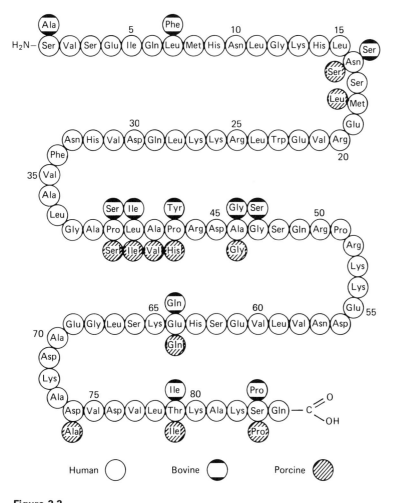

Figure 3.2

Complete amino acid sequence of human PTH.

Effects of PTH on bone

Parathyroid hormone has a direct effect on bone turnover, increasing rates of both bone resorption and new bone formation. All cells in bone (osteoclasts, osteoblasts, and osteocytes) are affected. The precise mechanisms by which PTH increases bone turnover and how it interacts with bone cells remain to be clarified. The major effects appear to be stimulation of osteoclast activity, followed by an increase in new bone formation. However, it is possible that neither of these effects is mediated directly by PTH. It is more likely that complex interactions between other cells and the bone cells, or even between PTH and other mediators, are necessary for these effects to be observed.

Effects of PTH on osteoclastic bone resorption

Parathyroid hormone clearly stimulates osteoclastic bone resorption both in vivo and in vitro.

The first observations in vivo were made by Barnicot,[5] who found that when parathyroid glands were grafted into the subcutaneous tissue adjacent to rodent calvariae there was a marked increase in adjacent osteoclast activity. Similar increases in osteoclast activity may be found in patients with primary and secondary hyperparathyroidism. These results were confirmed by the observations made by Gaillard[6] and later by Raisz[7] in organ culture systems. In organ cultures of fetal rat long bones or neonatal mouse calvariae, incubation of the organ cultures with PTH leads to a marked increase in osteoclast number and activity, an increase in the ruffled border area of the osteoclast, release of calcium from the bone, and the degradation of the bone matrix associated with a release of lysosomal enzymes and collagenase.[8]

The precise mechanism by which PTH stimulates osteoclast activity is not known. It appears likely that the effects on mature multinucleate cells and cells late in the osteoclast lineage are mediated through an intermediate cell, possibly a cell in the osteoblast lineage,[9] although Perry et al[10] have suggested that an immune cell may also be involved. Presumably PTH acts on this intermediate cell to cause the release of a soluble mediator which, in turn, is responsible for osteoclastic resorption. Using the modified Dexter marrow culture system, PTH has been found also to cause increased formation and differentiation of mature multinucleated cells with osteoclast characteristics from progenitors.[11] Kurihara et al[12] have found that these effects of PTH are direct, and not mediated through intermediary cells. Similar observations have been reported by Kumegawa's group.[13] The effects of PTH on bone resorption are presumably different from those of growth factors such as epidermal growth factor, transforming growth factor α (TGFα) and the osteotropic cytokines (see below), because they are not dependent on cell replication. This was demonstrated by Lorenzo et al,[14] who found that agents which block cell replication have no effects on bone resorption mediated by PTH.

The relationship between PTH, cAMP generation, and bone resorption is still unclear.[15] When bone cultures are incubated with PTH, there is an increase in adenylate cyclase activity and cAMP content of the bone cultures. However, it is possible that this increase in cAMP is not related to the resorption process, since similar changes

occur with calcitonin. The precise molecular mechanism by which osteoclasts stimulated by PTH resorb bone remains uncertain, but it is probably similar to the molecular mechanisms involved in osteoclastic bone resorption mediated by the other known bone resorbing factors. These responses include the production of acid and proteolytic enzymes under the ruffled border of the osteoclast to cause release of mineral from the bone and degradation of the bone matrix.

Effects of PTH on cells in the osteoblast lineage

Parathyroid hormone stimulates bone turnover and causes an increase in bone formation in vivo. In patients with primary hyperparathyroidism there is a marked increase in rates of bone turnover, in bone formation, and in marrow fibrosis. However, the effects on cells with the osteoblast phenotype in vitro are different from those seen in vivo. Parathyroid hormone causes cells lining bone surfaces to retract, and Jones and Boyde[16] have suggested that this allows access of the osteoclast to the bone surface so that it can resorb bone. This change in shape of bone cells in response to PTH has been noted by a number of workers. With the use of labeled PTH and autoradiography, it has been shown that parathyroid hormone receptors are present on spindle-shaped cells in the intertrabecular space of the metaphyseal region of rat long bones.[17] These cells were distinct from mature osteoblasts. These cells possibly represent a cellular target for the effects of PTH on bone resorption at this site. When PTH is added to osteosarcoma cells with the osteoblast phenotype, it causes a decrease in the synthesis or content of proteins which represent differentiated function of the osteoblasts. For example, when PTH is incubated with cultured rat osteosarcoma cells, there is a decrease in alkaline phosphatase content and a decrease in Type I collagen synthesis. Similarly, when PTH is added to organ cultures of fetal rat calvariae, there is a decrease in bone collagen synthesis.[18] This is in contrast with what is observed in vivo. The explanations for these differences are unclear. It may mean that the time course required for in vitro experiments is insufficiently long to observe the effects that are seen in the in vivo situation. Alternatively, the overall

Table 3.1   Effects of PTH on bone.

Osteoclasts
   stimulates osteoclastic bone resorption
   causes differentiation of committed precursors
   activates mature multinucleated osteoclasts (indirectly)

Osteoblasts
   causes retraction of bone lining cells
   inhibits differentiated osteoblast function in vitro (acutely)
   enhances rates of bone turnover in vivo

effects of PTH on the bone remodeling sequence in vivo may be mediated by secondary coupling factors released during the resorption process, which override the direct suppressive effects of PTH on cells with the osteoblast phenotype. The generation of coupling factors could lead to chronic stimulation of differentiated osteoblast function and subsequent new bone formation.

The effects of PTH on bone may be dependent on other hormones. It has long been realized that PTH is less effective on bone in the absence of 1,25-dihydroxyvitamin D. Thus, patients with secondary hyperparathyroidism and chronic renal failure remain hypocalcemic. It has been shown in rats that PTH and 1,25-dihydroxyvitamin D work in concert to elevate the plasma calcium by stimulating bone resorption.[19] Other factors such as epidermal growth factor and interleukin-1 may also enhance the effects of PTH on bone.

In summary, the effects of PTH on bone turnover are complex (Table 3.1) and still not completely understood. It is clear that PTH increases osteoclast number and activity and in vivo causes an increase in new bone formation, but the precise mechanisms by which these effects are mediated and the relationship between the effects of PTH on osteoblasts and osteoclasts remains to be elucidated.

Effects of PTH on the kidney

The major effect of PTH on the kidney is to increase the distal tubular reabsorption of calcium.[20] About 10 per cent of the filtered load of calcium is reabsorbed in the distal tubules, and

this is tightly controlled by PTH. However, in patients with primary hyperparathyroidism, there is an increased filtered load of calcium so that, although distal tubular calcium reabsorption is enhanced, urinary calcium excretion may still be above normal. However, it is not as high as it would be in the absence of PTH. Current data suggests that the effects of PTH on the renal tubules are enhanced by 1,25-dihydroxyvitamin D.[21] The relationship of the effects of PTH on renal tubular calcium reabsorption to the generation of cyclic nucleotides remains unclear. Parathyroid hormone clearly has other important effects on the renal tubules, including the inhibition of renal tubular reabsorption of phosphate and inhibition of bicarbonate reabsorption in the proximal tubules. These effects are responsible for increased fractional excretion of bicarbonate and increased fractional excretion of phosphate in patients with primary hyperparathyroidism.

Effects of PTH on the gut

Parathyroid hormone probably has no direct effect on the gastrointestinal tract. Nevertheless, patients with primary hyperparathyroidism have an increase in the absorption of calcium from the gut. This effect is mediated by 1,25-dihydroxyvitamin D which is synthesized in the kidneys in response to increased circulating PTH concentrations.

Active vitamin D metabolites

The vitamin D metabolites are steroid-like compounds which are derived from the plant ergosterol and ingested in the diet, or are synthesized in the skin by exposure of epidermal cells to ultraviolet light.[22–24] The structure of the vitamin D metabolites is shown in Figure 3.3 and the metabolic pathways in Figure 3.4. The precursor in the skin is 7-dehydrocholesterol which, on exposure to ultraviolet light, forms previtamin D_3 through steps which are not enzyme controlled. The dietary form is absorbed in the small intestine and, like the form synthesized in the skin, is transported in the plasma bound to carrier proteins.

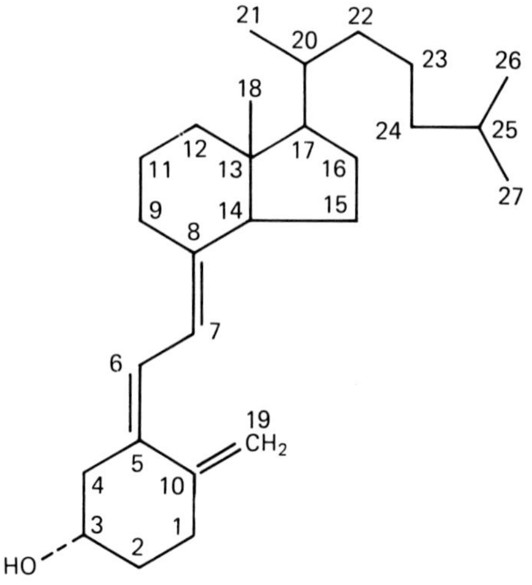

Figure 3.3

Structure of vitamin D_3 (cholecalciferol). Cholecalciferol is formed in the skin by irradiation of 7-dihydrocholesterol. Ergocalciferol is formed by irradiation of ergosterol, which is derived from plants and is ingested in the diet. Ergocalciferol (vitamin D_2) differs from cholecalciferol (vitamin D_3) by having a double bond between the C22 and 23 positions and a methyl group at C24. Dihydrotachysterol does not have a double bond between C10 and C19. This results in rotation of the ring so that the hydroxyl group at C3 corresponds to the C1 position, and dihydrotachysterol does not require renal hydroxylation for activity.

In the plasma, vitamin D is transported by lipoproteins, albumin, and the specific D-binding protein (DBP).[25] This DBP is related to the Gc-complex and has recently been molecularly cloned.[26–28] The DBP has high affinity and high capacity for the sterol, and binds it preferentially. It is still not known whether DBP is merely a transport protein or whether it is also involved with entry of the vitamin D metabolites into the cell. It is now possible to measure free-serum levels of 1,25-dihydroxyvitamin D, which are in the range of less than 0.05 per cent of the total circulating concentration.[29]

As circulating vitamin D reaches the liver, it is hydroxylated in the 25-position by an enzyme system which is located mainly in the liver microsomes, although there is also nonspecific enzyme activity in the mitochondria. 25-Hydroxylase activity is not confined solely to the liver, but is also found in the gut and kidney. The 25-hydroxylation reaction in the microsomes requires reduced nicotinamide–adenine dinucleotide phosphate (NADPH), molecular oxygen, and magnesium, and is probably mediated by the cytochrome p450 system. It does not appear to be tightly controlled, and it is not influenced by the ambient calcium, phosphate, or PTH concentrations, or by the vitamin D status. However, it is influenced by anticonvulsants, which presumably decrease 25-hydroxylase activity by means which are still not entirely clear. Serum 25-hydroxyvitamin D concentrations do depend to some extent on the availability of the substrate, and there is some degree of substrate inhibition. 25-Hydroxyvitamin D is the major circulating form of the hormone and possibly also the major storage form of vitamin D. It is weakly active in its own right, and it has been shown both to resorb bone and to increase gut absorption of calcium. It is about three orders of magnitude less potent than 1,25-dihydroxyvitamin D.[30]

25-Hydroxyvitamin D is metabolized (and activated) by hydroxylation in the 1-position in the kidney. There has been some debate over whether extrarenal hydroxylation of 25-hydroxyvitamin D occurs, although it is clear from in vitro systems that extrarenal synthesis of 1,25-dihydroxyvitamin D can occur in the immune cells of patients with sarcoidosis, with certain lymphomas, and also in the normal placenta.[31–34] The 1-hydroxylase system in the kidney is present in the mitochondria, and it is found in the proximal convoluted tubules and proximal straight tubules. The enzyme in the proximal convoluted tubules responds to PTH and cAMP, whereas the enzyme in the proximal straight tubule responds to calcitonin. This enzyme system has been difficult to study in mammals, and the precise regulation of 1-hydroxylase activity in the kidney still remains controversial. However, the kidney is clearly the major site of regulation of vitamin D activity in normals. The major affectors appear to be the ambient PTH concentration and low phosphate concentrations. Other stimuli are low calcium, vitamin D deficiency, calcitonin, growth hormone, prolactin, and estrogen. 1-Hydroxylase activity is inhibited by 1,25-dihydroxyvitamin D, hypercalcemia, and parathyroidectomy, and increases in the ambient phosphate concentration.

The receptor for 1,25-dihydroxyvitamin D has been fully characterized.[35] It is a 3.2 S protein which is present in the nucleus and which binds 1,25-dihydroxyvitamin D with high affinity and low capacity. It is approximately 55 000 daltons in size. The binding of 1,25-dihydroxyvitamin D to this receptor causes a conformational change which leads to binding to specific DNA sites, and to subsequent production of specific calcium-transport proteins. The receptor for vitamin D has considerable homology to the v-erb-A oncogene, as well as to the other steroid hormone receptors, including the glucocorticoid receptor, the estrogen receptor, the progesterone receptor, the thyroid hormone receptor and the retinoic acid receptor.[36,37,38] Studies of the receptor are of clinical interest because there is now evidence that the receptor is important in a number of disease states. In vitamin D-dependent rickets type-II, in which there is target tissue resistance

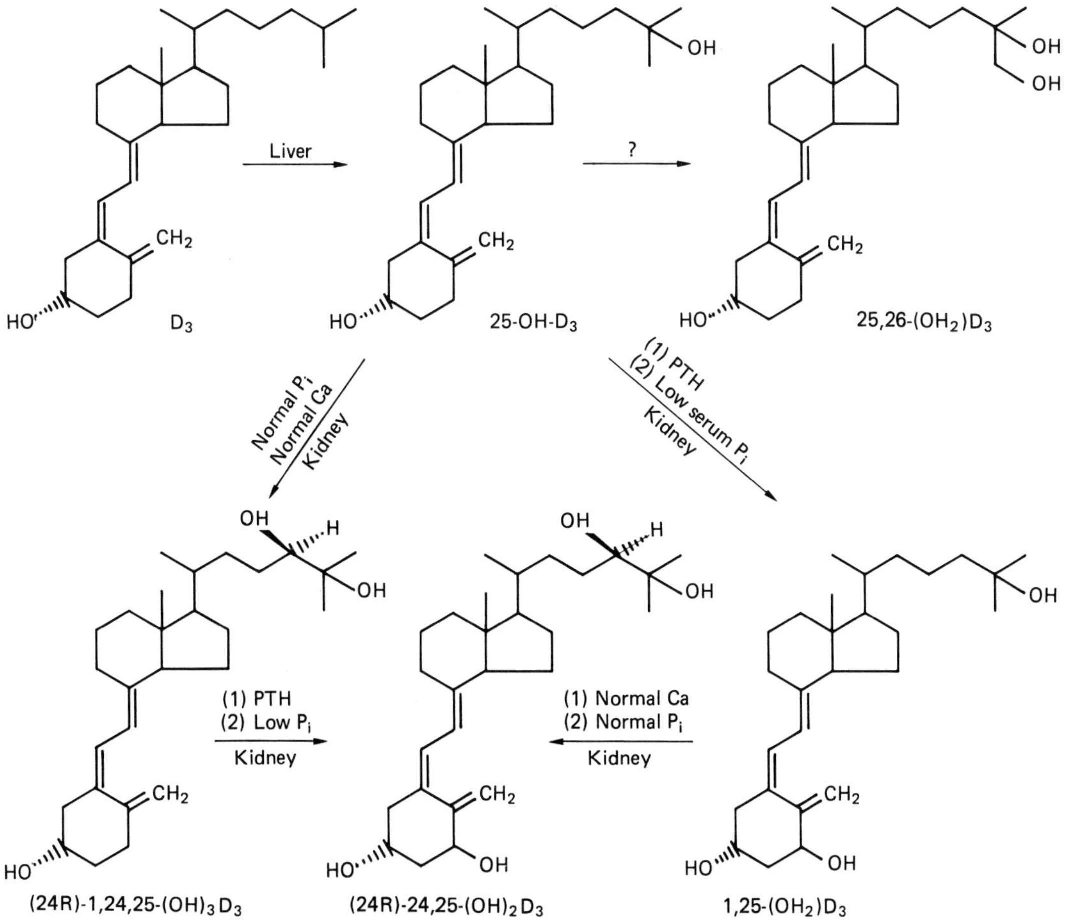

Figure 3.4

Metabolic pathway of vitamin D metabolism and activation.

to 1,25-dihydroxyvitamin D and the serum levels of this hormone are increased,[39] there are at least several varieties showing defects in the receptor protein and subsequent 1,25-dihydroxyvitamin D resistance. In some cases, these abnormalities have been identified as single amino acid substitutions in the receptor protein in the critical regions where the receptor binds to DNA (the zinc fingers).[40]

Biological effects of vitamin D metabolites (Table 3.2)

The major biologically active metabolite of vitamin D is 1,25-dihydroxyvitamin D. 25-Hydroxyvitamin D is also biologically active, but is three orders of magnitude less potent than 1,25-dihydroxyvitamin D. There has been an argument for some years over whether 24,25-dihydroxyvitamin D has any direct biological activity, and this issue still has to be resolved. 1,25-Dihydroxyvitamin D increases the absorption of calcium and phosphorus from the gut by active transport. It also increases bone resorption, although its effects on bone resorption appear to be relatively unique, since it causes differentiation of committed progenitors into mature cells (Table 3.2).[41] The relative significance of the effects on bone compared with the effects on gut as far as calcium homeostasis is concerned remain unclear. The increased synthesis of 1,25-dihydroxyvitamin D leads to increases in the serum concentrations of calcium and phosphorus, whereas lack of vitamin D causes failure

Table 3.2 Biological effects of 1,25-dihydroxyvitamin D on bone.

Osteoclasts
  stimulates osteoclastic bone resorption
  increases osteoclast formation
  causes fusion of committed osteoclast precursors
  activates mature multinucleated osteoclasts in vitro
Osteoblasts
  enhances bone mineralization
  modulates expression of osteoblast collagen synthesis

of mineralization of newly formed bone matrix, a disease which in adults is known as osteomalacia and in children as rickets, and which is characterized by impaired mineralization of the growth plate. It is likely that the major effects of 1,25-dihydroxyvitamin D on the gut and on bone are to increase the supply of mineral which is available for the mineralization process.[42] Thus, increased absorption of calcium from the gut and increased bone resorption will lead to an increase in calcium and other minerals at the remodeling site, which are then available for the normal mineralization of the newly formed bone matrix.

Recently there have been some exciting new advances in understanding the actions of 1,25-dihydroxyvitamin D on cells in the bone marrow micro-environment. 1,25-Dihydroxyvitamin D clearly acts as a fusogen for cells of the monocyte–macrophage lineage, and as a factor which aids differentiation of certain leukemic cell lines which also belong to the monocyte–macrophage lineage.[43,44] It has been found that 1,25-dihydroxyvitamin D stimulates the fusion and differentiation of cells with the characteristics of osteoclast progenitors into mature cells.[41] However, these effects are not specific for cells of the osteoclast lineage, since similar effects are seen with peripheral blood monocytes. Support for a potential role for 1,25-dihydroxyvitamin D as an osteoclast differentiation agent was provided by studies in an infant with malignant osteopetrosis, a disorder where osteoclasts failed to form normally, and where treatment with 1,25-dihydroxyvitamin D led to the production of active bone-resorbing osteoclasts and formation of a normal bone marrow cavity.[45]

1,25-Dihydroxyvitamin D also has effects on immune cell function, which could influence local factors which regulate osteoclast activity. It inhibits the production of interleukin-2 by normal activated lymphocytes, which causes impairment of mitogenesis.[46] Moreover, production of interleukin-1 by cells in the monocyte–macrophage lineage is enhanced by 1,25-dihydroxyvitamin D.[47] Recently it has been found that immune cells stimulated with phytohemagglutinin produce 1,25-dihydroxyvitamin D,[48] and others have shown that these cells develop 1,25-dihydroxyvitamin D receptors.[49]

1,25-Dihydroxyvitamin D probably works on bone in concert with PTH as discussed above, in the section on PTH. It also has direct effects on the kidney, although its renal effects are almost certainly less important than those on gut and the bone. In the kidney it increases 24-hydroxylase activity, which decreases the production of 1,25-dihydroxyvitamin D.[22] It also works in concert with PTH to enhance calcium reabsorption, and in opposition to PTH to enhance phosphate reabsorption. However, its effects on phosphate reabsorption are relatively weak and are overridden by the direct effects of PTH.

The effects of 1,25-dihydroxyvitamin D on the gut are to stimulate the absorption of calcium and phosphate, as discussed in more detail in Chapter 2.

Recently, Ishizuka et al,[50–57] have described a vitamin D metabolite which is present in human serum at concentrations of 4×10^{-10} M and has different effects from the other vitamin D metabolites. This metabolite, the 23(S)25(R)-1,25-dihydroxyvitamin D$_3$-lactone, opposes the effects of 1,25-dihydroxyvitamin D on bone. Thus, this metabolite inhibits bone resorption in organ culture and inhibits the formation of cells with osteoclast characteristics. It appears also to inhibit bone resorption and osteoclast formation stimulated by parathyroid hormone and interleukin-1, suggesting that its effects are not mediated through the vitamin D receptor. Not all investigators believe this metabolite is present under physiological conditions, and determination of the importance of this very interesting metabolite in bone metabolism and calcium homeostasis awaits further investigation.

Calcitonin

Calcitonin is a 32 amino acid peptide which is synthesized and secreted by the parafollicular cells of the thyroid gland.[58] These cells are found in the ultimo-branchial gland of birds, fish, and reptiles, where the parafollicular cells form a separate organ. Calcitonin has a particularly complex gene structure.[59,60] The gene is composed of six exons and five introns, and there is evidence that in different tissues which express this gene, there is differential splicing of the transcribed gene to produce different peptides (Figure 3.5). In the thyroid, the major product of the calcitonin gene is calcitonin, whereas in the brain the major

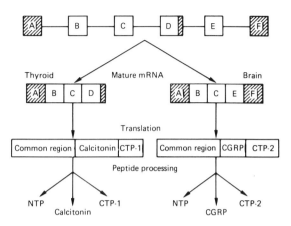

Figure 3.5

Structure of the calcitonin gene. This gene can undergo alternative tissue-specific ribonucleic acid (RNA) processing. In the thyroid, calcitonin is preferentially expressed. In the brain CGRP is preferentially expressed. A, B, C, D, E and F represent exons. Exons which are noncoding are hatched. NTP = N-terminal peptide; CTP = C-terminal peptide.[60]

product is the calcitonin gene-related peptide (CGRP). Calcitonin is secreted as a larger precursor called procalcitonin. This is of 17 500 daltons in man, and it contains a classical leader sequence. The 32 amino acid secreted peptide has a proline-amide group at the C-terminus, which for some years made the chemical synthesis of biologically active calcitonin extremely difficult. There is also a disulfide bridge at the N-terminal end of the molecule, between amino acids 1 and 7. Both the disulfide bridge and the proline-amide group are necessary for biological activity.

The circulating form of calcitonin is a single-chain polypeptide of 32 amino acids.[58] Metabolism occurs predominantly in the kidney, and the plasma half-life is approximately 5 min. The major regulator of calcitonin secretion is the extracellular fluid calcium concentration. Calcitonin is released from the parafollicular cells in response to an increase in the extracellular fluid calcium concentration, and is inhibited by a fall in the ambient calcium concentration. Calcitonin is also released in response to gastric hormones such as gastrin, glucagon, secretin, and cholecystokinin-pancreozymin (CCK-PZZ). The most potent secretogogue is gastrin. Calcium and pentagastrin are used, either alone or together, as provocative tests to determine the capacity of a patient to secrete calcitonin.

The precise biological role of calcitonin remains unclear. It inhibits osteoclastic bone resorption acutely, rapidly, and effectively.[61] It acts directly on the osteoclasts to cause an increase in cAMP generation and contraction of the osteoclast cell membrane.[62,63] It probably causes the dissolution of osteoclasts into mononuclear cells.[64] It may also decrease the formation and replication of preformed osteoclasts. However, its effects are probably transient, and it may have little effect in the control of trabecular bone volume or in calcium homeostasis chronically (Figure 3.6). Patients who are athyroid or patients who have tumors of calcitonin-secreting cells (medullary thyroid carcinomas) have no major abnormalities in either bone histomorphometry or calcium homeostasis.

It remains unclear why osteoclasts escape from the effects of calcitonin. The effect is seen both in vitro and in vivo,[65,66] and it could be due to downregulation of calcitonin receptor number or binding in the continued presence of calcitonin or, alternatively, to the emergence of a population of osteoclasts which are not sensitive to calcitonin. The teleological role for a hormone with such a transient effect has confounded bone physiologists. The most fashionable hypothesis is that calcitonin is necessary during growth, to protect the body against postprandial hypercalcemia and to reduce bone turnover following a calcium-containing meal. Calcitonin has no major effects on bone formation, vitamin D metabolism, or absorption of calcium from the gut. Recently, there have been suggestions that it may stimulate proliferation in osteoblast-like cells[67] and Burns and colleagues[68] have suggested that procalcitonin also has the capacity to enhance osteoblast-like cell proliferation. Whether these effects seen in vitro have relevance to bone formation in vivo remains to be demonstrated. Receptors for calci-

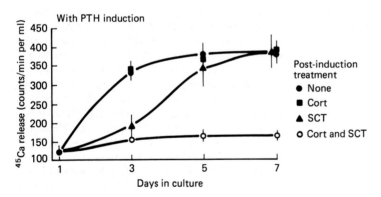

Figure 3.6

Effects of calcitonin and corticosteroids on calcium release from bone in organ culture. Parathyroid hormone stimulates previously incorporated ^{45}Ca from fetal rat long bone cultures. This stimulation of bone resorption is inhibited by salmon calcitonin (100 mU/ml). However, bone resorption is inhibited only transiently and the bones soon escape from the effects of calcitonin and the resorption increases again. When small doses of cortisol (10^{-6}M) are added with calcitonin, this escape from inhibition of bone resorption is prevented.[65] Similar observations are seen in vivo when patients with hypercalcemia are treated with calcitonin. (Reproduced from Raisz et al, Effect of glucocorticoids on bone resorption in tissue culture, *Endocrinology* (1972), **90**: 961–7.)

tonin are widespread in many tissues, but it is probable that the presence of calcitonin receptors in most of these tissues has no physiological significance. Calcitonin receptors are also found on many tumor cells, and may reflect dedifferentiation.

CGRP is produced from the calcitonin gene by alternative splicing.[59] Its major source is the brain. It does have an effect on bone which is similar to that of calcitonin, although it appears to be about three orders of magnitude less potent. There does appear to be some overlap of effects with calcitonin, although there clearly are also effects which are independent of the calcitonin receptor. Thus, its effects cannot be explained simply by binding to and activating the calcitonin receptor. Its major effects appear to be as a vasodilator which inhibits arterial contraction, in contrast with calcitonin. It is effective when given either intra- or extraluminally. The vasodilator effects are probably mediated by CGRP released from nerves around the vasculature. In vivo it causes a tachycardia and a fall in blood pressure. Like calcitonin, it can be produced by nonthyroid tumors and has some potential as a tumor marker. It is a weak inhibitor of bone resorption in vitro.[69]

Other hormones which affect calcium homeostasis

There is a series of other hormones which affect calcium homeostasis, but which do not appear to be under feedback regulation by the circulating concentration of calcium. These are considered in more detail with the appropriate disease states. Most of these factors are active on bone. They may be more important in the regulation and control of bone volume than in calcium homeostasis.

Osteoclast activating factor

Osteoclast activating factor (OAF) was the term used in the past to refer to the bone-resorbing activity produced by normal lymphocytes and monocytes when they are stimulated by an anti-

gen to which they have previously been exposed, or by phytohemagglutinin. It is not one molecule. The OAF activity represents the net effects of a family of molecules released by immune cells (also called cytokines), which can either stimulate or inhibit osteoclastic bone resorption. The cytokines which are currently known to influence bone resorption are interleukin-1, lymphotoxin, and tumor necrosis factor, all of which have recently been purified and molecularly cloned.

Interleukin-1

Interleukin-1 is the biological activity produced by monocytes which stimulates thymocyte proliferation. Its biological activity is composed of at least two separate proteins encoded by separate single copy genes in man. Interleukin-1 was first shown to stimulate bone resorption by Gowen et al.[70] This work was later confirmed by Heath et al[71] who showed that homogenously pure porcine interleukin-1 could resorb bone and then again by Gowen and Mundy,[72] who showed that recombinant murine interleukin-1α and β resorb bone in vitro. Interleukin-1 is capable of resorbing bone in very low concentrations, as low as 10^{-12} M (Figure 3.7). Interleukin-1 probably represents the major form of OAF present in short-term cultures of phytohemagglutinin-stimulated peripheral blood leukocytes.[73] Its effects on mature osteoclasts are not direct, but rather are mediated by other cells, including cells in the osteoblast lineage. However, in addition to these effects interleukin-1 also acts as a growth factor on osteoclast precursors. The effects of interleukin-1 on bone turnover and on calcium homeostasis have also been examined using intermittent injections in normal mice.[74,75] Following daily injections of interleukin-1α or β for 3 days, there is a marked increase in osteoclast activity which persists for 10 days.[74] This occurs in bones adjacent to the site of injection, and also in distant bones. However, there is a difference between the local effect and the distant effects. Locally, there is accumulation of chronic inflammatory cells and this accumulation of chronic inflammatory cells is inhibited by treatment of the mice with indomethacin. There is no accumulation of chronic inflammatory cells at distant sites. Thus, it appears that this accumulation of chronic inflammatory cells is related to prostaglandin synthesis.

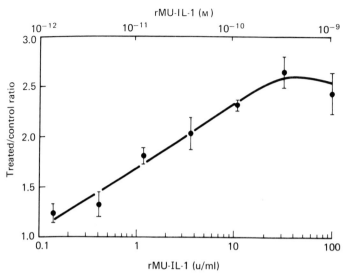

Figure 3.7

Stimulation of bone resorption in organ cultures in neonatal mouse calvariae by recombinant murine interleukin-1. These data represent means and standard errors for quadruplicate determinations. Bone resorption is determined by treated to control ratios with release of previously incorporated ^{45}Ca.[72]

There is also an increase in the plasma calcium. However, despite this progressive increase in the plasma calcium, there is a transient fall in plasma calcium for several hours after the bolus injection.[75] This fall in plasma calcium is inhibited by indomethacin, and therefore appears likely to be prostaglandin mediated. It causes a marked increase in the serum calcium when infused in vivo,[76] and has been linked to hypercalcemia associated with certain forms of malignancy.

Interleukin-1 also has effects on bone-forming cells. It stimulates the proliferation of human bone cells, some of which have the osteoblast phenotype (Figure 3.8), but inhibits collagen synthesis in fetal rat bones cultured for 48 h.[77] However, low-dose pulse injections may lead to stimulation of bone collagen synthesis in longer cultures. When given by intermittent injections in vivo, as noted above, initially there is an increase in osteoclastic bone resorption.[74] However, after a number of days, osteoclasts disappear and there is an influx of bone-forming cells and repair of the resorbed bone by the formation of new bone. This new bone is initially woven and over a period of 28 days completely replaces the bone removed by the activity of osteoclasts. It is likely that this effect is related to withdrawal of interleukin-1 and may be due to other factors released as a consequence of the resorption process. When interleukin-1 is continuously infused, osteoblast activity is suppressed.[74,76]

Tumor necrosis factor

Tumor necrosis factor is an 18 000 dalton peptide which is released by activated macrophages. It has recently been purified, cloned and expressed in *Escherichia coli*.[78] It is a powerful stimulator of osteoclastic bone resorption, having similar effects to those of interleukin-1 (Figure 3.9).[79] In fact, the biological activity of tumor necrosis factor in many systems parallels that of interleukin-1, and it has been shown that tumor

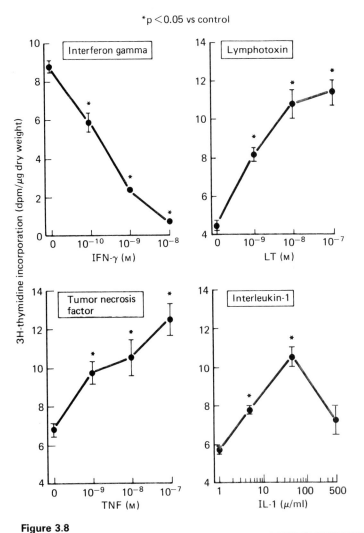

Figure 3.8

Effects of the osteotropic cytokines interferon gamma, lymphotoxin, tumor necrosis factor and interleukin-1 on DNA synthesis in cultured fetal rat calvariae. Each point is the mean ± standard error of the six half-calvariae.[77]

necrosis factor can induce interleukin-1 secretion from monocytes. Tumor necrosis factor has other important biological activities. It inhibits lipoprotein lipase in vitro, and has been linked to the cachexia which is associated with malignancy and chronic infections.[80] It also appears to be the mediator of endotoxic shock.[81] Like interleukin-1, it will cause an increase in the plasma calcium when infused in vivo.

Recently, it has been found that when the complementary DNA for human tumor necrosis factor (TNF) is transfected into Chinese hamster ovarian cells and these TNF-secreting cells are inoculated in nude mice, the nude mice develop hypercalcemia and show evidence of markedly increased osteoclastic bone resorption.[82] This indicates that increased chronic production of TNF leads to osteoclastic bone resorption and

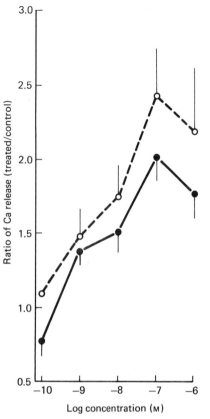

Figure 3.9

Effects of recombinant human TNFα (open circles) and lymphotoxin (TNFβ) (closed circles) on bone resorption from fetal rat long bones previously incorporated with ^{45}Ca in vitro. Treated to control ratios are from ^{45}Ca released from the bones. Each value represents the mean ± standard error for four pairs of bone cultures.[79]

hypercalcemia in vivo. To date, solid tumors have not been shown to produce TNF and it is not likely that direct production of TNF by solid tumors is a common mechanism of tumor-mediated hypercalcemia. However, there is evidence that normal immune cells can be stimulated by the presence of a tumor to produce TNF excessively, and TNF-induced hypercalcemia could be important in this situation.[83]

Lymphotoxin

Lymphotoxin is a multifunctional lymphokine which has cytolytic and cytostatic effects on neoplastic tumor cells.[84] Human lymphotoxin has been purified to homogeneity, a cDNA sequence isolated, the gene cloned and sequenced, and recombinant human lymphotoxin has been expressed in *E. coli*.[85] Lymphotoxin is a lymphocyte product which has biological properties essentially identical to those of tumor necrosis factor (Figure 3.9). It binds to the tumor necrosis factor receptor and has considerable amino acid sequence homology, but is encoded by a separate gene. Like tumor necrosis factor and interleukin-1, it acts on preformed mature osteoclasts indirectly through an intermediate cell, and probably also acts as a growth factor to increase the formation of osteoclasts by stimulating replication of progenitors (Figure 3.10).

Gamma-interferon

Gamma-interferon is a multifunctional lymphokine produced by activated lymphocytes. It has a wide range of biological effects on normal and transformed cells, including the inhibition of cell replication, alteration of cell surface characteristics, effects on B-cell activation and monocyte maturation, and induction of other lymphokines.[86-89] In some systems it has been shown to enhance both the production and cytotoxic effects of lymphotoxin and tumor necrosis factor.[90] Results show that gamma-interferon inhibits osteoclastic bone resorption stimulated by interleukin-1, lymphotoxin, and tumor necrosis factor in the mouse calvarial organ culture system (Figure 3.11).[70,72] Its effects on osteoclasts appear to be as an antidifferentiation or antifusion agent, rather than as an antiproliferative agent (Figure 3.12). This is in contrast with its interactions with lymphotoxin and tumor necrosis factor in other systems, where it may enhance or mimic the effects of these cytokines. More studies are required to determine whether gamma-interferon can be an effective agent for the treatment of hypercalcemic states, or an effective inhibitor of osteoclastic bone resorption.

It is almost certain that there are additional cytokines yet to be discovered which both stimulate and inhibit bone resorption. The first clue has

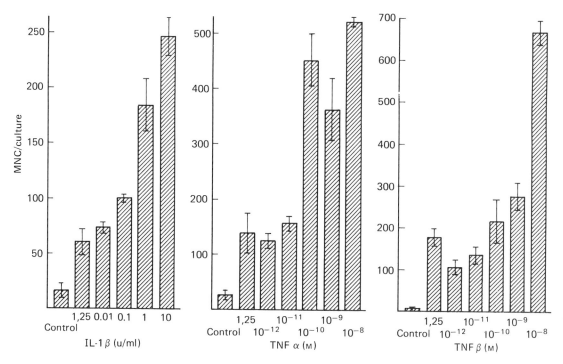

Figure 3.10

Effects of interleukin-1, tumor necrosis factor α and tumor necrosis factor β (lymphotoxin) on the capacity of human marrow mononuclear cells to form multinucleated cells with osteoclast characteristics. These cultures were performed for 3 weeks. (Reproduced from Pfeilschifter et al, Interleukin-1 and tumor necrosis factor stimulate the formation of human osteoclast-like cells in vitro, *J Bone Min Res* (1989) **4**: 113–18.)

come recently from the studies of Abe et al,[90,91] who have shown that activated cell cultures produce a differentiation factor that appears to be different from any of the known bone-resorbing cytokines.

Other cytokines and bone resorption

Recently it has been found that there are several other related cytokines which may affect osteoclastic bone resorption. These cytokines are characterized by their capacity to induce differentiation of myeloid precursors into monocyte-macrophages. These activities have been called

differentiation inducing factor (DIF), leukemia inhibiting factor (LIF) and interleukin-6 (IL-6).[92] It is now apparent that DIF and LIF probably represent the same glycoprotein.

Differentiation inducing factor (DIF)

DIF is a heavily glycosylated protein which is present in the conditioned media harvested from mouse L929 cells and Ehrlich's ascites tumor cells.[93,94] Using highly purified preparations of murine origin, Abe et al[95] showed that this factor stimulated osteoclastic bone resorption in cultures of neonatal mouse calvariae. This effect was inhibited by indomethacin. In separate experiments, the same activity increased osteoclast

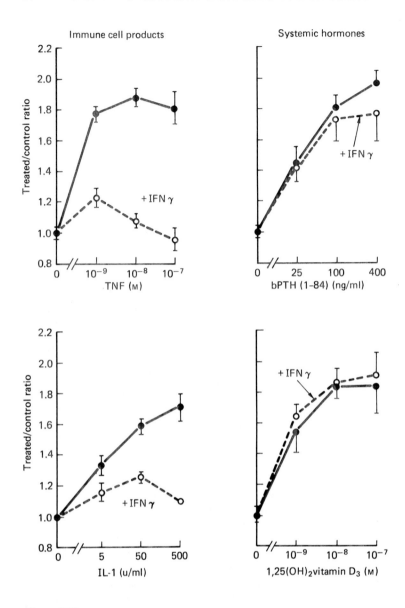

Figure 3.11

Effects of interferon on bone resorption stimulated by immune cell products interleukin-1 (IL-1) and tumor necrosis factor, and the systemic hormones PTH and 1,25-dihydroxyvitamin D on bone resorption in organ culture. Interferon is a more effective inhibitor of the cytokines than the systemic hormones.[70] TNF = Tumour necrosis factor; IFN γ = interferon gamma.

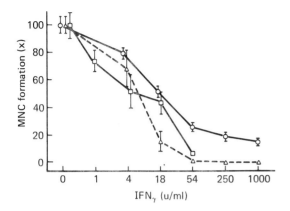

Figure 3.12

Effects of increasing concentrations of interferon gamma on formation of cells with osteoclast characteristics in cultures treated with 1,25-dihydroxyvitamin D. (Reproduced from Takahashi et al, Recombinant human gamma interferon inhibits formation of human osteoclast-like cells, *J Immunol* (1986) **137**: 3541–9.)

formation in cultures of murine marrow mononuclear cells.[96] The same activity is also released by mouse osteoblastic clonal MC3T3-E1 cells.[96,97] This activity has recently been purified by Lowe et al,[98] and was found to be a glycoprotein comprised of 180 amino acids.

Leukemia inhibitory factor (LIF)

Recently, Gearing et al[99] cloned a glycoprotein from a T-lymphocyte cell line which caused differentiation of myeloid cells into mature monocyte-macrophages and which they called leukemia inhibitory factor (LIF). LIF appears to be identical to DIF. Although less is known of the effects of LIF on osteoclastic bone resorption, preliminary studies suggest that it also stimulates osteoclastic bone resorption in organ culture.[100] Recently, Metcalf and Gearing[101] have suggested that LIF may have important effects on bone formation. They transfected murine hematopoietic cells with the cDNA encoding LIF and reported that when these cells were injected into syngeneic mice there was a marked increase in serum LIF

and a striking increase in bone formation, as well as bone resorption. However, these findings will require careful histomorphometry to assess their significance.

Interleukin-6 (IL-6)

IL-6 is a B-cell stimulatory factor which is present in the supernatants of mitogen-stimulated leukocytes.[102] IL-6 has a molecular weight of 21,000 and comprises 184 amino acids.[103] IL-6 is a ubiquitous factor which is produced by many replicating cells in addition to activated leukocytes. It has multiple functions, one of which is to enhance differentiation of myeloid cells into monocyte-macrophages. In this regard, its biological activity is similar to that of DIF and LIF.

IL-6 is produced in myeloma and it appears likely that it is important as a growth factor for myeloma cells. This myeloma cell growth factor is present in the serum of patients with advanced myeloma, and serum concentrations correlate with disease severity.[104] Kawano et al[105] have suggested that IL-6 may be an autocrine growth factor in myeloma. A number of groups have looked for effects of IL-6 on osteoclastic bone resorption. Most workers have found that IL-6 has no effect on organ cultures of fetal rat long bones or neonatal mouse calvariae, the standard screening assays for detecting bone-resorbing factors. However, there are other suggestions that IL-6 may still be involved in osteoclast activity. Kurihara et al[106] have shown that IL-6 increases osteoclast formation in cultures of human marrow mononuclear cells. Moreover, Lowik et al[107] showed that in an organ culture system using embryonic murine metacarpal bones, IL-6 stimulated osteoclast formation and bone resorption. Feyen et al[108] have found that IL-6 production is increased in cultures of rodent calvarial cells and organ cultures of mouse calvariae. Production of IL-6 in both isolated bone cells as well as the organ cultures is enhanced by treatment with PTH. Suda et al[92] also found that IL-6 was produced by cells with the osteoblast phenotype, but could find no increase in production in response to systemic bone resorbing agents such as PTH. However, they found IL-6 production by these murine osteoblast-like cells was enhanced by IL-1α and TNFα.

Recently, our group has found that IL-6 may be involved in the bone resorption induced by interleukin-1. Specific monoclonal antibodies to IL-6 will block IL-1-mediated bone resorption. This occurs in an assay system where IL-6 alone has no effect. These data suggest that for IL-1 to exert its effects on osteoclastic bone resorption, IL-6 needs to be present in the microenvironment of the osteoclasts.[109]

Prostaglandins

Prostaglandins of the E-series stimulate osteoclastic bone resorption in vitro, and have been linked to the hypercalcemia associated with some forms of cancer.[110,111] They have also been suggested as local mediators of bone destruction in chronic inflammatory diseases, since they are produced in large amounts at sites of chronic inflammation.[8] The effects of prostaglandins on osteoclasts in vitro appear to be unique. They stimulate osteoclasts directly, causing cytoplasmic contraction similar to that seen when osteoclasts are exposed to calcitonin.[112] How they cause an overall increase in osteoclast stimulation therefore remains unclear. Their effects on osteoblasts seem to be mediated through cAMP and mimic those of PTH.

Transforming growth factor α (TGF α)

This polypeptide stimulator of cell growth and replication, which has many characteristics in common with epidermal growth factor (EGF), is considered in more detail under the section on solid tumors and the hypercalcemia of malignancy. The first suggestion that TGF α may be responsible for increased bone resorption in some cases of malignancy was made by Mundy and Martin.[113] This peptide of 5600 daltons has similar chemical and biological characteristics to EGF, including marked conformational homology. All of its known effects appear to be mediated through the EGF receptor. In vitro studies suggest that it stimulates osteoclast precursors as well as activating the mature cell. It clearly resorbs bone in vitro,[114–117] and TGF α, or a similar factor working through the EGF receptor, is responsible for the bone-resorbing activity present in several tumor models of the hypercalcemia of malignancy.[118] Recently, it has been shown that systemic injections or infusions of TGF α cause increases in the plasma calcium.[119]

Recently, our group has performed similar experiments with TGF α to those performed with TNF. Chinese hamster ovarian cells were transfected with human TGF α cDNA and nude mice were inoculated with these human TGF α-secreting cells. It was found that there is a marked increase in plasma calcium in these animals, associated with increased osteoclastic bone resorption. There is also impaired osteoblast activity and bone formation. WE have also recently shown that injections of PTH-rP cause a marked increase in the plasma calcium, much greater than that produced by injections of similar doses of PTH-rP alone (Yates and Mundy, unpublished).

TGF β and PDGF

Transforming growth factor β and platelet-derived growth factor (PDGF) are other growth factors which could conceivably have an effect on bone volume, bone resorption, and possibly the plasma calcium. The former is a growth factor which is rich in platelets, bone, and placenta, and which is also produced by many replicating and transformed cells. It is a multifunctional polypeptide which was first described by Sporn et al.[120] The molecule is a homodimer of 24 000 daltons which has a precursor which consists of 391 amino acids.[121] The receptor has been characterized and appears to be extremely ubiquitous.[122] All cells so far tested seem to have a small number of high-affinity TGF β receptors. Surprisingly, the receptor does not have tyrosine kinase activity. The name TGF β is unfortunate, because it is more often a growth inhibitor than a growth stimulator, and it is produced by many normal tissues, including platelets and the placenta, as well as by transformed cells.[120] It has now been shown to be very abundant in normal bone.[123] There are many closely related molecules. Mullerian inhibitory substance, an inhibitor of cell growth which is described by Tucker et al[124] and

CIF-A, a cartilage-inducing factor from the bone matrix which is described by Seyedin et al,[123] are all identical to platelet-derived TGF β. In addition, there are structurally related molecules, such as inhibin[125] and CIF-B,[123] another cartilage-inducing factor found in the demineralized bone matrix. Transforming growth factor β is extraordinarily ubiquitous. Although its role in normal bone remodeling and turnover has still to be determined, the data which are currently available indicate that it is extraordinarily abundant in the bone matrix. Since TGF β activity is produced by resorbing bone cultures, its production by these bones is modulated by osteotropic hormones and it stimulates cells with the osteoblast phenotype, it is likely that it plays a pivotal role in

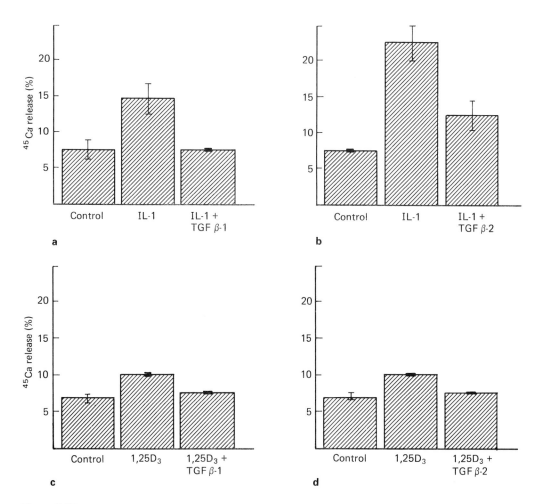

Figure 3.13

Inhibitory effects of TGF β-I and TGF β-II at 50 ng/ml on bone resorption in fetal rat long bones stimulated by interleukin-1 (a,b) and by 1,25-dihydroxyvitamin D (c,d). Values are means ± standard error of the mean for five bones per group. (Reproduced from Pfeilschifter et al, Transforming growth factor beta inhibits bone resorption in fetal rat long bone cultures, *J Clin Invest* (1988), **82**: 680–5.)

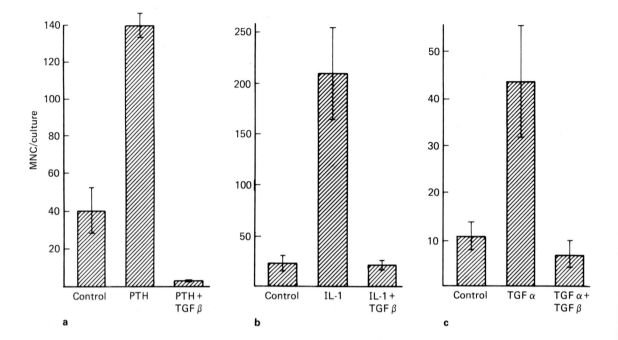

Figure 3.14

Effects of TGF β on formation of cells with osteoclast characteristics from human bone marrow cultures. TGF β inhibited the effects of PTH (a), IL-1 (b) and TGF α (c) to stimulate osteoclast formation. (Reproduced from Chenu et al, Transforming growth factor beta inhibits formation of osteoclast-like cells in long term human marrow cultures, *Proc Natl Acad Sci USA* (1988) **85**: 5683–7.)

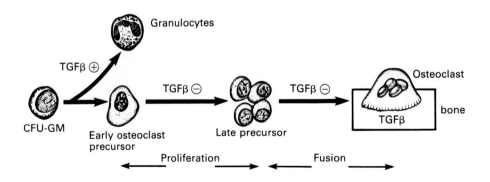

Figure 3.15

Proposed sites of action of TGF β on inhibition of osteoclast formation and activity. TGF β acts on proliferation, on differentiation and on the mature preformed osteoclast.

the bone-remodeling process. It has bidirectional effects on bone cells. It stimulates bone resorption in neonatal mouse calvariae, an effect which is mediated through prostaglandin synthesis,[116] and it has no detectable stimulatory effect on bone resorption in fetal rat long bones, but rather is inhibitory (Figures 3.13–3.15).[126]

Moreover, TGFβ inhibits the formation of cells with the osteoclast phenotype when incubated with human marrow mononuclear cells.[127,128] These effects on osteoclast formation are similar to those of hydroxyurea, which acts predominantly as an inhibitor of osteoclast precursor replication.

Recently, our group has examined the effects of TGFβ on bone cell function in vivo (Marcelli, Yates and Mundy, unpublished). TGFβ was injected and infused into the subcutaneous tissue over the calvariae of normal mice, and was also injected and infused into the peritoneal cavity of normal mice. TGFβ does not appear to have any systemic effects, either on calcium homeostasis or bone histology. However, at the sites of local injections, TGFβ has profound effects on bone formation. It increases osteoblast activity causing a marked stimulation of proliferation of the precursors, as well as stimulating the formation of mineralized bone. However, the effects of TGFβ on bone formation may depend on the doses used and the tissues to which it is exposed. TGFβ can also affect cartilage cells to stimulate proliferation and differentiated function and may enhance cartilage formation in vivo.[129] It has been proposed as a factor which may play a role in the stimulation of ectopic bone formation, as well as enhancing the closure of bone defects.

There are other properties of TGFβ which suggest it may play an important role in bone remodeling. It is present in relatively abundant amounts in the bone matrix as noted above,[130] its production by bone cells and bone cultures seems to be regulated by parathyroid hormone and other osteotropic hormones,[131] and it has powerful effects on both osteoclasts and osteoblasts.[132] Moreover, recent data suggest that it is chemotactic for cells with the osteoblast phenotype,[133] and it is released from bone as a latent complex which may be activated either by exposure to an acid microenvironment such as exists under the osteoclast, or by the action of stimulated osteoclasts independent of acid production.[134] The author has examined the

latent complexes for TGFβ and reports that it appears likely that there are two latent complexes, one which comprises the TGFβ precursor and a second which represents binding of the active TGFβ moiety to α₂-macroglobulin. In other tissues, the latent complex for TGFβ appears to comprise the precursor plus a second modulator protein. However, this modulator protein has not been detected in the bone tissues examined.

Platelet-derived growth factor is the major mitogen which is present in serum for cells of mesenchymal origin. Like TGF β, it is sequestered within the α-granules of the platelet and is released when platelets are activated by contact with thrombin, or collagen surfaces on subendothelial basement membranes at sites where blood vessels are injured.[135,136] In macrophages, PDGF-like activity has also recently been described.[137,138] In addition to its mitogenic effects, it has chemotactic activity for monocytes, neutrophils, and fibroblasts. Determination of its N-terminal amino acid sequence has shown that it consists of two polypeptide chains which are linked by disulfide bonds and are encoded by two separate genes. The PDGF-B polypeptide shares extensive sequence homology with P28 sis, the oncogene product of the simian sarcoma virus, an acute transforming primate retrovirus.[139,140] Platelet-derived growth factor has an effect on mesenchymal cells which appears to be different from most other growth factors such as EGF, TGF α, and the somatomedins. It is a competence growth factor, which means that cells exposed for 30 min or less to PDGF become competent to replicate their DNA and divide, even if PDGF is then removed from the medium. In contrast, other growth factors such as EGF and the somatomedins are required for prolonged intervals (nearly until the onset of the S-phase) to sustain a mitogenic response.

Platelet-derived growth factor is encoded by the c-sis proto-oncogene. Studies on expression of this proto-oncogene as determined with DNA probes to the v-sis gene show that this oncogene is frequently expressed in human osteosarcoma cells. Heldin et al[141] found that a human osteosarcoma cell line (U-2 0S) secreted a heat- and acid-stable mitogenic polypeptide with chromatographic behavior identical to that of human PDGF. This labeled polypeptide could be immunoprecipitated by anti-PDGF antiserum, and PDGF competed for binding to cellular receptors with this factor. The mitogenic activity produced by these cells can be blocked with antiserum to PDGF.[142] Moreover, recently a PDGF-like peptide has been found to be enriched in the bovine bone matrix.[143] These studies showed that this peptide was present at a level of about 50 ng/g dry bone. As further evidence that PDGF-like peptides may be important in bone cell function, there is now a small literature suggesting that platelet-derived PDGF acts on bone cells in vitro. Tashjian et al[144] showed that prostaglandin synthesis in cultured mouse calvariae was stimulated by platelet-derived PDGF, and Canalis[145] showed that PDGF could stimulate bone collagen synthesis in vitro. Graves et al[146,147] have shown that PDGF has a mitogenic effect on osteosarcoma cell lines which have the osteoblast phenotype.

The effects of these factors on normal human bone cell function and calcium homeostasis remain to be determined. They have potential

roles as coupling factors linking bone formation with previous bone resorption, since TGF β has been shown to be produced by resorbing bone cultures, and both TGF β and PDGF are probably released when the bone matrix (which contains them) is degraded by resorbing osteoclasts.

Thyroid hormones

The thyroid hormones thyroxine and triiodothyronine stimulate osteoclastic bone resorption in vitro (Figure 3.16),[148] and patients with hyperthyroidism may have markedly increased rates of bone turnover, and may occasionally develop hypercalcemia. However, normal thyroid hormone activity is not necessary for calcium homeostasis, nor is it required for normal bone structure, since patients with myxedema do not develop hypocalcemia or decreased bone volume. Conversely, their rates of bone turnover are markedly diminished. The effects of the thyroid hormones on bone cell function and calcium homeostasis are discussed in more detail in Chapter 9.

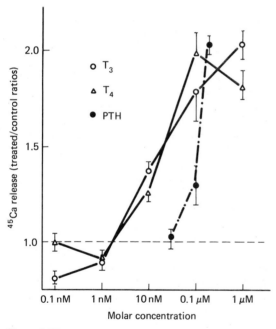

Figure 3.16

Dose response effects for thyroxine, triiodothyronine and parathyroid hormone on bone resorption using fetal rat long bone cultures. Values are means ± standard errors for four bones per point for 4 days in culture. (Reproduced with permission, copyright 1976 by the American Society for Clinical Investigation.)[148]

Glucocorticoids

Glucocorticoids have complex and multiple effects on bone and calcium homeostasis.[149] Glucocorticoids inhibit both bone formation and bone resorption when used in pharmacologic doses. The inhibition of bone formation is probably the major mechanism responsible for the osteopenia associated with chronic glucocorticoid use. Inhibition of bone resorption is probably the mechanism by which glucocorticoids mediate beneficial effects on lowering the serum calcium in occasional patients with the hypercalcemia of malignancy (Figure 3.17). When used chronically, glucocorticoids lead to bone loss, possibly as a result of their effects to inhibit the formation of mature osteoblasts from precursors. They may also inhibit the formation of osteoclasts from less mature cells. However, their effect on the gut to impair the gut absorption of calcium leads to secondary hyperparathyroidism and subsequent increased osteoclastic bone resorption, so they

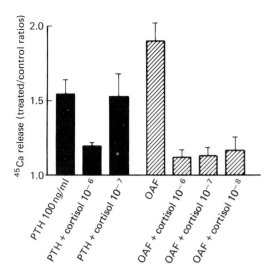

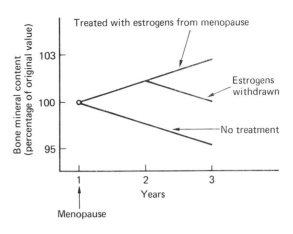

Figure 3.17

Effects of cortisol on bone resorption stimulated by parathyroid hormone and by bone resorbing activity present in supernatant media harvested from normal activated leukocyte cultures. Cortisol is more effective in inhibiting bone resorption stimulated by bone resorbing cytokines than by parathyroid hormone. OAF = Osteoclast activating factor. (Reproduced from Mundy et al, Pathogenesis of hypercalcemia in lymphosarcoma cell leukemia – role of an osteoclast activating factor-like substance and mechanism of action for glucocorticoid therapy, *Am J Med* (1978) **65**: 600–606.)

Figure 3.18

Effects of estrogens to slow the rate of bone loss after the menopause. In normal women, bone mineral content begins to decrease following the menopause. This loss in bone mineral content is prevented by estrogen therapy.

may have bidirectional effects on bone resorption: a direct inhibitory effect and an indirect stimulatory effect. Whether they have direct effects on the gut or whether their effects are mediated through impaired metabolism of vitamin D is not clear. It is possible that both mechanisms are involved.

Estrogens and androgens

Estrogens are important inhibitors of bone resorption in vivo. They clearly inhibit the enhanced bone resorption which is associated with the immediate postmenopausal period (Figure 3.18).[150,151] The mechanism by which

estrogen lack leads to an enhancement of bone resorption, and how this is inhibited by estrogen replacement therapy, is entirely unknown. Estrogen receptors have been difficult to demonstrate in mammalian bone cells, and estrogens have not been shown to have any direct effects on organ culture systems.[152] There have been several recent reports of estrogen receptors in cells with the osteoblast phenotype,[153,154] but the relationship of this to bone resorption is not yet apparent. However, one possibility is that production of TGFβ by cells with the osteoblast phenotype may be regulated by estrogens.[155] Estrogens may enhance TGFβ expression by cells with the osteoblast phenotype. If this were so, then removal of this estrogen effect at the time of the menopause may lead to removal of a normal inhibitory factor produced in the bone microenvironment which inhibits osteoclast formation and osteoclast activity.[127,128] Thus, under these circumstances, bone resorption may be enhanced. Whether this explanation is responsible for the enhanced osteoclastic bone resorption associated with estrogen deprivation requires further study. In birds and mice estrogens cause sclerosis of the marrow, increased bone formation, and a form of osteopetrosis. It is possible that the effects of estrogen lack to enhance bone resorption in humans are mediated indirectly by other hormones, such as enhanced production of cytokines, effects on PTH release, or effects on vitamin D metabolism. When used before maturity, estrogens and androgens stimulate skeletal maturation and closure of the epiphyses, resulting in short stature.

Other factors

Insulin and insulin-like growth factor 1

There are a series of other factors which affect skeletal homeostasis. Their role in bone metabolism and calcium homeostasis remains unknown. These include growth hormone, insulin-like growth factor 1, and insulin. Growth hormone derived from the anterior pituitary stimulates the production of somatomedin C (insulin-like growth factor 1) in peripheral tissues, and particularly in the liver. It is also likely that growth hormone leads to enhanced production of somatomedin C in bone cells. In vivo, growth hormone is necessary for normal skeletal growth, but not in utero, since anencephalic individuals are of normal birth size. Somatomedin C stimulates protein synthesis and sulfation in cartilage, and it affects differentiated function in osteoblasts including collagen synthesis. Similar effects are mediated by insulin. In vitro, insulin is a powerful stimulator of bone collagen synthesis.[156] Whether this effect has physiological significance remains unknown. Diabetics with insulin deficiency may have osteopenia, although there are multiple reasons why chronically ill individuals, such as those who have severe Type 1 diabetes mellitus, could develop impaired bone formation.

Hematopoietic colony stimulating factors

Normal hematopoiesis is controlled by a number of well-characterized growth regulatory factors. These factors have distinct biological specificities, but their effects may overlap and it is now apparent they may influence cells which are precursors of osteoclasts. However, despite this overlap there are specific growth factors for each of the cell lineages involved in hematopoiesis. (It also appears likely to the author that a specific growth factor or factors exists for the osteoclast lineage.) These hematopoietic growth factors have usually been called colony stimulating factors, although this is a misleading term since it is derived from the capacity of these factors to support single or multilineage colony formation in semisolid media in vitro or in the spleens of irradiated mice in vivo.[157–159] The term is unsatisfactory because committed progenitors for some types of hematopoietic cells (for example, lymphocytes) are not suited to this type of colony formation assay. A better term would be hematopoietic growth regulatory factors. All of these factors are necessary for committed precursor cells of individual cell types to survive, since without the presence of the hematopoietic growth factor the committed precursor cell will die.[159] In the presence of the growth regulatory factor (and possibly other differentiation factors) the precursor cell will proliferate and go on to differentiate into a mature cell. For the earliest multipotential precursors several factors acting synergistically may be required.[160] This differentiation has been shown recently to be a stochastic process.[161]

In order to understand the actions of these hematopoietic growth regulatory factors and the influence they have on hematopoietic cell (and possibly osteoclast) development, the steps involved in the formation of hematopoietic cells need to be considered. The hematopoietic cells arise from a small population of pluripotent stem cells which appear during a short interval of early embryonic life, and thereafter self-renew and differentiate into precursors of at least nine different hematopoietic lineages throughout the lifetime of the animal. Under the influence of these growth factors primitive stem cells form committed progenitor cells which retain a more limited capacity for self-renewal, less than their pluripotential forebears. In the presence of the appropriate environment and growth factors the committed progenitors will then go on to form mature cells which usually have no capacity for self-renewal, and which are therefore terminally differentiated. In order to proliferate, the pluripotent stem cells require hematopoietic growth factors. To differentiate, the committed progenitor cells usually need growth factors, but perhaps in higher concentrations. Without growth factors it is now clear that the committed stem cells will not survive.[159] The colony stimulating factors therefore have multiple functions on hematopoietic cells including: (a) cell survival; (b) support of differentiation; and (c) stimulation of functional activity in terminal cells.

There are currently four clearly defined hematopoietic growth regulatory molecules for cells of the granulocyte–monocyte lineages. It is likely that other growth regulatory molecules may be defined in the future. The four molecules that are currently well characterized are interleukin-3 (or multi-CSF), CSF-GM, CSF-M (or CSF-1), and-CSF-G.

Interleukin-3 (molecular weight = 28 000)

This factor appears to be involved in the self-renewal of pluripotent stem cells, the production of hematopoietic progenitor cells, and the production of mature myeloid, eosinophil, megakaryocyte, erythroid cells, and mast cells. Rodent forms of this molecule have been cloned molecularly.[162] It is still not entirely clear whether this factor is present in humans, although a factor with this biological activity has recently been purified from a human hepatoma cell line and so this appears likely.[163] A similar multilineage growth factor has been identified in a human urinary bladder cell line 5637, which requires CSF-1 for it to have an effect on developmentally early bone marrow cells.[160]

CSF-GM (molecular weight = 20 000)

CSF-GM is required for all of the steps involved in myelopoiesis from the pluripotent stem cell to the mature granulocyte and, in the absence of CSF-GM, committed granulocyte–macrophage progenitor cells die rapidly. CSF-GM from both mouse and human sources has recently been molecularly cloned and expressed.[164,165] In addition to its growth regulatory effects, CSF-GM also has direct effects on mature neutrophils.[166] It has been shown to be equivalent to neutrophil migration inhibition factor, and is also a neutrophil activator.

CSF-1 or CSF-M (molecular weight = 26 000)

CSF-1 has been purified by Stanley and Heard[167] and others,[168] the N-terminus sequenced[169] and

human CSF-1 molecularly cloned.[170] CSF-1 is a mononuclear phagocyte lineage-specific growth factor whose receptor has been characterized and localized to mononuclear phagocytic precursors.[171]

CSF-G (molecular weight = 19 600)

CSF-G is a hemopoietic growth factor for granulocytes, but like GM-CSF it will also support multilineage colony formation in culture. It will also induce various human leukemic cells to undergo terminal differentiation to macrophages and granulocytes. Human G-CSF has been purified[172] and molecularly cloned and sequenced.[173,174]

There is now a mounting body of evidence that these colony stimulating factors (which are produced by immune cells) are involved in osteoclast formation.

(a) Association with hypercalcemia. Several tumors of both animal and human origin that are associated with hypercalcemia have also been accompanied by leukocytosis.[175–177] In some cases, the tumors have been shown to produce a colony stimulating factor for cells of the monocyte–granulocyte series, as well as bone-resorbing activity.[178–181] In these tumors it remains possible that both of these biological activities are due to the same factor. It is an attractive hypothesis that a colony stimulating factor could also affect osteoclasts (cells that probably originate from the monocyte–macrophage lineage).

(b) Burger and colleagues have developed a system for osteoclast formation in which they add exogenous mononuclear cells to organ cultures of 17-day mouse metatarsal bones which contain a mineralized core and from which the periosteum has been stripped by enzyme treatment.[182] The cells invade the mineralized core, resorb the mineralized matrix, and mature morphologically into osteoclasts. This system is potentiated by culturing the cells with a source which is rich in colony stimulating activity (derived from fibroblast conditioned medium).

(c) Recent work has been presented suggesting that colony stimulating factors may play a role in the regulation of bone resorption in one murine variant of osteopetrosis, the op/op mouse.[183] These workers showed that there was a failure of normal monocyte–macrophage differentiation in these mice, which was probably caused by decreased production of colony stimulating activity by stromal marrow fibroblasts. When the stem cells were removed from these animals, they responded normally to colony stimulating factor in vitro. However, colony stimulating factor production by the stromal cells of the affected animals was decreased. These data suggested that deficient colony stimulating factor production in vivo was responsible for impaired osteoclast formation, rather than a deficiency in the osteoclast progenitors themselves.

(d) Recent studies have shown that colony stimulating factors influence the formation of osteoclasts in vitro. In a long-term human marrow culture system developed by Roodman and co-workers, multinucleated cells with the characteristics of osteoclasts (including the capacity to form resorption lacunae) form after 3 weeks in culture.[41,184] This system is a modified form of the Dexter marrow culture system in which marrow cells are grown together with an adherent cell layer containing various cell types including macrophages, fibroblasts, and other stromal cells. In this system recombinant human CSF-GM and highly purified CSF-1, both of which stimulate the proliferation of monocyte–macrophage precursors, increase the number of cells with osteoclast characteristics.[185] This effect is seen after 1 week of incubation with a source of colony stimulating factor, and is potentiated by the subsequent addition of 1,25-dihydroxy-vitamin D. Autoradiographic studies indicate that these forms of colony stimulating factor stimulate osteoclast cell formation by increasing the proliferation of the precursor cells, rather than by differentiation of the committed progenitors.

The rapid development in the understanding of the hematopoietic growth-regulatory molecules raises several important questions that have intrigued investigators working in the field of osteoclast biology for some time. These questions include the following.

(a) Does the osteoclast arise from similar pluripotent stem cells as white cells, and particularly macrophages? Could the osteoclast and macrophage arise from a common precursor cell? The bulk of the current evidence suggests that they do, although none of this evidence is definitive, and doubt has been thrown on this concept recently by Horton and Chambers, who argue that the failure of the osteoclast to express common leukocyte antigenic markers suggests that they may arise from a bone-marrow precursor that is separate from the multipotential precursor of the granulocyte and macrophage lineages.[186]

(b) How do the currently described and available hematopoietic growth-regulatory molecules (colony stimulating factors) influence osteoclast formation and activation? These molecules have been shown to have effects on white cells other than their primary targets, and preliminary evidence suggests that they have effects on cells with the characteristics of osteoclast progenitors.

(c) What are the growth-regulatory molecules for the osteoclast (that is, the 'osteoclast colony stimulating factors')? What factors influence the formation, localization, stimulation, and inhibition of osteoclasts and their precursors?

Clinical features and differential diagnosis of hypercalcemia

Clinical features

The clinical features of hypercalcemia are protean and vary considerably from patient to patient. They tend to be most prominent in patients who have a rapid increase in the serum calcium, as well as in those with the highest levels of serum calcium. They also tend to be more exaggerated in the elderly than they are in the young. Increases in serum calcium affect all organ systems, but most prominently the neurologic system, the gastrointestinal tract, the cardiovascular system, and the kidney. They are summarized in Table 4.1.

Table 4.1   Clinical features of hypercalcemia.

Neurological and psychiatric
  lethargy, drowsiness
  confusion, disorientation
  disturbed sleep, nightmares
  irritability, emotional problems, depression
  hypotonia, decreased deep tendon reflexes
  stupor, coma

Gastrointestinal
  anorexia, vomiting
  constipation
  ? peptic ulceration
  acute pancreatitis

Cardiovascular
  ? hypertension
  inotropic, chronotropic
  arrhythmias
  synergism with digoxin

Renal
  polyuria, polydipsia
  hypercalciuria
  nephrocalcinosis
  impaired glomerular filtration

Neurological and psychiatric features

The most prominent neurological features of hypercalcemia are drowsiness and lethargy. Other common features in the early stages include headache, irritability, and nightmares. Disturbed sleep is common, and some patients relate vivid and terrifying nightmares that occur while they are hypercalcemic. As the serum calcium increases, these neurologic features may be followed by progressive confusion and disorientation. Eventually the patients become stuporous and then comatose, and this comatose state usually heralds imminent death. Some patients show generalized muscle weakness, hypotonia, and absent deep-tendon reflexes.

Psychiatric problems have frequently been described in patients with hypercalcemia. These problems have been found most commonly in patients with primary hyperparathyroidism, with marked improvement when hypercalcemia is corrected by successful parathyroid gland surgery. Psychiatric complications may be mild, and include depression, or emotional problems, with irritability and inability to cope with stressful life situations. Less frequently patients may suffer from delusions[1] or hallucinations.[2] Fifty per cent of patients with primary hyperparathyroidism demonstrate some form of psychopathology.[3] The evidence that milder disturbances are related to the serum calcium is quite convincing, since a lowering of the serum calcium often leads to improvement. However, it is less clear whether lowering the serum calcium improves frank psychosis.

Patients with hypercalcemia may develop localizing neurological signs, but this is very rare. Other neurological features which have been reported include nerve deafness,[4] and ataxia and visual scotomas.[5] Determination of the cerebrospinal fluid protein reveals concentrations which are frequently >60 mg/dl.[6] In most patients who have an electroencephalogram (EEG) taken, there may be characteristic changes. These are related to the absolute concentration of serum calcium and may last for several weeks after the serum calcium returns to normal. However, the EEG may be normal with the serum calcium as high as 15 mg/dl.[7] Characteristic changes include paroxysms of frontal dominant, moderately high-voltage, 2–4 Hz activity which may be misinterpreted as diffuse cerebral metastases.[8,9]

Patients with hypercalcemia frequently have muscle weakness and are easily fatigued. Primary hyperparathyroidism is associated with a specific form of atrophy of Type II muscle fibers, described in more detail in Chapter 8.

Gastrointestinal features

The most frequently described gastrointestinal symptom associated with hypercalcemia is constipation. This is seen most commonly in patients with primary hyperparathyroidism, or with any form of hypercalcemia of long standing. Even then it is not as prominent a symptom as the literature would suggest. More common in the patient with an acute increase in serum calcium is anorexia, followed by uncontrolled vomiting. Occasionally patients may have gastric atony.[10]

The relationship between hypercalcemia, hyperparathyroidism and peptic ulceration remains obscure. In some patients an increase in the serum calcium increases gastric acid and pepsin secretion.[11–13] However, whether this induces peptic ulceration is not certain. Hypercalcemia does not increase stimulated gastric secretory responses.[14] The increase in gastric acidity associated with hypercalcemia is probably mediated by neural mechanisms via acetylcholine, since atropine and pentolinium block the effects of calcium to stimulate gastric hyperacidity. In contrast, magnesium opposes the effects of calcium on gastric acid output.[15]

Rarely, when patients develop a serum calcium of >18 mg/dl (4.5 mmol/l), metastatic gastric calcification may occur.[12] Like other forms of metastatic calcification, this is most likely to occur when both the serum calcium and the serum phosphorus are elevated, such as occurs in the patient with impaired renal function.

Occasionally patients with hypercalcemia develop acute pancreatitis. This is seen most frequently in patients with primary hyperparathyroidism, and it has not to the author's knowledge been clearly described with malignancy. This topic is discussed in more detail in Chapter 8. Acute pancreatitis does not occur only in patients with hypercalcemia due to primary hyperparathyroidism. There have been reports of a few patients with familial hypocalciuric hypocalcemia (FHH) who have developed acute pancreatitis (Chapter 9), and the author has seen a patient with FHH who had this association. The mechanisms are not clear. It has been suggested that hypercalcemia may cause activation of trypsin and resultant destructive changes in the pancreatic parenchyma, or that hypercalcemia may provoke intravascular coagulation since some of the clotting factors may be calcium-activated. It has also been suggested that calcium precipitates may obstruct the pancreatic ducts. The precise mechanism is not known.

Cardiovascular features

It is still widely debated whether hypertension is directly related to serum calcium. This issue is also discussed in Chapter 8. Calcium infusions often cause a significant increase in blood pressure, which returns to baseline after the infusion.[16,17] Blood pressure is increased in patients who are intoxicated with vitamin D.[18] Agents which lower the serum calcium, such as intravenous phosphates and EDTA,[19,20] may decrease blood pressure. Why calcium should have an effect on blood pressure is not known, but it may be due to the vasoconstrictive effects of calcium on smooth muscle.

The serum calcium also affects myocardial function. Hypercalcemia increases myocardial contractility (an inotropic effect) and shortens ventricular systole (a chronotropic effect). There are characteristic electrocardiographic (EKG) changes which include shortening of the QT-interval,[21] coving of the ST-segment and broadening of the T-wave[22] when the serum calcium is >13 mg/dl (Figure 4.1). Reversible first-degree heart block or Wenckeback phenomenon can occur.[23] Patients with severe hypercalcemia may develop asystole or ventricular arrhythmias, usually as a terminal event.

Calcium and digoxin have an apparent synergistic effect on heart muscle, but it is not known if this is deleterious to myocardial function. It is clear that digoxin causes a net uptake of calcium by myocardial cells, and counteracts heart failure and hypertension associated with hypocalcemia.[24] Patients treated with digoxin who are also given intravenous calcium are particularly sensitive to digoxin and might die suddenly.[25] For these reasons caution should be exercised when digoxin is administered to patients with hypercalcemia.

Effects of hypercalcemia on the kidney (Table 4.1)

A characteristic feature of hypercalcemia is the inability to concentrate the urine and subsequent hyposthenuria. This is probably due to an inhibitory effect of calcium on the actions of antidiuretic hormone on the collecting tubules.[26] It does not appear to be due to changes in the glomerular filtration rate,[27] although it may in part be due to inhibition of sodium transport in Henle's loop.[28]

Most patients with hypercalcemia unrelated to primary hyperparathyroidism have metabolic alkalosis.[29] This is in contrast to patients with primary hyperparathyroidism, who often have a mild hyperchloremic acidosis (Chapter 8). The metabolic alkalosis in patients with hypercalcemia associated with malignancy may be due to bone destruction, which increases the plasma buffering capacity. Another possibility is that hypercalcemia itself causes increased acidification of the urine. Glomerular function may be impaired in patients with hypercalcemia of malignancy, particularly malignancies such as myeloma associated with Bence Jones nephropathy, or lymphomas associated with obstructive uropathy. Patients who have increased intestinal calcium absorption often develop nephrocalcinosis and impaired renal function, since phosphate absorption is also increased. Nephrocalcinosis, when it occurs, usually appears at sites of dying tissue or where there is a local increase in the pH.

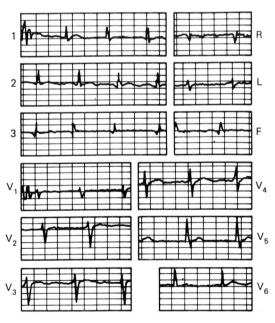

Figure 4.1

Electrocardiogram of a patient with hypercalcemia, showing the characteristic changes, namely coving of the ST-segment, shortening of the QT interval and flattening of the T wave.

It occurs particularly commonly in the collecting ducts. The ascending limb of Henle's loop and the proximal tubules are usually spared.[30–32] In patients with chronic hypercalcemia, chronic interstitial nephritis may occur and lead to progressive impairment in glomerular filtration.

Differential diagnosis

Clinicians are frequently confronted with hypercalcemia as an isolated laboratory abnormality. As indicated in Table 4.2, the list of potential diagnostic possibilities is very large. All of these causes must be considered in the differential diagnosis, although in practice 90 per cent of cases are due either to primary hyperparathyroidism or to malignancy (Table 4.3). Thus, the first two conditions that the clinician considers when faced with the differential diagnosis of hypercalcemia is whether the underlying disorder responsible is malignancy or primary hyperparathyroidism. If neither of these appears likely, then the less frequent causes of hypercalcemia must be considered.

Table 4.2   Differential diagnosis of hypercalcemia.

Primary hyperparathyroidism

Malignant disease

Hyperthyroidism

Immobilization

Vitamin D intoxication

Vitamin A intoxication

Familial hypocalciuric hypercalcemia (FHH)

Diuretic phase of acute renal failure

Chronic renal failure

Thiazide diuretics

Sarcoidosis and other granulomatous diseases

Milk-alkali syndrome

Addison's disease

Paget's disease

Table 4.3   Relative frequency of causes of hypercalcemia.

	No. of patients	Percentage of cases
Primary hyperparathyroidism	111	54
Malignant disease	72	25
Lung	25	35
Breast	18	25
Hematologic (myeloma 5, lymphoma 4)	10	14
Head and neck	4	8
Renal	2	3
Prostate	2	3
Unknown primary	5	7
Others (gastrointestinal 4)	8	8

The data presented in Table 4.3 are not biased by patient selection. It was obtained from a survey of 207 consecutive patients with hypercalcemia monitored in an urban population of a major British city of 1 million people. In this population it was possible to trace every patient in this city who presented with hypercalcemia over 6 months, because of the means by which all laboratory tests are performed under the National Health Service. All laboratory tests for this population were performed in four hospital clinical chemistry laboratories. Some of these patients were ambulant outpatients and some were hospitalized, so there was no bias due to causes of hypercalcemia which are more likely to be present because the hospitalized population was over-represented.

In many situations there is no absolute test to distinguish the conditions likely to be responsible for hypercalcemia, so the clinician must seek all of the relevant information and then make a judgment based on the available pieces of evidence.

In the following pages, tests that are frequently used in the differential diagnosis of hypercalcemia are considered, followed by the individual causes and what the author considers to be the most useful diagnostic points.

Tests used in the differential diagnosis of hypercalcemia

Serum calcium

Most hospital laboratories measure the total serum calcium by automated or autoanalyzer technology. However, the total serum calcium is composed of the ionized calcium, which is metabolically active and regulated by the homeostatic mechanisms, and biologically inert calcium, most of which is bound to protein, and some of which is complexed or chelated to citrate and other anions (Figure 4.2). About 47 per cent of the total serum calcium is free (also called ionized) and biologically active. Since the total serum calcium is only an indirect measurement of the free calcium, its measurement may be misleading. Total serum calcium may give an inaccurate assessment of the free calcium in patients with abnormalities in the plasma proteins, or in those with abnormalities in acid–base status. Of the circulating calcium which is bound to plasma proteins, most is bound to albumin (about 80 per cent) and the rest to globulins. Abnormalities in plasma proteins are likely to lead to spurious results in two situations. First, and most commonly, decreases in the serum albumin (frequently seen in patients with cancer, liver disease, or any other chronic illnesses or protein wasting states) may have a low circulating albumin concentration, and thereby a low total serum calcium. Alternatively, some patients have abnormalities in the circulating globulins, and may have an increase in a globulin which avidly binds calcium. This may happen in myeloma, but it can also occur in patients with other types of monoclonal gammopathies. In this situation the total serum calcium will be increased, whereas the ionized or free serum calcium may be normal. The other major problem that may make the total serum calcium difficult to interpret is a change in the acid–base status, which changes calcium binding to plasma proteins. In the presence of alkalosis, the free (or ionized) calcium is relatively decreased due to increased binding to plasma proteins, but in the presence of acidosis the free calcium will be relatively increased. However, changes in the free calcium due to changes in the blood pH are usually relatively small and rarely significant.

Realizing that total serum calcium may be a misleading index of the ionized or free calcium, McLean and Hastings[33] developed a nomogram to allow clinicians to calculate the ionized calcium, based on knowledge of the total serum proteins. Although this nomogram has been modified by many investigators over the years, the general concept does have considerable practical utility, particularly in patients who are markedly hypoalbuminemic. The correction factor favored by the author is to allow 0.8 mg/dl of corrected total serum calcium for every gram of albumin that the serum albumin is less than 4 g/dl (or 0.1 mmol/l for every 5 g/l when the serum albumin is less than 40 g/l). When expressed as a formula, the corrected total serum calcium will therefore equal the observed total serum calcium − 0.8(4.0 − the serum albumin in g/dl). The normal range for the corrected total serum calcium is 8.5–10.2 mg/dl (2.1–2.55 mmol/l).

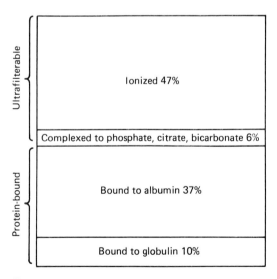

Figure 4.2

Forms of calcium which comprise the total serum calcium. The total serum calcium is partly composed of ionized or free calcium, partly of calcium bound to circulating proteins, and partly of calcium chelated or complexed with anions.

The absolute measurement of the total serum calcium gives little differential diagnostic information. Most patients with hypercalcemia of malignancy and primary hyperparathyroidism have a total serum calcium of <13 mg/dl (3.25 mmol/l). It is rare to find a patient with hypercalcemia due to malignancy with a serum calcium of >17 mg/dl (4.25 mmol/l), but this does occasionally occur. The only patients that the author has seen with serum calcium of >22 mg/dl (5.5 mmol/l) have been patients with primary hyperparathyroidism. Patients with a total serum calcium of >18 mg/dl (4.5 mmol/l) have been seen who have been surprisingly asymptomatic, but all of these patients have had primary hyperparathyroidism. Possibly the increase in the serum calcium in these patients has not occurred acutely, and this may be why symptoms are not as prominent as in those patients in whom the serum calcium has risen more rapidly.

Free (or ionized) calcium

Measurement of the serum free calcium is more readily available now than it was a few years ago, but it is still not a routine measurement in most hospital laboratories. Measurement of free calcium in the serum with ion-selective electrodes has improved dramatically over the past 10 years. It is clear when this measurement is performed that many patients with subtle hypercalcemia or hypocalcemia will be recognized who would not be recognized if total serum calcium alone was measured. However, once the diagnosis of hypercalcemia or hypocalcemia is made, it is not clear that in routine clinical situations the measurement of free calcium offers a significant advantage over the measurement of the total serum calcium for following the progress or response of a patient to therapy.

Serum phosphorus

Most autoanalyzers will give a measurement of the serum phosphorus. Serum phosphorus is not as tightly regulated as the serum calcium, and variations of up to 50 per cent may be seen in the normal individual following changes in the diet, fasting, vomiting, or renal function. The serum phosphorus may fall by one-third within 2 h of the

patient digesting a meal that is rich in carbohydrates. Serum phosphorus also varies with age and sex, and is clearly higher in children and adolescents with growing bones. It is also slightly higher at any age in females than in males. Patients with primary hyperparathyroidism have a decrease in the serum phosphorus, due to the effects of PTH to inhibit renal tubular phosphate reabsorption, unless they have impaired glomerular filtration. Patients with hypercalcemia due to malignancy may also have a decrease in the serum phosphorus due to renal phosphate wasting, so in this situation the measurement does not give much useful diagnostic information. Patients who have hypercalcemia due to sarcoidosis, vitamin D intoxication, or the milk-alkali syndrome usually have an increase in the serum phosphorus due to increased gut absorption and impaired renal function. Patients with hypercalcemia due to myeloma also have a normal serum phosphorus or a phosphorus which is increased, since renal function is often poor in these patients.

Tubular reabsorption of phosphorus

Patients with primary hyperparathyroidism and some patients with malignancy have renal phosphate wasting, associated with impaired renal tubular phosphate reabsorption. As a consequence the measurement of phosphate excretion and its expression as the tubular reabsorption of phosphate (TRP), fractional excretion of phosphate or tubular maximal reabsorption of phosphate as a fraction of glomerular filtration rate (TMP/GFR) may be abnormal. However, as a differential diagnostic test it adds little useful information, in the opinion of the author.

Serum chloride and bicarbonate

Patients with primary hyperparathyroidism frequently have a mild metabolic acidosis. This is usually reflected by a serum chloride which is >103 mmol/l. Patients with most other types of hypercalcemia, and in particular hypercalcemia of malignancy, almost always have a serum chloride which is <100 mmol/l. The use of the serum chloride as a differential diagnostic point in the diagnosis of hypercalcemia was first emphasized by Wills and McGowan,[34] and in the opinion of the author it remains a useful index. The reason that the serum chloride is increased in patients with primary hyperparathyroidism is that PTH acts on the renal tubules to prevent bicarbonate reabsorption, thereby leading to a mild form of hyperchloremic renal tubular acidosis.

Urine calcium

Measurement of renal calcium excretion is most useful in the diagnosis of FHH, and should be performed in all patients with presumed asymptomatic primary hyperparathyroidism, in order to exclude FHH. Urine is collected for 24 h in containers with an acid preservative. Creatinine should be determined on the same specimen to ensure that the specimen collection has been adequate. In patients with FHH, the urine calcium excretion is almost always <100 mg/g creatine, whereas in most patients with primary hyperparathyroidism urine calcium is usually >200 mg/g creatinine. In patients with the hypercalcemia of malignancy there is also frequently an increase in renal tubular calcium reabsorption mimicking the situation in primary hyperparathyroidism, so that urine calcium measurements may overlap those of patients with primary hyperparathyroidism. The usual urinary calcium excretion is 200–600 mg/g creatinine in patients with both primary hyperparathyroidism and malignant disease. As discussed in Chapter 1, fasting urine calcium provides an indirect assessment of net bone resorption.

Serum alkaline phosphatase

Alkaline phosphatase is a marker enzyme of the mature osteoblast. It is usually increased in patients with increased rates of bone resorption, because formation of new bone is usually coupled to previous bone resorption. Consequently, osteoblast activity is an indirect assessment of previous osteoclastic resorption. Exceptions to this rule occur in myeloma and some lymphomas, where there is frequently little evidence of an increase in new bone formation. Serum alkaline phosphatase is usually increased in patients with primary hyperparathyroidism and in patients with most types of malignancy associated with hyper-

calcemia. The hematologic malignancies are an exception. The alkaline phosphatase present in the serum consists of a number of isoenzymes which may come not only from bone, but also from liver, tumor, or the placenta. These isoenzymes can be distinguished by electrophoresis or by their relative heat-lability. The bone isoenzyme is unstable at 56°C for 15 min ('bone alkaline phosphatase burns').

Urine cAMP

The urinary cAMP corrected for creatinine or for glomerular filtration rate gives similar information to assays of PTH. The concentration of cAMP in the urine depends mainly on circulating biologically active PTH. Urinary cAMP is increased in patients with primary hyperparathyroidism and in some patients with malignancy, presumably due to the actions of PTH-like peptides on the renal tubules. It is therefore of limited use in the differential diagnosis of primary hyperparathyroidism from hypercalcemia due to malignancy. The calculation of nephrogenous cAMP from measurements of cAMP in both the blood and the urine has been used frequently in investigative studies, but for the practising clinician offers no advantage over the simpler measurement of urinary cAMP. Measurements of urinary cAMP are difficult to interpret in patients with renal failure. Urine collections for cAMP should be made under the same conditions as those for urine calcium. The measurement is best expressed in terms of the glomerular filtration rate, that is in nM 100 ml glomerular filtrate.

Serum immunoreactive PTH

Radioimmunoassays for PTH have now been available for 20 years and have been improved considerably in recent years. However, the measurements are still far from being ideal. The major problems are that PTH circulates as fragments, and these multiple circulating fragments vary both in biological activity and immunoreactivity. The PTH assay is discussed in more detail in Chapter 8. Unfortunately, the fragments of the molecule which have been the most readily measured by immunoassay are those which are

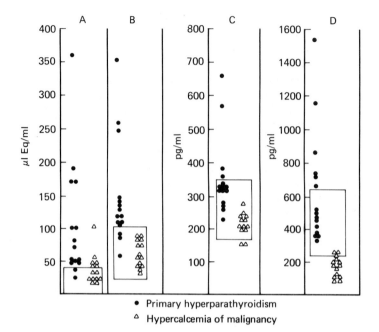

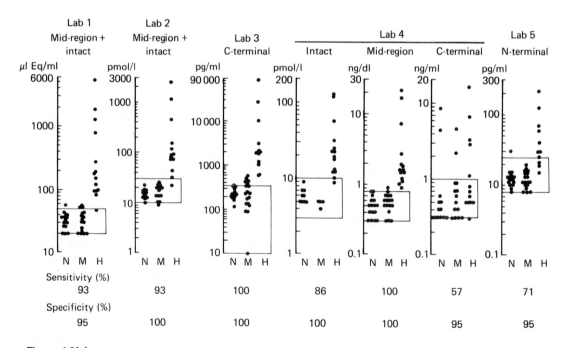

Figure 4.3(a)

Upper panel represents a 1979 report of four different PTH assays on patients with hypercalcemia (reproduced from Raisz et al, Comparison of commercially available parathyroid hormone immunoassays in the differential diagnosis of hypercalcemia due to primary hyperparathyroidism or malignancy, *Ann Intern Med* (1979), **91**, 739–40). Lower panel is a 1987 comparison of different types of PTH (lanes A–D) assay on aliquots of blood derived from patients with primary hyperparathyroidism and the hypercalcemia of malignancy (N = Normal; M = malignancy, H = hyperparathyroidism).[35]

least active biologically and those which are most dependent on renal function for their clearance. The newer N-terminal assays more closely reflect minute-to-minute secretion by the parathyroid gland, and they measure the biologically active portion of the molecule, but in practice they seem to be even less satisfactory than assays directed towards the C-terminal or mid-region of the molecule (Figure 4.3(a)).[35]

Improved results with assays for circulating PTH have been claimed recently with the introduction of techniques utilizing antibodies directed at two different sites on the PTH molecule. These include the immunoradiometric assay or IRMA[36,37,38] and the two-site immunochemiluminometric assay.[39] These techniques limit cross-reactivity with inactive PTH fragments, as they utilize two separate antisera directed at the N-terminal and C-terminal ends of the PTH molecule. Thus, these assays measure intact PTH and should not detect inactive PTH fragments. These assays probably improve discrimination between patients with primary hyperparathyroidism and patients with the hypercalcemia of malignancy. Blind et al[37] found no overlap between primary hyperparathyroidism and malignancy-associated hypercalcemia in 70 patients with primary hyperparathyroidism and 39 patients with hypercalcemia of malignancy (Figure 4.3(b)). In only one of their patients with primary hyperparathyroidism was serum PTH in the normal range. In 37 out of 39 patients with hypercalcemia of malignancy,

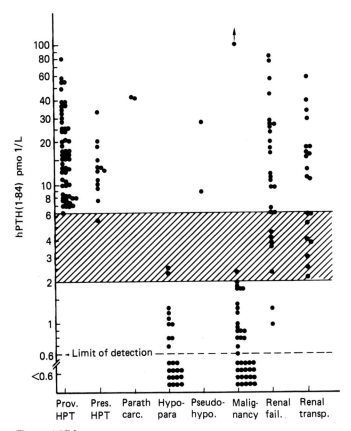

Serum intact hPTH-(1–84) concentrations in 70 patients with presumed (Pres. HPT; n = 13) or proven (Prov. HPT; n = 57) primary hyperparathyroidism, 2 patients with parathyroid carcinoma (Parath. carc.), 25 patients with hypoparathyroidism (Hypopara.), 2 patients with pseudohypoparathyroidism (Pseudohypo), 40 patients with malignancy-associated hypercalcemia, 27 patients with chronic renal failure (Renal Fail.; among them 17 patients receiving hemodialysis), and 20 patients after renal transplantation (Renal transp.). The *hatched area* represents the normal range, and the *arrow* indicates a value beyond the scale.

Figure 4.3(b)

Parathyroid hormone concentrations in patients with the hypercalcemia of malignancy, primary hyperparathyroidism and other disorders of calcium homeostasis (redrawn from Blind et al[37])

serum intact PTH was below the upper limit of normal. In comparison with a standard mid-region PTH radioimmunoassay, the IRMA was found to provide much better discrimination between patients with primary hyperparathyroidism and patients with hypercalcemia of malignancy.

In practical terms the interpretation of the PTH assay should be made with some consideration of these technical difficulties. Patients who are hypercalcemic and have a PTH concentration which is >30 per cent above normal in any assay in the presence of normal renal function are almost certain to have primary hyperparathyroidism. However, there are still many patients with primary hyperparathyroidism who have measurements which are in the normal range or only slightly increased when measured with many of the available commercial assays. Similarly, in patients with nonparathyroid malignant disease and hypercalcemia PTH measurements may be suppressed, in the normal range, or even slightly increased. However, the measurements will not be markedly increased unless the patient has renal failure. In some cases at least, the reason that they are slightly increased or in the normal range could reflect cross-reactivity of antisera with PTH-like factors present in the circulation of some of these individuals, although it is more likely to be caused by the release of biologically inert PTH fragments from the glands suppressed by hypercalcemia. In the past some investigators have used nomograms in patients with hypercalcemia due to malignancy, in order to interpret a normal PTH as being 'inappropriately increased'. However, these nomograms are of very limited use in the differential diagnosis of hypercalcemia, and they have almost certainly been overinterpreted.

Vitamin D metabolites

Two biologically active vitamin D metabolites are frequently measured; serum 25-hydroxyvitamin D and 1,25-dihydroxyvitamin D. Serum 25-hydroxyvitamin D is the major circulating form of the vitamin, and is a reflection of vitamin D intake. It is increased in patients who ingest excessive vitamin D and may be increased 10–20-fold above the normal range, which is 10–80 ng/ml. It tends to be higher in patients living in sunny climates

Table 4.4 Types of hypercalcemia associated with increased 1,25-dihydroxyvitamin D.

Sarcoidosis and other granulomatous diseases

Malignant lymphomas

Primary hyperparathyroidism

and in the months when patients are exposed to more sunshine, such as spring and summer.

Serum 1,25-dihydroxyvitamin D has limited use in the differential diagnosis of hypercalcemia (Table 4.4). It is increased mildly in some patients with primary hyperparathyroidism, although there is some disagreement as to how frequently this occurs. Serum 1,25-dihydroxyvitamin D may also be increased in patients with hypercalcemia due to sarcoidosis or other granulomatous diseases. In this situation the 1,25-dihydroxyvitamin D is presumably produced in the granulomatous tissue. Serum 1,25-dihydroxyvitamin D will also be increased in occasional patients with hypercalcemia due to T-cell lymphomas, Hodgkin's disease, or even B-cell lymphomas. However, this too is probably an unusual occurrence, and most hypercalcemic patients with lymphomas do not have increased serum 1,25-dihydroxyvitamin D.

Bone Gla protein

The bone Gla protein (also called osteocalcin) is the major noncollagen protein in the bone matrix. The function of this protein still remains unknown more than 10 years after its discovery. It is now readily measurable in the serum by radioimmunoassay, and it is a reliable parameter of osteoblast function. However, it may not be produced by the same cells in the osteoblast lineage as alkaline phosphatase, and its measurement does not entirely parallel that of serum alkaline phosphatase. It is usually increased in patients with primary hyperparathyroidism, but it is decreased in some patients with hypercalcemia of malignancy. Why it should be decreased in this situation remains unclear, although there are frequently differences in rates of bone formation between patients with primary hyperparathyroidism and patients with malignant disease. At present the measurement of serum bone Gla protein is still an investigative tool, and in the opinion of the author it provides little useful or interpretable information in the differential diagnosis of hypercalcemia.

Skeletal radiology

Skeletal X-rays may be useful in the differential diagnosis of hypercalcemia in special situations. A few patients with primary hyperparathyroidism have the characteristic radiologic changes of osteitis fibrosa which can be seen on radiography. These include subperiosteal resorption of the phalanges and the 'salt-and-pepper' skull, as well as the other changes noted in Chapter 8. These features are essentially pathognomonic of primary hyperparathyroidism in patients with normal renal function. The author has seen only one hypercalcemic patient with malignant disease (and without primary hyperparathyroidism) who had these changes—an elderly woman with a bronchial carcinoid. In patients with malignancy, skeletal X-rays may indicate the presence of osteolytic bone lesions. Of course, skeletal radiology may be essentially diagnostic for myeloma, and sometimes the pattern of osteolytic metastases can be highly suggestive of a diagnosis of breast cancer.

Bone scanning

Although the bone scan should not be used as a routine test in the differential diagnosis of hypercalcemia, it will often be useful and particularly in the patient with suspected metastatic disease. It may direct the clinician to sites where localized skeletal X-rays should be performed. The bone scan may be negative in a patient with myeloma who has obvious radiologic lesions, due to the suppression of osteoblast activity.

Corticosteroid suppression test

The corticosteroid suppression test was introduced many years ago by Dent[40] and still has limited use in the differential diagnosis of hypercalcemia. This test is most useful in the differential diagnosis of mild hypercalcemia. In such patients who have primary hyperparathyroidism, the serum calcium will almost never fall in response to 10 days of corticosteroid therapy (prednisone 30 mg/day, or hydrocortisone 100 mg/day), whereas all patients with hypercalcemia due to sarcoid or other granulomatous diseases, or vitamin D intoxication, and some patients with hypercalcemia due to malignancy will respond with a decrease in the serum calcium.

Bone histomorphometry

Quantitative bone histomorphometry performed on nondecalcified biopsy specimens taken from the iliac crest may occasionally give useful information in the differential diagnosis of hypercalcemia (Figure 4.4). The specimen consists of a through-and-through section of bone which is taken from the iliac crest by percutaneous needle biopsy after the patient takes two short courses of tetracycline, separated by approximately 10 days. Patients with hypercalcemia due to primary hyperparathyroidism usually have distinctive changes in rates of bone resorption and bone formation. Most patients show evidence of increased osteoclastic bone resorption associated with increased rates of bone turnover, increases in osteoid tissue, and increases in the mineral apposition rate. In contrast, patients with malignant disease may show similar increases in osteoclastic bone resorption, but they usually show impairment of bone formation, with little evidence of osteoblast activity, decreased osteoid tissue, and decreases in mineral appositional rate. The reason for these differences is unknown. It is possible that in some patients with cancer, factors such as immobilization or poor nutrition may be involved. However, it is also likely that there are tumor products which specifically inhibit osteoblast activity. Transforming growth factor α, interleukin-1, TNF, and lymphotoxin have all been found to have this effect (Chapter 3).

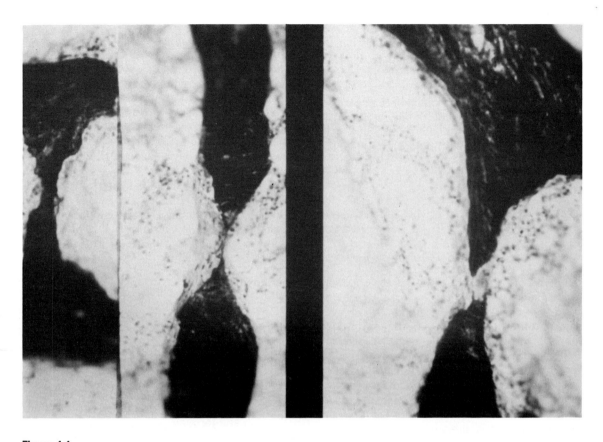

Figure 4.4

Iliac crest biopsy specimen taken from a patient with hypercalcemia of malignancy. The two panels show evidence of increased osteoclastic bone resorption and decreased osteoblast activity.

Other tests

Other tests should occasionally be used in the differential diagnosis of hypercalcemia. Parathyroid gland localization should be reserved for special situations such as before re-exploration of the neck after a failed attempt to locate a parathyroid adenoma, because of the cost and limited usefulness. However, with the advent of useful and noninvasive techniques such as real-time high resolution ultrasonography, this practice is changing. This is discussed in more detail in Chapter 8. All patients with hypercalcemia of uncertain cause should have thyroid function tests to exclude thyrotoxicosis. Whenever myeloma is suspected, serum protein electrophoresis, measurement of serum and urine monoclonal proteins, and bone marrow biopsy and aspirate should be performed to exclude this disorder. Photon absorptiometry (single beam or dual beam) is of no value in the differential diagnosis of hypercalcemia, but measurements of cortical bone density may be helpful in following patients with primary hyperparathyroidism to monitor rates of bone loss. Measurements of urine hydroxyproline excretion have been used to monitor changes in rates of bone resorption in patients with osteolytic metastases, but it is not a direct measurement of bone collagen degradation, and it has no current clinical use in this situation.

Watson et al,[41] apparently dissatisfied with the available PTH radioimmunoassays, used discriminate analysis (the discriminate functions including plasma phosphate, chloride, bicarbonates, urea, alkaline phosphatase, and erythrocyte sedimentation rate), and found this approach to be inexpensive and usually accurate in the differential diagnosis of hypercalcemia assessed retrospectively.

Important clinical points in the differential diagnosis of hypercalcemia

Points which favor a diagnosis of primary hyperparathyroidism

Age and sex

Primary hyperparathyroidism occurs most frequently in elderly females. In these patients hypercalcemia is usually asymptomatic and mild. Thus, when hypercalcemia occurs in an elderly woman who is asymptomatic, and if there is no obvious underlying malignancy, primary hyperparathyroidism is by far the most likely diagnosis.

Duration of hypercalcemia

The length of history is important in the differential diagnosis of hypercalcemia. If a patient has been mildly hypercalcemic for more than 1 year, then primary hyperparathyroidism is the most likely cause, particularly if the patient is hyperchloremic and hypophosphatemic. For this reason it is important for the clinician to check back as far into the past as possible for any previous determinations of serum calcium.

Hyperchloremia

Hyperchloremia is particularly helpful in the differential diagnosis of primary hyperparathyroidism. In most patients with primary hyperparathy-

roidism the serum chloride will be >103 mmol/l provided the patient has normal renal function. In the opinion of the author, the chloride/phosphorus ratio, advocated by some, is not useful because the serum phosphorus is so variable and is influenced by so many factors that have nothing to do with the cause of the hypercalcemia. Since the denominator fluctuates so much, the ratio shows too wide a range to be helpful.

Urinary calcium excretion

The urinary calcium excretion is an important measurement, since it can exclude FHH (Chapter 9). The urine calcium excretion is usually <100 mg/g creatinine in this disorder, but is usually >200 mg/g creatinine in primary hyperparathyroidism.

Urinary or nephrogenous cAMP

Measurement of urinary cAMP excretion or nephrogenous cAMP is not particularly helpful in the differential diagnosis of hypercalcemia. In primary hyperparathyroidism both measurements should always be increased. If they are not increased, then the diagnosis of primary hyperparathyroidism should be reconsidered. However, as indicated above, many patients with hypercalcemia of malignancy also have increases in this parameter.

Serum PTH

Measurement of the serum PTH is only helpful if it is greater than 30–50 per cent above normal (depending on the assay). It is not helpful in any way if it is within the normal range. A measurement in the normal range does not exclude primary hyperparathyroidism in any of the known assays.

Other points

There are some laboratory findings which point strongly towards the diagnosis of primary hyperparathyroidism. One of these is the radiological demonstration of subperiosteal resorption of the fingers. Another is a history of renal stones, which is much more common in primary hyperparathyroidism than in any of the other causes of hypercalcemia.

An important point to consider in the differential diagnosis of hypercalcemia, and in particular the diagnosis of primary hyperparathyroidism, is that primary hyperparathyroidism is a common disease which frequently occurs in patients with cancer. Therefore, simply because a patient has cancer it does not mean that primary hyperparathyroidism is not the cause of the hypercalcemia.

Points which favor a diagnosis of cancer

There are several important points to bear in mind in making a diagnosis of hypercalcemia due to underlying malignancy.

The first point is the pattern of malignancies which are associated with hypercalcemia. Hypercalcemia is rarely associated with some common malignancies such as cancer of the colon or cancer of the uterine cervix. If hypercalcemia occurs in patients with these disorders, then another cause should be sought carefully. Similarly, hypercalcemia rarely occurs in breast cancer, except late in the course of the disease and when there is widespread osteolytic bone destruction. Thus, when hypercalcemia occurs in a patient with breast cancer without obvious metastatic bone disease, another cause of hypercalcemia (such as primary hyperparathyroidism) should be diligently sought.

When a patient with apparent malignant disease presents with renal failure and hypercalcemia the most likely cause by far is myeloma.

When Hodgkin's disease is complicated by hypercalcemia, it occurs most frequently in elderly men, in association with bulky abdominal disease, and often with increased serum 1,25-dihydroxyvitamin D.

Occult malignancy as a cause of hypercalcemia is quite unusual. Thus, when hypercalcemia occurs and there is no obvious underlying malignancy, it is more likely that another cause is responsible.

Hypercalcemia is fairly frequently associated with some uncommon malignancies. Awareness of this point may help in the differential diagnosis of hypercalcemia. Thus, hypercalcemia is common in patients with rare malignancies such as cholangiocarcinoma and VIPomas (neoplasms of vasoactive intestinal peptide-secreting cells). It is also relatively common in patients with squamous cell tumors of the lung, head, and neck, and occurs frequently with adenocarcinomas of the kidney and ovary.

Serum 1,25-dihydroxyvitamin D measurements are suppressed in most patients with the hypercalcemia of malignancy, except in a few patients who have a T-cell lymphoma/leukemia, a B-cell lymphoma, or Hodgkin's disease associated with the hypercalcemia.

Other types of hypercalcemia

There are other useful diagnostic points which should be considered in the differential diagnosis of hypercalcemia related to the other less common causes.

In the granulomatous diseases both the serum calcium and the serum phosphorus are increased together, and the patients frequently have nephrocalcinosis. These patients have increases in the serum 1,25-dihydroxyvitamin D concentration, and the hypercalcemia is frequently associated with impaired renal function. These patients will respond to 10 days of corticosteroid therapy (the corticosteroid suppression test).

In vitamin D intoxication the serum 25-hydroxyvitamin D concentration is always markedly increased. These patients also have an increase in the serum calcium and phosphorus, and they develop nephrocalcinosis with impaired renal function.

In FHH renal calcium clearance is always low, so urine calcium excretion is less than 100 mg/g creatinine. These individuals are usually young, frequently asymptomatic, and have a family history of hypercalcemia. Because this is an autosomal dominant disorder with complete penetrance, half of the siblings or first-degree relatives will be affected. Another important differential diagnostic for this condition is that the hypercalcemia has been present from early life, unlike the multiple endocrine neoplasia syndromes (Chapter 9).

Malignancy and hypercalcemia— humoral hypercalcemia of malignancy, hypercalcemia associated with osteolytic metastases

Humoral hypercalcemia of malignancy

Epidemiology

Since determinations of the serum calcium have become routine, the recognition of patients with hypercalcemia associated with malignancy has increased, as has the recognition of other causes of hypercalcemia. It is estimated that the annual incidence rate of hypercalcemia of malignancy is about 150 new cases per million population per year. These figures come from an epidemiologic study performed in Birmingham in 1979.[1] In that special location it was possible to detect all of the new patients with hypercalcemia in a population of 1 million people over a 5-month period. Seventy-two patients with hypercalcemia of malignancy were recognized in a total of 207 consecutive patients with hypercalcemia.[2] Although hypercalcemia due to malignancy is not as frequent as primary hyperparathyroidism in the general population, it occurs relatively more frequently in the hospital population. Patients with primary hyperparathyroidism are frequently asymptomatic and are ambulant, whereas patients with the hypercalcemia of malignancy are usually hospitalized when hypercalcemia is first detected.

The frequency of the different types of malignancy which are associated with hypercalcemia are outlined in Table 5.1. It is important to make

Table 5.1 Malignancies associated with hypercalcemia (%).

Lung	35
Breast	25
Hematologic (myeloma, lymphoma)	14
Head and neck	6
Renal	3
Prostate	3
Unknown primary	7
Others	8

Stewart et al, Synthetic human parathyroid hormone-like protein stimulates bone resorption and causes hypercalcemia in rats, *J Clin Invest* (1988) **81**: 596–600.

several points. First, the most common malignancies which occur in patients are frequently associated with hypercalcemia. Thus, lung cancer and breast cancer are the two most common solid tumors, and these are also the two most common malignancies associated with hypercalcemia. However, there are some relatively common causes of malignancy which are not frequently associated with hypercalcemia. Examples are carcinomas of the female genital tract and carcinomas of the colon.[2] However, some rare malignancies are frequently associated with hypercalcemia. For example, cholangiocarcinoma and

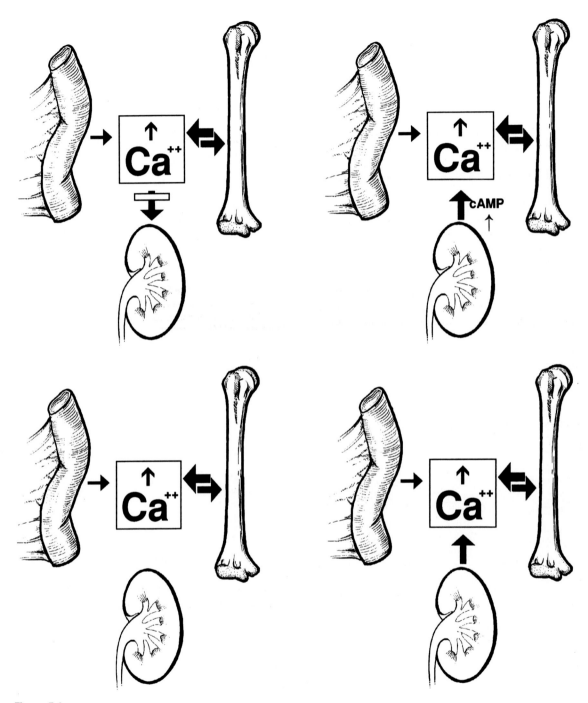

Figure 5.1

Patterns of different abnormalities in calcium transport between the extracellular fluid and gut, kidney and bone in different types of hypercalcemia of malignancy. In myeloma, there is an increase in bone resorption and impaired glomerular filtration. In lung cancer, there is increased bone resorption and increased renal tubular calcium reabsorption, associated with increased nephrogenous cAMP. In breast cancer there is increased bone resorption and increased renal tubular calcium reabsorption. In some lymphomas, there is increased bone resorption and increased gut absorption of calcium.

VIPomas of the pancreas occur infrequently, but when they do occur they are often associated with hypercalcemia. It is now clear that there is no single unifying mechanism for hypercalcemia associated with malignancy. In the following chapters hypercalcemia of malignancy is arbitrarily considered in four separate groups (Figure 5.1). In the first group, hypercalcemia is associated with solid tumors which may or may not be accompanied by bone metastases. This is the type of hypercalcemia associated with lung cancer, renal cancer, and ovarian cancer, and is usually called the humoral hypercalcemia of malignancy (HHM). The second group are the solid tumors which are associated with extensive bone metastases. The most frequent example in this group is breast cancer. In the third group is myeloma, where hypercalcemia seems to be due to a combination of impaired glomerular filtration and increased bone resorption. In the final group are the lymphomas, where hypercalcemia seems to be due, at least in some cases, to increased absorption of calcium from the gut, together with increased bone resorption.

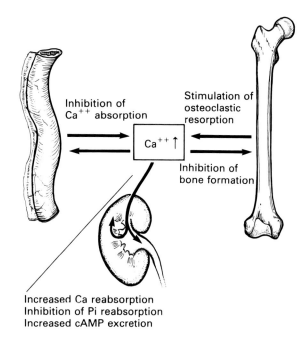

Increased Ca reabsorption
Inhibition of Pi reabsorption
Increased cAMP excretion

Figure 5.2

Patterns of abnormalities in patients with solid tumors associated with the humoral hypercalcemia.

Humoral hypercalcemia of malignancy (HHM)

The humoral hypercalcemia of malignancy is the name given to the hypercalcemic syndrome which occurs in patients with solid tumors, and which is due to the production of systemic or circulating humoral factors that cause hypercalcemia. These patients may or may not have bone metastases. The important point is that the hypercalcemic factor or factors produced by the tumor (or in response to the tumor) acts systemically.

metabolic alkalosis, increased renal tubular calcium reabsorption, and morphologic parameters of increased bone resorption associated with decreased bone formation (Figure 5.2).

Features of HHM

The HHM syndrome is often associated with an increase in the fractional excretion of phosphate, an increase in nephrogenous cAMP and decrease in plasma immunoreactive PTH.[3,4] Other features which are common include suppressed or low normal serum 1,25-dihydroxyvitamin D, mild

Types of malignancies associated with HHM

The malignancies frequently associated with HHM include squamous cell carcinoma of the lung, head and neck, carcinoma of the kidney, carcinoma of the ovary, and carcinoma of the pancreas. Of these the most frequent are squamous cell carcinomas of the lung, head, and neck.[2]

Pathophysiology of the hypercalcemia in HHM

In those cases of hypercalcemia associated with solid tumors that have been studied thoroughly, there appears to be an increase in osteoclastic bone resorption, an increase in renal tubular calcium reabsorption, and impaired calcium absorption from the gut (Figure 5.1). The mechanisms responsible for these changes are still not entirely clear, but it appears certain that they are due to circulating hypercalcemic factors released by the tumor cells, or by normal cells provoked by the presence of a tumor. The recent dramatic progress in this area has been summarized by Mundy.[5]

Evidence for a role for increased bone resorption

The evidence that bone resorption is increased in patients with the hypercalcemia of malignancy is overwhelming. Histologic studies have revealed evidence of increases in osteoclast activity,[6,7] there is increased hydroxyproline in the urine,

Table 5.2   Tumor factors and bone.

TGFs

PTH-like factors

Prostaglandins

Leukocyte cytokines (OAFs)

1,25 D

indicating breakdown of bone matrix, and hypercalcemia is frequently improved by drugs which inhibit osteoclast activity, such as bisphosphonates, calcitonin, corticosteroids, and plicamycin (mithramycin). Moreover, factors which stimulate osteoclastic bone resorption in vitro have frequently been detected in cell culture media harvested from these tumors or in tumor extracts (reviewed in Mundy[4]). The major interest in this area is identification of the factors produced by the tumor cells which are responsible for this increase in bone resorption. Factors potentially responsible include PTH, PTH-like factors, transforming growth factors, colony stimulating factors, and leukocyte cytokines (Table 5.2). There are almost certainly additional novel bone-resorbing factors produced by solid tumors associated with hypercalcemia which are not in the above list.[8]

Parathyroid hormone (PTH)

Since PTH is a major regulator of osteoclastic bone resorption as well as calcium homeostasis, and since many of the features of the hypercalcemia of malignancy are similar to those of primary hyperparathyroidism, it is not surprising that PTH was long suspected to be responsible for the hypercalcemia of malignancy. This mechanism was first suggested by Albright[9] in a case presentation from the Massachusetts General Hospital presented in the *New England Journal of Medicine* (Case records, 1941). In that case a renal tumor occurring in a patient with hypercalcemia was treated with radiation therapy. The tumor decreased in size and the hypercalcemia improved. The patient had renal phosphate wasting in addition to hypercalcemia before therapy. The renal tumor extract was assayed for PTH-like activity, but none was found. In 1966 Lafferty[10] described the clinical features of the syndrome which is now known as HHM or the humoral hypercalcemia of malignancy. He called it 'pseudohyperparathyroidism', and for the next 5–10 years the concept held sway that the hypercalcemic syndrome associated with cancer was due to PTH and it was called either the ectopic PTH syndrome or pseudohyperparathyroidism. In 1964, Tashjian et al,[11] using an early immunologic technique for the detection of PTH, claimed PTH-like activity in tumor extracts. Then, in 1967,

Sherwood et al[12] found PTH-like immunoreactivity in tumor extracts. Buckle et al[13] described a gradient of immunoreactive PTH across a renal tumor bed and, in a larger series, Benson et al[14] found PTH-like immunoreactivity in the serum of many patients with malignancy with or without hypercalcemia.

Despite all of these indications that PTH may be responsible for tumor hypercalcemia, some doubts on the role of PTH in nonparathyroid tumor hypercalcemia became apparent with the report by Powell et al[15] that in a series of thirteen tumors associated with hypercalcemia, none contained PTH immunoreactivity although bone resorbing activity was present in the tumor extracts. Following this report, and the development of more sophisticated PTH radioimmunoassays, it was realized that PTH immunoreactivity was more frequently low than elevated in patients with hypercalcemia of malignancy. Finally, Simpson et al,[16] using a cDNA probe for PTH, which detects PTH messenger-RNA (mRNA) transcripts in tumor tissue, could find no evidence of PTH gene transcripts in a series of hypercalcemic tumors associated with non-parathyroid malignancies. There is currently no well-documented case of a non-parathyroid tumor producing PTH.

What then could be the explanation for the finding of PTH immunoreactivity in patients with malignancy? There are several possibilities. The first is that there could be immunoreactive but biologically inert PTH fragments released from the suppressed parathyroid glands of patients with the hypercalcemia of malignancy. This could explain the small elevations in PTH immunoreactivity present in some patients where an alternative mechanism was responsible for causing hypercalcemia. The second explanation is that a tumor product related to PTH could cross-react with some PTH radioimmunoassays. This is not far-fetched, since a tumor factor which binds to the PTH receptor has now been identified which shares some amino acid sequence homology with PTH,[17] and could therefore conceivably be recognized by PTH antisera. However, these results were also obtained in patients where PTH-like factors are not produced, such as myeloma, and the similarities in the tumor factor and PTH are at the N-terminal end of the molecule, a region which those older assays did not recognize well. Whatever the explanation for these spurious PTH radioimmunoassay results, it is clear that PTH itself is rarely the mediator which causes hypercalcemia of malignancy.

However, there has been one recent case report which shows that ectopic PTH production by nonparathyroid tumors does occur. Yoshimoto et al[18] studied a patient with small cell carcinoma of the lung who developed severe hypercalcemia. In this patient, the plasma immunoreactive PTH concentration was markedly increased in three different radioimmunoassays, but the parathyroid glands were histologically normal at post-mortem. A tumor extract from a liver metastasis was examined, and immunoreactive PTH was identified which eluted with synthetic human PTH (1–84) on gel filtration chromatography. Northern analysis showed PTH mRNA in the tumor extract. PTH-rP mRNA was not detected. Thus, this patient is a rare but well documented example of ectopic PTH production in a non-parathyroid cancer.

Parathyroid hormone-like factors

As already noted, it has been postulated for many years that PTH or a PTH-like factor could be involved in the hypercalcemia of malignancy. Albright could not find PTH-like activity in the tumor extracts from his patient with renal cell carcinoma and hypercalcemia, but it is likely that his bioassay was too insensitive. The clinical observations reported by Lafferty[10] on patients with hypercalcemia of malignancy showed that the features of hypercalcemia of malignancy were similar to those of primary hyperparathyroidism, and included renal phosphate wasting, renal tubular calcium reabsorption, osteoclastic bone resorption, and hypercalcemia. Later observations also revealed increased nephrogenous cAMP.[19] In 1981, Stewart et al[20] reported that tumor extracts from patients with the hypercalcemia of malignancy contained biological activity which stimulated adenylate cyclase in renal tubular membranes. This effect was inhibited by antagonists to PTH, suggesting that these factors were interacting with the PTH receptor. At about the same time Goltzman et al[21] showed that tumor extracts and the serum of patients with the hypercalcemia of malignancy contained a factor which stimulated renal cell glucose-6-phosphate dehydrogenase (G6PD) content, an effect also produced by PTH and presumably mediated through the PTH receptor on renal

```
           -36         -30              -20               -10             1              10
hPTHrP   M Q R R L V Q Q W S V A V F L L S Y A D P S C G R S V E G L S R R L K R A V S E H Q L L H D K G K S
hPTH         M I P A K D M A K V M I V M L A I C F L T K S D G K S V K K R S V S E I Q L M H N L E K H


             20              30               40              50              60
hPTHrP   I Q D L R R R F F L H H L I A E I H T A E I R A T S E V S P – N S K P S P N T K N H P V R F G S D D
hPTH     L N S M E R V E W L R K K L Q D V H N F V A L G A P L A P R D A G S Q R P R K K E D N V L V E S H E


             70              80               90              100             110
hPTHrP   E G R Y L T Q E T N K V E T Y K E Q P L K T P G K K K K G K P G K R K E Q E K K K R R T R S A W L D
hPTH     K S L G E A D K A D V N V L T K A K S Q
```

Figure 5.3

Amino acid structure of the human PTH-related protein (hPTH-rP) compared with human PTH (hPTH).

cells. These results were confirmed in other tumors and using slightly different systems.[22,23] Since then this PTH-like factor or factors have been more fully characterized. Martin and his colleagues reported complete purification of the factor, and have shown considerable amino acid sequence homology with authentic PTH in the first thirteen amino acids—eight of the first thirteen amino acids are identical—which is the portion of PTH responsible for binding to the PTH receptor (Figure 5.3).[17] This group has also reported the molecular cloning of the complementary DNA for the gene which transcribes this factor.[24] Others also purified identical proteins in a breast cancer extract and in a cultured renal cancer cell line.[25,26] These factors apparently activate the PTH receptor in a slightly different manner from that of PTH. For example, according to the studies of Rodan et al,[22] they have a different time-course and produce a more pronounced effect on adenylate cyclase in these systems. In several tumor models studied,[27] the tumor cells produce bone-resorbing activity, but the bone-resorbing activity is not dependent on the PTH receptor because it is not blocked by PTH antagonists which inhibit the bone-resorbing effects of PTH. Moreover, patients who are pre-

sumably producing these factors do not have the same spectrum of clinical features that is seen in patients with primary hyperparathyroidism. Similar activity has been described in other tissues; for example, in keratinocyte cultures.[28] Similar biological activity has also been described in tumor extracts from patients with oncogenic osteomalacia.[29]

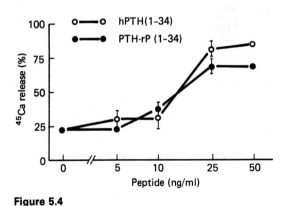

Figure 5.4

Effects of hPTH and hPTH-rP on bone resorption in organ culture.[32]

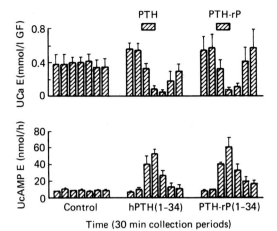

Figure 5.5

Effects of PTH and PTH-rP on urinary calcium excretion and urinary excretion of cAMP.[32]

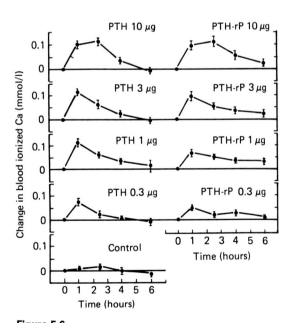

Figure 5.6

Effects of PTH-rP on blood ionized calcium in intact mice following a single subcutaneous injection.[32]

Descriptions of the effects of synthetic peptides representing the N-terminal end of this protein are now being reported, and they appear to be essentially identical to those of PTH on the kidney and bone.[30–32] Synthetic peptides or analogs of PTH-rP (PTH-related protein) stimulate osteoclastic bone resorption in vitro (Figure 5.4),[30–33] stimulate renal tubular calcium reabsorption (Figure 5.5)[32] and cause hypercalcemia in vivo (Figure 5.6 and 5.7).[31–33] PTH-rP also stimulates osteoclastic bone resorption in vivo (Figures 5.8 and 5.9). Differences in potency with PTH have been suggested,[30] although these have not been described by other workers who have compared the effects with those of PTH (Figure 5.10).[31–33]

These findings give rise to obvious questions, such as what is the normal physiologic role of this protein, and what is its phylogenetic relationship to PTH and the PTH receptor. The immediate issue is that this protein does not explain all of the manifestations of humoral hypercalcemia of malignancy. The syndrome of humoral hypercalcemia of malignancy associated with solid tumors is quite distinct from that of primary hyperparathyroidism, and cannot be explained by what is currently known of the effects of synthetic peptides of PTH-rP on kidney, bone and gut (Figure 5.2). For example, osteoclastic bone resorption is much more prominent in humoral hypercalcemia of malignancy than in primary hyperparathyroidism and, in most cases, rates of bone formation are reduced in malignancy-associated hypercalcemia. There are also differences in renal tubular function between the two syndromes. Patients with primary hyperparathyroidism have normal or increased 1,25-dihydroxyvitamin D production which is possibly a consequence of increased PTH and lowered ambient phosphate concentrations. However, in the hypercalcemia of malignancy, even when PTH-rP is produced by the tumor, 1,25-dihydroxyvitamin D production is often suppressed,[34] and patients tend to be alkalotic, rather than hyperchloremic and mildly acidotic, as is the case in primary hyperparathyroidism.[35] Why do these differences exist when a peptide which activates PTH receptors is involved in both syndromes? One possibility is that the full length PTH-rP molecule behaves differently from the synthetic peptide. However, our group has tested the intact PTH-rP molecule and found that it behaves identically to synthetic PTH-rP(1–34) on bone resorption in vitro, renal cell and bone cell

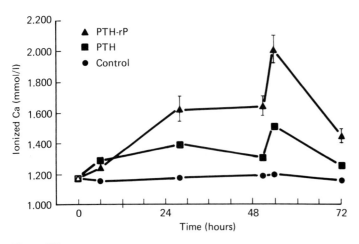

Figure 5.7

Repeated injection protocol, with effects of single injection of PTH and PTH-rP on blood calcium 2 h later. Note brisk rise after injection given at 50 h followed by decline, most likely due to increased renal tubular calcium reabsorption.

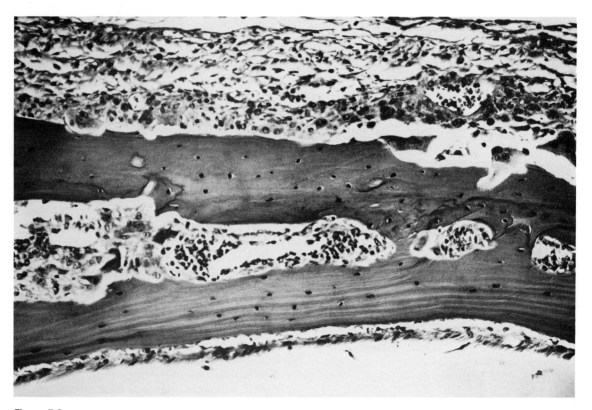

Figure 5.8

Effects of repeated subcutaneous injections of PTH-rP on calvarial bones of intact mice, showing increase in osteoclastic activity.[32]

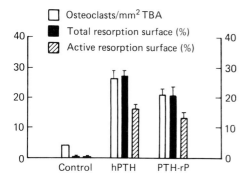

Figure 5.9

Effects of PTH and PTH-rP on bone resorption assessed quantitatively in intact mice.[32]

target organs are modified by other factors, such as interleukin-1, transforming growth factor α, and tumor necrosis factor. These factors also stimulate osteoclastic bone resorption[37] and increase plasma calcium.[38,39] They are produced by tumors associated with hypercalcemia, and often together with the PTH-rP. All of these are more powerful bone resorbing factors than PTH in vivo[32,38] and in vitro.[40] These factors act synergistically with PTH (and probably also with PTH-rP) to stimulate bone resorption in vitro[41] possibly because their major effect is to stimulate proliferation of osteoclast progenitors.[42] Recently, our group has observed that TGFα acts synergistically with PTH-rP to increase plasma calcium,[43] and Sato et al[44] have shown that PTH-rP acts synergistically with interleukin-1α to increase plasma calcium in vivo. However, the effects of PTH (and presumably PTH-rP) on some target cells are also substantially impaired by these factors. For example, TGFα, interleukin-1 and TNF all decrease adenylate cyclase responsivity to PTH in cultured osteoblastic cells,[45] and it is likely they will have the same effects on these responses to PTH-rP. A third possibility is that PTH-rP is produced by tumors in a constant manner, whereas PTH is produced by parathyroid

adenylate cyclase activity and produces a similar hypercalcemic response following single injections. It also causes similar effects on 1-hydroxylase activity in vitro.[36] A more likely explanation is that the effects of the PTH-rP on

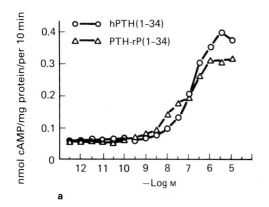

a

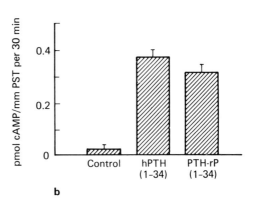

b

Figure 5.10

Effects of human PTH (1–34) and human PTH-rP (1–34) on adenylate cyclase activity in kidney cortical homogenates (a) and isolated renal tubules (b).[32]

adenomas as pulsatile spikes, and the 1-hydroxylase enzyme system responds differently to pulsatile spikes than it does to a constant infusion of PTH or PTH-rP.

Further studies may prove that the complete PTH-rP molecule has additional actions to PTH, possibly by interacting with another distinct receptor. PTH-rP is homologous to PTH in the first 16 amino acids, but thereafter there is no sequence homology. Although PTH (1–84) behaves identically to PTH (1–34), PTH-rP has a basic amino acid carboxyl terminus of 57 residues which may be responsible for other biological effects.

PTH-rP is produced very commonly by many solid tumors.[46] Most squamous cell skin cancers contain PTH-rP, detected by immunofluorescent techniques. Normal skin cells also express the protein. Since it is not associated with hypercalcemia in these tumors, presumably the burden of PTH-rP secreting cells is not sufficient to disrupt calcium homeostasis. One question of immediate interest is whether PTH-rP has a normal physiological function on calcium or phosphate homeostasis. Patients with parathyroid gland disease or ablation have hypocalcemia and hyperphosphatemia, suggesting that a normal tissue source outside the parathyroids does not produce enough PTH-rP to prevent the features of hypoparathyroidism. Thus, it is possible that PTH-rP is an alternative form of PTH produced in the parathyroid glands themselves. If so, it might be the cause of hypercalcemia in some patients with primary hyperparathyroidism. There are a few patients with this disorder who do not have an increase in serum immunoreactive PTH. PTH-rP is expressed in some parathyroid adenomas. However, it is not known as yet whether it is secreted by parathyroid adenomas.[47]

Little is yet known of the regulation of PTH-rP production, although preliminary reports suggest that it may be enhanced by epidermal growth factor, transforming growth factor α, 1,25-dihydroxyvitamin D, prolactin and suppressed by tumor necrosis factor.[48,49,50] Recently, Rizzoli and Bonjour[51] have shown that changes in extracellular calcium concentration influence the production of PTH-rP by cultured Leydig tumor cells without affecting cell replication. Increasing concentrations of extracellular fluid calcium from 0.5–3 mM enhanced the production of PTH-rP. The number of Leydig cells was not affected by the differences in calcium concentration. The significance of these findings is not entirely clear although Kato and Yoneda (unpublished) have found similar results in the VX-2 rabbit carcinoma model of hypercalcemia. Since Rizzoli and Bonjour found effects between 1–2 mM calcium, which is in the range which can be seen in vivo, it is possible that the tumor factor leads to increased bone resorption which leads to an environment which provides a survival advantage for PTH-rP secreting cells. This hypothesis requires further confirmation with in vivo studies.

Different forms of PTH-rP may be produced by alternative RNA splicing.[52,53] Alternative splicing has led to differences in the structure and length of the C-terminal end of the molecule. The N-terminal end, which is the critical region responsible for binding to the PTH-rP receptor and for the biological effects which resemble those of PTH, is apparently not affected. In our experience, synthetic PTH-rP (1–34) and native PTH-rP appear to have identical effects on bone, kidney and calcium homeostasis.[33,36]

With the identification of PTH-rP, there has been great interest in using the techniques of molecular biology to characterize the gene encoding this protrein and comparing it with the gene for PTH. A number of studies have been reported, and these are summarized in recent reviews.[54,55,56] The human PTH-rP gene is a complex gene, more complex than that for PTH, although it is likely that both proteins arose from a single ancestral gene. This single ancestral gene probably duplicated and separated during evolution, as also occurred with proteins such as epidermal growth factor and transforming growth factor α. PTH-rP is encoded by a gene on the short arm of human chromosome 12, and PTH has been mapped previously to the short arm of chromosome 11. These chromosomes probably arose from a single ancestral chromosome. Human PTH-rP comprises three isoforms of 139, 141 and 173 amino acids, and these are encoded by a single gene which undergoes alternate splicing. The rat gene has also recently been characterized. Unlike the human gene, as yet there has been no evidence for isoforms or alternative splicing, although there is considerable homology between human and rat PTH-rP. The complexity of human PTH-rP and the multiplicity of exons suggest that this may provide a mechanism for tissue-specific regulation of PTH-rP expression.

There is great current interest in determining the physiological function and normal source for PTH-rP. Recent studies in pregnant ewes suggest that PTH-rP may play a role in maintaining active transport of calcium from the mother to the fetus across the placenta.[57] PTH-rP is produced in the fetal parathyroids. This biological activity cannot be mimicked by PTH, or by PTH-rP (1–34), suggesting that this biological response is mediated by the C-terminal end of the molecule and probably occurs independently of the PTH receptor. PTH-rP is also present in normal keratinocytes,[28] in lactating breast tissue,[58] some parathyroid adenomas[47] and in the placenta. Whether any of these sources have any physiological significance remains unknown. It is present in large amounts in human milk,[59] but whether this has any role in neonatal calcium homeostasis has yet to be shown.

Recently, Insogna et al[60] have suggested that purified native PTH-rP has similar activity in vitro to transforming growth factor β. However, TGFβ does not have PTH-like activity either in terms of its biological effects on plasma calcium and bone cells or on adenylate cyclase activity,[61] and the significance of these observations remains unclear.

Kukreja et al[62] used neutralizing antibodies to PTH-rP in a model of the hypercalcemia of malignancy. In this human tumour which was carried in nude mice, there was a significant fall in the plasma calcium when the animals were injected with neutralizing antibodies to PTH-rP. However, plasma calcium was not suppressed as would be expected if PTH-rP was the only stimulus for the increase in the plasma calcium.

Recently, Mehdizadeh et al[63] examined the plasma PTH-like bioactivity in nude mice bearing a human renal cell carcinoma using a sensitive cytochemical assay. They found that although the PTH-like biological activity was increased in hypercalcemic tumor-bearing mice, there was poor correlation between the plasma calcium and the PTH-like activity in different mice. Together with findings that PTH-rP expression is common in cancers not associated with hypercalcemia, it appears likely that other factors in addition to PTH-rP may be involved in many of these models of the hypercalcemia of malignancy.

Radioimmunoassays for PTH-rP are only just being described and are still in their developmental stages. Early data suggests that 50 per cent of patients with hypercalcemia of malignancy have an increase in PTH-rP,[64] although there are clearly discrepancies at the present time from assay to assay. The sensitive cytochemical assay, which is not a practical clinical tool, also detects circulating PTH-rP.[65]

Transforming growth factor α (TGFα)

It has long been recognized that tumor cells have the capacity to continue proliferating in culture indefinitely, whereas normal cells tend to die out after a number of passages. The concept first proposed by Todaro and his colleagues was that tumors produce their own endogenous growth factors which are responsible for this replication. This concept, which was expanded in the 1970s with the suggestion that tumors produce factors with characteristics in common with EGF, evolved into the endocrine–autocrine–paracrine hypothesis of tumor cell growth presented by Sporn and Todaro.[66] The first endogenous growth factor described was the sarcoma growth factor, which was found by the early 1980s to consist of at least two peptides, one with similarities to EGF and the other distinct from EGF.[67–69] The peptide which was similar in some ways to EGF has since been renamed TGFα. The other peptide, which depends on EGF for its effects on some indicator cells, has been called TGFβ.[68] The TGFs are operationally defined by their capacity to stimulate the anchorage independent growth of indicator cells (normal fibroblasts) in soft agar.

Transforming growth factor α is a 5000 dalton polypeptide which has many homologies to EGF. It has approximately 40 per cent amino acid homology to EGF, and has a similar conformation.[70,71] Like EGF, it is encoded by a much larger precursor which is about thirty times the size of the major secreted molecule.[72,73] It has no clear signal peptide, but a hydrophobic region suggesting that it may be inserted into the cell membrane and then clipped by a specific and novel protease.[74] Larger forms are present in culture media of cells expressing TGFα. Like EGF, TGFα has three disulfide bridges, and there is considerable amino acid homology between these two molecules in the region between the fifth and sixth cysteines.

Biological effects   Apart from being a factor which stimulates cell growth and replication, it is now clear that TGFα has powerful effects on bone cells. It stimulates osteoclastic bone resorption in vitro.[75–78] This has been shown in organ cultures of fetal rat long bones using rat synthetic TGFα and in neonatal mouse calvariae using human recombinant TGFα (Figures 5.11 and 5.12). Like EGF, its resorbing effects in calvariae seem to be dependent on prostaglandin synthesis.[77] It may stimulate osteoclastic bone resorption by increasing the proliferation of osteoclast precursors,[42] thus acting as a regulatory growth factor or colony stimulating factor for cells in the osteoclast lineage. This effect is seen at extremely low concentrations—as low as 0.01 ng/ml. It is expressed in large amounts during embryonic development at about the same time that the marrow cavity is forming, so it could be important in the bone resorption necessary for the formation of the normal marrow cavity. It is not known which normal cells produce TGFα.

Evidence for a role in the hypercalcemia of malignancy   There is good evidence that TGFα is involved in increased bone resorption which is associated with some solid tumors. First, it is clear that natural, synthetic, and recombinant TGFα resorbs bone in vitro.[75,76,79] Second, in several tumor models of the hypercalcemia of malignancy, the increased bone-resorbing activity produced by the cultured tumor cells is inhibited by antisera to the EGF receptor (Figures 5.13 and 5.14).[80,81] Third, bone-resorbing activity produced by some tumors co-purifies with large-molecular weight species of TGFα (Figure 5.15).[80] Finally, TGFα administration causes an increase in the plasma calcium in mice. This has been shown both by injections of TGFα,[82] and by using Chinese hamster ovarian cells transfected with TGFα and injected in nude mice. In the latter circumstance, there is a marked increase in plasma calcium and an increase in osteoclastic bone resorption, but a suppression of osteoblast activity.[43] Whether TGFα production alone is frequently responsible for hypercalcemia or whether it usually works in concert with other factors such as PTH-like factors remains to be determined. Recent studies using PTH-rP injections in nude mice bearing CHO tumors transfected with the TGFα gene and producing TGFα

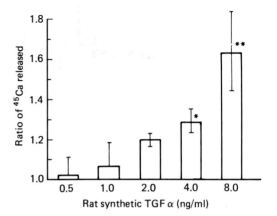

Figure 5.11

Stimulation of bone resorption by synthetic TGFα assessed by measuring the release of ^{45}Ca from previously labeled fetal rat long bones. (*p < 0.05; **p < 0.025). (Reproduced with permission, copyright 1985 by the AAAS.)[75]

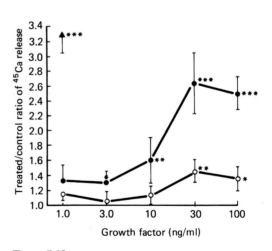

Figure 5.12

Comparison of dose–response curves of human recombinant TGFα and EGF (n = four bones). ▲ Bovine parathyroid hormone 1–84 at 400 ng/ml; ● TGFα; ○ EGF; ***p < 0.005; **p < 0.025; *p < 0.05.[76]

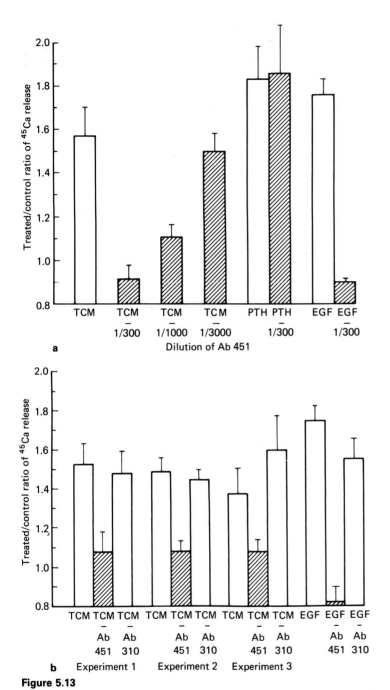

Figure 5.13

Effects of antisera to the EGF receptor on the bone-resorbing activity produced by Leydig tumor cell media (TCM). These tumor cells produce both TGF α and PTH-related peptide. Antisera 451 to the EGF receptor, which block EGF and TGF α biological activity, inhibit bone-resorbing activity produced by the Leydig cell tumor (a). Antisera 310 which bind to the EGF receptor but do not block EGF or TGF α biological effects have no effect on the bone-resorbing activity produced by the rat Leydig tumor (b).[80]

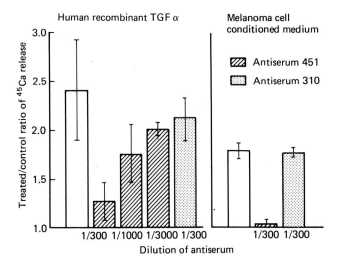

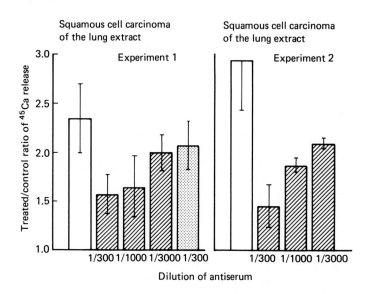

Figure 5.14

Effects of EGF receptor antisera against recombinant TGF α and the bone resorbing activity produced by A375 human melanoma cells (from which TGF α was purified)—upper panels—and against extracts of this lung tumor—lower panels. Antisera 310 inhibit TGF α biological effects, whereas antisera 451, although raised to the receptor, do not inhibit either TGF α biological effects or binding.

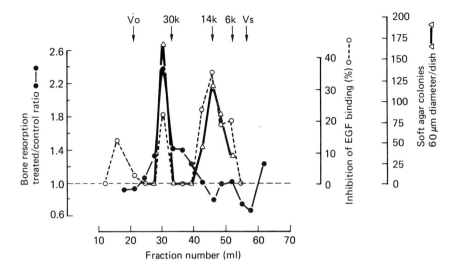

Figure 5.15

Coelution of bone-resorbing activity and TGFα [stimulating activity growth in soft agar and competition with epidermal growth factor (EGF) for receptor binding] in tumor extracts from the rat Leydig tumor associated with humoral hypercalcemia of cancer. (Reproduced with permission, copyright 1983 by the AAAS.)[79]

show a markedly enhanced increase in plasma calcium.[43]

Related peptides Epidermal growth factor is closely related to TGFα, both in its conformation and in its biological effects. However, EGF appears to be less potent than TGFα in its effects on bone resorption and bone formation.[42,78] Similar differences in the biological potency of TGFα and EGF have been described in other systems.[83] Recently, a vaccinia virus protein was described which also binds to the EGF receptor. It has both amino acid and conformational homology to EGF and TGFα, and causes biological effects similar to those of EGF and TGFα.[84] Epidermal growth factor is not known to be produced by any tumors, although many have been surveyed (G. Todaro, pers. commun.). Less is known of the vaccinia virus protein.

Colony stimulating activity

Clinical syndrome A number of tumors have now been reported in which the patients develop a humoral syndrome manifested by extreme leukocytosis.[85–87] This leukocytosis is characterized by a neutrophilia and granulocytosis. This syndrome is particularly common in Japan, and it has been mostly associated with squamous cell carcinomas of the head and neck, including the upper and lower jaw, the tongue, and the lung. Several tumors have also been associated with hypercalcemia, which is in turn associated with a marked increase in osteoclastic bone resorption. When these tumors are inoculated into athymic nude mice, the mice develop a similar syndrome of leukocytosis and hypercalcemia.[88–91] Moreover, there are several examples in which the tumor cells have been shown to produce both colony stimulating activity and bone-resorbing activity in vitro. The association of hypercalcemia with leukocytosis is not invariable, since there are several cases reported which develop leukocytosis alone. Although the syndrome has been reported most frequently in Japan, several examples have been described in the USA[92,93] and there is one animal model of this syndrome.[94]

The association between leukocytosis and hypercalcemia in certain solid tumors is clearly not unusual. The papers citing this association have been reviewed by Mundy et al.[3] The relationship between the colony stimulating activity and the bone-resorbing activity still remains unclear, but it has been studied in most detail by Sato et al.[95] The colony stimulating activity produced by tumors was studied by Sato et al in a squamous cell carcinoma of the lower jaw derived from a 33-year-old man with profound leukocytosis (granulocytes > 150 000/ml) and hypercalcemia (>15 mg/dl). In this particular patient the human tumor cells were inoculated into nude mice, which developed granulocytosis of 300 000/ml and hypercalcemia of 20 mg/dl. The leukocytosis in the nude mice resolved with the removal of the tumor. The tumor cells were established in culture, and the clonal line derived from these cultures produced colony stimulating activity. In this tumor, it was found that the malignant cells expressed PTH-rP, granulocyte colony stimulating factor (G-CSF) and interleukin-1α (IL-1α). The authors postulated that all of these factors were acting in concert to produce the paraneoplastic syndrome. They found that the leukocytosis could in part be accounted for by the combined effects of IL-1α and G-CSF. They had previously shown that IL-1α potentiates the PTH-rP induced increase in plasma calcium.

Lee and Baylink[94] extensively studied an animal model with this syndrome. This was the CE-transplantable mouse mammary tumor, which develops marked granulocytosis (>60 000/ml), serum calcium concentrations of >22 mg/dl, with corresponding increases in the urine calcium, urine hydroxyproline, and urine cAMP. Bone histomorphometry was performed in tumor-bearing nude mice, and this showed an increase in marrow medullary area and in the number of endosteal osteoclasts. The changes in serum calcium, granulocytosis, and urine calcium all improved with removal of the tumor, suggesting that the tumor was producing a humoral factor and was nonmetastatic. The increases in the serum and urine calcium coincided perfectly with the increase in granulocytes in tumor-bearing animals.

Clearly, the relationship between colony stimulating factor production and production of bone-resorbing activity is of great interest to the bone resorption field. There are several possibilities. It is possible that the bone-resorbing factor and the colony stimulating factor are the same material.

However, the data of Sato et al[95] suggest that the biological activities can be separated, and it has not been possible to show a direct effect of colony stimulating factors (CSF-GM, CSF-M, or interleukin-3) on osteoclastic bone resorption in organ culture. A second possibility is that the tumors are coincidently releasing related proteins, one being a colony stimulating factor and the other a bone-resorbing factor. It is conceivable that these factors are separate, but that they act in concert on bone. For example, it has been found that colony stimulating factors can work in concert with 1,25-dihydroxyvitamin D on the formation of cells with osteoclast characteristics in vitro.[96] It is also possible that the bone-resorbing factor produced by the tumor cells has coincident effects on the formation of granulocytes from a common precursor. For example, it has been found that TGFα stimulates the replication of osteoclast precursors (although it has no effect on granulocyte or macrophage colony formation).[45] A third possibility is that, despite the evidence of Sato et al, the growth factors are not produced by the tumor. For example, it is possible that production by the tumor of a factor which resorbs bone leads to the production by other cells such as fibroblasts, monocytes, or lymphocytes, of factors which induce colony formation.

Recently, more light has been shed on the relationship between leukocytosis and hypercalcemia. Yoneda et al[97] and Sabatini et al[98] studied different tumors associated with hypercalcemia and leukocytosis. In the tumor studied by Sabatini (the A375 human melanoma), the tumor cells in experimental animals produced GM-CSF. Plasma TNF was also increased in the animals, but was not produced by the tumor cells. Rather, the tumor cells stimulated host-immune cells to release TNF. Sabatini and colleagues purified the factor produced by the A375 melanoma cells responsible for this effect and found that it was GM-CSF. In another remarkable tumor studied by Yoneda et al, tumor-bearing nude mice developed leukocytosis, hypercalcemia, cachexia and splenomegaly. Plasma TNF was increased four-fold. Yoneda found that the tumor cells produced G-CSF, GM-CSF, M-CSF and interleukin-6, but not TNF. He also found that the host-immune cells produced excessive TNF, and that antibodies to TNF reduced hypercalcemia and cachexia in tumor-bearing nude mice and lowered the white cell count. Yoneda found that these MH-85 tumor cells also produced a factor which stimulated TNF production by host-immune cells. However, this activity could not be blocked by antibodies to GM-CSF. Yoneda has purified this factor and shown that it is a macromolecular protein of 40 kilodaltons which is distinct from GM-CSF. Its identification may have important implications not only for the pathophysiology of these paraneoplastic syndromes, but also for the interactions which occur between tumor cells and host immune defense.

Other bone-resorbing factors

It appears very likely that tumors produce other factors which stimulate osteoclastic bone resorption, and which have not yet been clearly delineated. In the Walker rat model of the hypercalcemia of malignancy, the bone-resorbing factor is distinct from either PTH-like factors or TGFα.[99] The nature of this macromolecular protein is still not known. In a human tumor associated with hypercalcemia, Bringhurst et al[8] partially characterized the bone-resorbing factor and distinguished it from both TGFα or PTH-like activity. This factor also appeared to be a macromolecular protein.

Recently there have been several preliminary reports implicating interleukin-1 in the hypercalcemia associated with solid tumors. Sammon et al[100] and Sato et al[101] have both described solid tumors associated with hypercalcemia in which the primary tumor expresses interleukin-1. In both cases bone-resorbing activity was produced by the tumor cells, and this bone-resorbing activity was inhibited by neutralizing antisera to interleukin-1. In the tumor described by Sato et al[101] the form of interleukin-1 expressed by the tumor appears to be interleukin-1α, since the bone-resorbing factor was suppressed by the antibody to interleukin-1α and expressed interleukin-1α mRNA. In this tumor PTH-like activity was also present, but this appeared to be distinct from the bone-resorbing activity due to interleukin-1. Since both interleukin-1α and β are potent hypercalcemic agents when infused or injected in vivo,[102] they could be responsible for hypercalcemia in these tumors either by acting alone or in concert with PTH-rP. In a recent report, Sato et al[44] have put forward evidence for a synergistic effect between PTH-rP and interleukin-1α on plasma calcium in vivo, at least at one dose.

Tumors may, however, produce interleukin-1α in amounts which are too small solely to cause hypercalcemia. Fried et al[103] examined two human squamous cell carcinomas which were not associated with hypercalcemia. These tumors did not produce PTH-rP. However, the conditioned media from tumor cell cultures stimulated bone resorbing activity in vitro, and this activity was clearly due to interleukin-1α.

One of the earlier fashionable theories for the pathogenesis of hypercalcemia was the production of a bone-resorbing prostaglandin by the tumor cells. Some tumors do, in fact, produce large amounts of prostaglandins in vitro, prostaglandins can resorb bone,[104] and infusions of large amounts of prostaglandins can cause slight increases in the serum calcium.[105] Rarely, patients with the hypercalcemia of malignancy will respond to inhibitors of prostaglandin synthesis, such as aspirin or indomethacin,[106] and there are several animal models of hypercalcemia where these agents have caused a decrease in the serum calcium.[107] However, this situation is extremely uncommon. It seems unlikely that prostaglandins are commonly involved in the HHM syndrome as the systemic humoral factor, and even then their role may be as subsidiary factors rather than as the primary agents responsible for the increase in osteoclast activity.

One possibility for a role for prostaglandins might be their production in bone by normal cells in response to systemic factors. In some organ culture systems the cytokines and TGFs stimulate prostaglandin synthesis. This occurs particularly in organ cultures of mouse calvariae. It is not clear whether generation of prostaglandins within bone as a mechanism for bone resorption is relevant to human systems. The data of the Chambers group,[108] using human cells and examining bone resorption directly, suggest that drugs which inhibit prostaglandin synthesis have no effect on the formation of resorption lacunae by human osteoclasts stimulated by the cytokines.

There was an earlier observation implicating a role for prostaglandin metabolites in the hypercalcemia of malignancy which has still not been adequately explained. Seyberth et al[109] described a number of patients with solid tumors and hypercalcemia in whom the excretion of prostaglandin metabolites in the urine was markedly increased. In these patients, this excretion of prostaglandin metabolites was inhibited by drugs which inhibit prostaglandin synthesis, and there was also a fall in the serum calcium, although this fall was much less convincing than the decrease in the excretion of the prostaglandin metabolite. Whether in these patients prostaglandins were directly involved or whether this was an epi-phenomenon and the hypercalcemia was due primarily to other factors is not known. However, every investigator in this field agrees that drugs which inhibit prostaglandin synthesis are rarely effective in lowering the serum calcium.

Evidence for production of bone-resorbing factors by normal cells

Most workers looking for evidence of production of bone-resorbing activity in patients or animal models of the hypercalcemia of malignancy have concentrated on the tumor cells as the source of bone-resorbing activity. However, it is also con-ceivable that bone-resorbing activity could be produced by normal cells in response to the presence of the tumor (Figure 5.16). Recently, it

has been found that in the rat Leydig tumor model of the HHM syndrome, tumor culture medium stimulates normal rat or human immune cells to release tumor necrosis factor (TNF) and interleukin-1.[110] The evidence that TNF could be involved in the increased bone resorption and hypercalcemia in this model is provided by the observation that the serum of the animals is turbid, the animals have hypertriglyceridemia (a common feature of increased production of TNF), and the animals become extremely cachectic. Moreover, when TNF activity is measured in the serum, it is markedly increased. Since the tumors in these animals are rich sources of both TGFα and PTH-rP, these observations together suggest that both the tumor and the normal cells are releasing a series of bone-resorbing factors which possibly work in concert to stimulate osteoclastic bone resorption.

The concept that powerful hypercalcemia-inducing factors are produced by normal host-immune cells in cancer has been confirmed recently in a tumor model studied by Yoneda et al.[97] In this model, which was derived from a patient with a squamous cell carcinoma of the

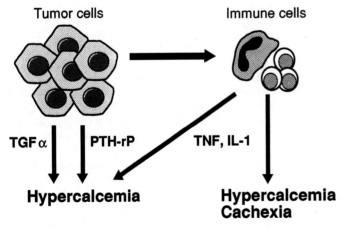

Figure 5.16

Hypothesis for production of bone resorbing cytokines by normal immune cells in hypercalcemia of malignancy.

maxilla, the tumor cells were established in culture and were transplanted into nude mice. Tumor-bearing nude mice exhibited the paraneoplastic syndromes of hypercalcemia, leukocytosis and cachexia. The same paraneoplastic syndromes were present in the patient from whom the tumor was removed. In tumor-bearing nude mice with hypercalcemia, circulating tumor necrosis factor was increased four-fold and hypercalcemia was abrogated by injecting the tumor-bearing nude mice with antibodies to tumor necrosis factor. Moreover, removal of the spleen, which is the site of the majority of host-immune cells in nude mice, led to a marked fall in the plasma calcium. Thus in this model, it is clear that the tumor cells were provoking normal host-immune cells to release TNF which in turn was responsible for the hypercalcemia.

Of major interest in this situation is the tumor product which is responsible for inducing release of TNF from host-immune cells. We have examined several models which exhibit this syndrome and have found that the production of CSF-GM by the tumor cells is responsible in some.[98] However, in the tumor studied by Yoneda et al,[97] conditioned media harvested from the tumor cells which stimulated TNF production by cells in the monocyte lineage was not neutralized by antisera to GM-CSF which blocked GM-CSF activity. The factor responsible appears to be a protein of approximately 40 kilodaltons molecular weight and which is still being characterized.

Evidence for renal mechanisms in the pathophysiology of humoral hypercalcemia

European investigators have long considered that the kidney may be central in the pathophysiology of hypercalcemia, but the renal contribution to hypercalcemia has been relatively ignored by American investigators. The kidney could potentiate hypercalcemia in two ways—by reducing glomerular filtration or by increasing renal tubular calcium reabsorption—both of which would impair renal calcium excretion. In fact, all investigators agree that glomerular filtration is reduced in many patients with hypercalcemia, and is usually related to volume depletion associated with emesis, impaired fluid intake, overuse of potent loop diuretics, or radiation therapy or cytotoxic drugs causing vomiting. In patients with

myeloma (as well as in some patients with lymphomas), glomerular filtration can be impaired because of complications such as Bence Jones nephropathy, uric acid nephropathy, pyelonephritis, or obstructive uropathy caused by retroperitoneal lymph-node enlargement. Volume depletion causes increased renal tubular sodium reabsorption in the proximal tubules, which is associated with increased renal tubular calcium reabsorption.

The first evidence that increased renal tubular calcium reabsorption is important in the pathophysiology of hypercalcemia came from clinical observations made by Peacock et al.[111] These investigators made careful observations on the serum calcium and renal calcium excretion in patients with hypercalcemia. In normal individuals there is a curvilinear relationship between the serum calcium and renal calcium excretion (Chapter 1). Patients with increased renal tubular calcium reabsorption show decreased urinary calcium excretion for any level of the serum calcium, whereas patients with decreased renal tubular calcium reabsorption show an increased renal calcium excretion for any level of the serum calcium (Chapter 1). Using the particular relationship between renal calcium excretion and serum calcium, patients with increased renal tubular calcium reabsorption lie to the right of the curvilinear line for any level of the serum calcium. However, there are some limitations to these observations. These limitations are: (a) the relationship between calcium excretion and serum calcium is based on acute calcium infusions in normal individuals, and this relationship may be different in states of chronic hypercalcemia; (b) the relationship is dependent on sodium and volume status, and is altered when sodium excretion is increased; (c) usually the total serum calcium rather than the ultrafilterable calcium is measured; (d) glomerular filtration is often impaired in these patients, and this will upset the relationship between renal tubular calcium reabsorption and serum calcium; and (e) loading with saline will alter the relationship between renal tubular calcium handling and serum calcium, since excess natriuresis (sodium excretion greater than 4 mmol/l glomerular filtrate) markedly enhances distal delivery of calcium and exceeds reabsorptive capacity at that site.[112] Nevertheless, despite these drawbacks, the observations have now been made independently by so many investigators that they seem certain to be correct and, more recently (see below), Bonjour and co-workers have shown beyond question that increased renal tubular calcium reabsorption does play an important pathogenetic role in some cases. In a recent study designed to overcome some of the limitations of the earlier studies, Tuttle et al[113] examined the relationship between renal calcium excretion and serum ionized calcium in patients with the hypercalcemia of malignancy in whom careful measurements were made of volume status and glomerular filtration rate, and comparisons were made with patients with similar degrees of renal impairment. Glomerular filtration rate was measured by iothalamate clearance. Surprisingly, most of the patients had normal plasma volumes before rehydration. Eight patients were studied, including one with myeloma. All patients showed evidence of increased renal tubular calcium reabsorption compared with the appropriate controls, including the patient with myeloma. Several of the patients had no increase in circulating PTH-rP or urine cyclic AMP excretion. These findings indicate that increases in renal tubular reabsorption of calcium are common in the hypercalcemia of malignancy, that this occurs independent of volume status and that PTH-rP is not solely responsible for these increases in renal tubular calcium reabsorption.

The major question that arises is what the mechanism for increased renal tubular calcium reabsorption is. This is vehemently debated. Most workers in this area believe that a humoral factor produced by the tumor is responsible for increasing renal tubular calcium reabsorption. However, Bijvoet and co-workers, using observations made on patients treated with intravenous saline and the potent bisphosphonate aminohydroxypropylidene bisphosphate (APD), which completely inhibits bone resorption, believe that in many patients enhanced calcium reabsorption in the proximal tubule is linked to sodium reabsorption, which occurs as a consequence of prolonged hypercalcemia.[114] They argue that hypercalcemia itself causes a diuretic-type effect in the kidney, with calcium and sodium excretion being linked by the effects of the hypercalcemia (with a subsequent increased filtered load of calcium) on the renal tubules. However, even these workers agree that in some patients there is an additional specific effect on the renal tubule

which may be hormonally mediated. Moreover, the study by Tuttle et al[113] suggests that volume depletion and increased sodium excretion cannot account for the increases in renal tubular calcium reabsorption in the hypercalcemia of malignancy.

The best evidence for a humoral mechanism for increased renal tubular calcium reabsorption comes from observations in the rat Leydig tumor model of the HHM. Three separate observations indicate that increased renal tubular calcium reabsorption is in part responsible for hypercalcemia, and that this increase in renal tubular calcium reabsorption is mediated humorally. These observations are the following.

(a) In tumor-bearing rats treated with very large doses of the bisphosphonate dichlormethylene bisphosphonate (Cl_2MDP) (40 mg/kg body weight per day), which inhibits osteoclastic bone resorption, it was found that there was a slight decrease in the plasma calcium but a marked fall in the urine calcium excretion.[115] In animals treated with smaller doses (2.5 mg/kg body weight per day) there was no detectable effect on the plasma calcium, but the urine calcium was also lowered into the normal range. Both of these doses of the drug caused a decrease in urinary hydroxyproline excretion, indicating inhibition of bone resorption. This was confirmed by quantitative histology, which indicated a marked decrease in osteoclastic bone resorption and an increase in the percentage mineralized tissue. These results suggest that inhibition of bone resorption alone is insufficient to correct the plasma calcium in this model, and that increased renal tubular calcium reabsorption is present and unmasked when bone resorption is inhibited.

(b) Rizzoli et al[116] carried these results further in thyroparathyroidectomized rats bearing the Leydig cell tumor. These workers performed careful renal calcium clearance experiments in order to examine changes in renal tubular calcium handling as the tumor grew and the plasma calcium increased. They found that the urine calcium excretion was lower for the same plasma calcium in tumor-bearing animals than it was in animals that were thyroparathyroidectomized and treated with 1,25-dihydroxyvitamin D. The clearance studies showed that the urinary excretion of calcium per milliliter of glomerular filtrate was always less in the tumor-bearing rats than in control rats, no matter what the level of plasma calcium. They also examined renal phosphate clearance and urinary cAMP excretion. They confirmed evidence of decreased tubular reabsorption of phosphate and stimulation of urinary cAMP excretion. By also examining renal sodium handling, they were able to exclude a direct link between increased renal tubular sodium reabsorption and increased renal tubular calcium reabsorption in this situation.

(c) The same group performed further studies on rats bearing the Leydig cell tumor with the radiopharmaceutical WR-2721.[117] This compound impairs PTH secretion from the parathyroid gland, and has a PTH-independent effect on the inhibition of renal tubular calcium reabsorption. When they treated tumor-bearing animals with this agent, they found a marked fall in the plasma calcium at the same time as the urinary calcium excretion increased. The only reasonable explanation for this finding is that the agent was lowering plasma calcium by increasing renal calcium clearance, and that increased renal tubular calcium reabsorption plays an important pathogenetic role in this model of hypercalcemia. Other explanations such as a direct effect on tumor secretion of hypercalcemic factors seem unlikely, because there was no change in other parameters such as increased urinary cAMP excretion or increased renal phosphate excretion.

The identity of the hormonal factor responsible for stimulating renal tubular calcium reabsorption is not yet known. However, these studies suggest that a PTH-like factor is likely important. Studies with synthetic analogs of PTH-rP show that this peptide enhances renal tubular calcium reabsorption in vivo.[34] However, some of the renal changes seen in patients with HHM are dissimilar to those seen in primary hyperparathyroidism. For example, in the HHM syndrome there is usually no increase in 1,25-dihydroxyvitamin D production, nor is there inhibition of bicarbonate reabsorption. However, there are some changes which are similar to those seen with PTH excess, such as increased nephrogenous cAMP generation and increased renal phosphate wasting.

The relative importance of renal mechanisms in the pathophysiology of hypercalcemia of malignancy is still unknown. Kanis and co-workers have developed an algorithm to assess the relative contributions of impaired glomerular filtration, increased renal tubular calcium reabsorption, and increased bone resorption to hypercalcemia.[118] Kanis et al[119] believe that renal tubular calcium reabsorption is the major contributor in some patients.

Serum 1,25-dihydroxyvitamin D in HHM

It has generally been held that patients with the hypercalcemia of malignancy, with the rare exception of occasional patients with T- or B-cell lymphomas, have suppressed serum 1,25-dihydroxyvitamin D concentrations and decreased net calcium absorption from the gut. This is in contrast with some of the well-described animal or rodent models of hypercalcemia, where serum 1,25-dihydroxyvitamin D concentrations are usually normal or increased. This sometimes occurs in nude mice bearing human tumors, even in situations where the patients from whom the tumors were derived had suppressed serum 1,25-dihydroxyvitamin D concentrations before tumor removal. Recently, Yamamoto et al[120] reported studies in eighteen patients with renal cell carcinoma complicated by hypercalcemia, from a total of eighty-six hypercalcemic patients with various types of malignancy. Of eighteen patients with renal cell carcinoma and hypercalcemia, the majority of patients had serum 1,25-dihydroxyvitamin D concentrations in the normal range, although two had clearly elevated concentrations and two clearly suppressed concentrations. In patients with other types of malignancy, serum 1,25-dihydroxyvitamin D concentrations were clearly suppressed in the majority of patients with widespread bone metastases and with hematologic malignancies. In hypercalcemic patients with solid tumors but without bone metastases, the serum 1,25-dihydroxyvitamin D concentration was below the normal range in just over one-half of the patients. Although they postulated that decreased serum phosphorus may have been responsible for increasing the serum 1,25-dihydroxyvitamin D concentration in patients with renal cancer and hypercalcemia, this explanation is unconvincing since these patients were probably no more hypophosphatemic than patients with other types of solid tumors without extensive bone metastases. The difference in circulating serum 1,25-dihydroxyvitamin D concentrations in different types of malignancy-associated hypercalcemia remains unexplained.

However, there may be alterations in the responses of the 1-hydroxylase enzyme system in hypercalcemia of malignancy. Kukreja et al[121] used a nude mouse model of the hypercalcemia of malignancy and fed the mice a diet deficient in phosphorus. They found that the activity of the 1-hydroxylase enzyme was significantly greater in tumor-bearing hypercalcemic mice than could be expected for the degree of phosphate depletion. Possibly, the 1-hydroxylase enzyme in tumor-bearing nude mice was being stimulated by phosphorus depletion and PTH-rP, whereas in humans with hypercalcemic tumors the 1-hydroxylase activity is suppressed for reasons which are unknown.

A number of hypotheses have been proposed for the differences in vitamin D metabolism in the hypercalcemia of malignancy compared with primary hyperparathyroidism, and for the suppression of 1-hydroxylase activity in humans with HHM. One possibility is that a tumor product which inhibits the synthesis of 1,25-dihydroxyvitamin D in humans is not recognized when this tumor is implanted in nude mice or nude rats. Fukumoto et al[122] have provided evidence for such a factor in two tumors that they have studied. In one of these tumors, a cancer of the uterine cervix, they identified an activity with an apparent molecular weight of 25 kilodaltons which inhibited the effect of PTH or PTH-rP on 1-hydroxylase activity in primary cultures of rat kidney cells. Nude rats carrying this tumor had decreased serum 1,25-dihydroxyvitamin D. The nature of this inhibitory factor is unknown.

The possibility that other tumor products could interfere with the effects of PTH-rP on renal tubular cells is a reasonable notion. Clearly, TGFα, interleukin-1 and TNF can all inhibit the adenylate cyclase response in renal tubular cells to PTH or PTH-rP.[45] It is thus possible that these factors produced in conjunction with PTH-rP could impair the 1-hydroxylase response to PTH-rP in renal tubular cells. However, there are other possibilities. The differences in 1-hydroxylase activity between primary hyperparathyroidism and solid tumors associated with the hypercalcemia of malignancy may also be related to differences in pulsatile secretion of PTH-like activity in these conditions. For example, there is already evidence from animal studies that PTH delivered in a pulsatile manner may produce different responses in bone than when PTH is delivered by constant infusion. Thus, it is possible that PTH in primary hyperparathyroidism is produced in pulsatile spikes whereas PTH-rP in solid tumors associated with the hypercalcemia of malignancy is produced in a constant manner, and this causes a different response in the 1-hydroxylase system in renal tubular cells.

The possibility that increased production of sterols might be related to hypercalcemia was first suggested by Gordan et al,[123] although the presence of vitamin D-like sterols in tumor tissue was later refuted by Haddad et al.[124] However, a patient has recently been described who had hypercalcemia associated with small-cell cancer of the lung in which such a mechanism may have been applicable.[125] Both serum and tumor extracts from this patient contained 1,25(R)-dihydroxy-vitamin D. This was confirmed by high-performance liquid chromatography using straight-phase and reverse-phase columns. This metabolite competed with 1,25-dihydroxyvitamin D for binding to its receptor in intestinal cells, and it stimulated the release of previously incorporated ^{45}Ca from mouse calvariae in organ culture. Mass spectrometry was not performed, so definitive identification of this metabolite was not possible. The evidence that this metabolite was 1,24(R)-dihydroxyvitamin D came from comparing chromatographic mobility, receptor binding, and biological activity with synthetic 1,25(R)-dihydroxyvitamin D, and so remains likely but not confirmed.

Diagnosis of the humoral hypercalcemia of malignancy

There are several problems which may confront the physician when faced with a hypercalcemic patient. The first is the cause of hypercalcemia. In most patients with malignant disease, the presence of malignant disease is obvious and the cause of the hypercalcemia is clear. However, uncommonly a patient with hypercalcemia may have an occult malignancy. This is a fairly uncommon occurrence in the experience of the author (probably less than 10 per cent of all cases of hypercalcemia of malignancy). The more common situation is a patient with a known malignancy and hypercalcemia where it is unclear whether the hypercalcemia is due to the malignancy or whether it is due to some other coincidental disease process, such as primary hyperparathyroidism. In this situation the most useful diagnostic maneuvers after a careful history and physical examination are to look for positive evidence of primary hyperparathyroidism, such as an increased serum immunoreactive PTH concentration, hyperchloremic acidosis, and renal phosphate wasting. Although renal phosphate wasting is certainly not pathognomonic of primary hyperparathyroidism, in the author's experience hyperchloremic acidosis in a patient with normal renal function and clearly increased serum PTH clearly points to primary hyperparathyroidism as the cause of the hypercalcemia. In the future, measurements of the circulating factor of factors responsible will be helpful. For example, if circulating concentrations of PTH-rP, TGFα, interleukin-1α or TNF are increased, this will clearly point to a diagnosis of hypercalcemia of malignancy. Radioimmunoassays for these factors are currently being developed. A number of groups are now working on the development of PTH-rP assays, and reliable and precise assays should be available within the next few years. The differential diagnosis of hypercalcemia is discussed in more detail in Chapter 4.

Management of HHM

The best treatment for the HHM syndrome is complete ablation of the primary tumor itself. It is occasionally possible to excise the tumor surgically, and thereby successfully remove the source of the hypercalcemic factor or factors and alleviate the increased serum calcium. However, ablative therapy of the primary lesion is usually not possible, and the patient must be treated medically for hypercalcemia. The HHM syndrome is due to production by the tumor cells of a circulating factor which causes hypercalcemia, probably by its effects predominantly on bone resorption and renal tubular calcium reabsorption. Medical treatment has aimed at interrupting each of these pathogenetic mechanisms. This has involved the use of agents which increase renal calcium clearance, such as fluids, loop diuretics, and drugs which inhibit bone resorption, such as plicamycin (mithramycin), the bisphosphonates, calcitonin, and glucocorticoids. A new approach suggested recently has been the use of the agent WR-2721 which specifically inhibits renal tubular calcium reabsorption.[117] It is possible that in the future, with expanded knowledge of the mechanisms of renal tubular calcium reabsorption and better ways of assessing the relative importance of renal tubular calcium reabsorption and increased bone resorption in the pathophysiology of hypercalcemia in the individual case, it will be feasible to select appropriate forms of therapy to counteract the primary pathophysiologic mechanism responsible for the hypercalcemia. Medical management of hypercalcemia of malignancy is discussed in more detail in Chapter 7.

Osteolytic metastases and hypercalcemia

Some solid tumors cause hypercalcemia in association with disease which is widely metastatic to bone, causing extensive localized bone destruction. This is different from the situation in other solid tumors where a systemic humoral mechanism is solely responsible and the hypercalcemia is unrelated to the extent of bone metastasis. In the group with extensive localized bone metastasis, breast cancer is the most common cause. Despite occasional case reports to the contrary, the experience of the author is that breast cancer rarely causes hypercalcemia without the patient having extensive metastatic disease or being in the last few months of life. It seems likely that in these patients hypercalcemia is related directly to the extent of metastatic bone disease and bone destruction. This does not exclude the possibility that a humoral mediator acting on the kidneys could not also be involved in some of these cases. Although the majority of these patients do not have increased nephrogenous cAMP, some have increased renal tubular calcium reabsorption. This situation is complicated and, for this reason, the author likes to consider this group of malignancies separately from other solid tumors associated with hypercalcemia where only a systemic humoral mechanism is involved.

Frequency of hypercalcemia in breast cancer

Breast cancer is one of the more common tumors associated with hypercalcemia. When all patients with hypercalcemia are considered, breast cancer accounts for about 25 per cent of the total cases.[2] Breast cancer is second only to lung cancer as a cause of hypercalcemia. Galasko and Burn[126] found that, in a study of 127 patients with breast cancer, one-third had progressive hypercalcemia, one-third had asymptomatic hypercalcemia, and one-third had hypercalciuria with normocalcemia. Graham et al[127] reported that between 10 and 33 per cent of patients with breast cancer developed hypercalcemia. In a similar study Jessiman et al[128] reported fifty-nine episodes of hypercalcemia occurring in 145 patients with breast cancer. These episodes of hypercalcemia occurred spontaneously in two-thirds, and were related to treatment with hormones in one-third.

Although hypercalcemia is common in breast cancer, bone lesions are even more common. Approximately 90 per cent of patients dying with breast cancer have bone metastases.[129–131] The mean survival after diagnosis of hypercalcemia in these patients is 3 months[132] and is similar in patients with or without other metastases. Hypercalcemia and bone lesions are more common in estrogen-receptor positive tumors (see below, under hormone-mediated hypercalcemia), but liver metastases are less common than would be expected in this group of patients.[133–135] In those patients in whom hypercalcemia occurs, extensive bone metastases are almost always present (approximately 90 per cent in the large series of Coleman and Rubens[132]), the patient is usually late in the course of the disease and the cancer is

widespread. All of this evidence suggests empirically that hypercalcemia in breast cancer is usually related to extensive bone destruction.

Mechanism of hypercalcemia in breast cancer

Hypercalcemia in breast cancer almost certainly occurs as a consequence of increased bone resorption and increased renal tubular calcium reabsorption (Figure 5.16). From the small amount of evidence available, absorption of calcium from the gut seems to be decreased. Although there have not been many studies on renal tubular calcium reabsorption in this disease, Kanis and co-workers believe it is substantially increased in some cases[136] (Figure 5.17). To date, most investigative effort has focused on determining the cellular mechanisms involved with the increase in bone destruction, rather than identifying the mechanisms responsible for this renal tubular effect.

Mechanisms of bone destruction in breast cancer

Cellular mechanism

There has been some interest over the years in determining whether the cellular mechanism of bone destruction is due to the direct effects of tumor cells to cause bone resorption or whether it is indirectly mediated by the production of local osteoclast stimulating factors by the tumor cells or related cells which are responsible for bone resorption.

Evidence for a direct effect

Galasko[137] and Galasko and Bennett[138] were the first to show evidence that tumor cells could directly resorb bone. Their evidence was based on histologic studies, both in human material and in bone sections adjacent to tumor in the VX2 rabbit carcinoma which is associated with hypercalcemia.

If tumor cells could resorb bone directly, then presumably this would be due to similar mechanisms to those involved in osteoclastic bone

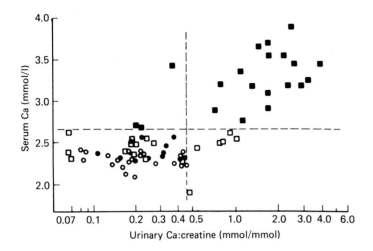

Figure 5.17

Relationship between serum calcium and urine calcium excretion in patients with breast cancer. Notice evidence for increased renal tubular calcium reabsorption in breast cancer.[136]

resorption, namely the production of acid and acid proteases (lysosomal enzymes) which would be responsible for removing mineral from bone and then causing resorption of the bone matrix. Eilon and Mundy[139] showed that cultured human breast tumor cells have the capacity to resorb bone directly in vitro, and to produce acid proteases. These experiments were performed using human breast cancer cells or their culture media and fetal rat long bones which had been devitalized to remove the endogenous osteoclasts. The bones were devitalized by exposing them to ultraviolet light overnight to kill the bone cells and leave the calcified acellular bone matrix. The capacity of the cancer cells to resorb the bones was assessed by measuring the release of previously incorporated ^{45}Ca or the release of ^{3}H-hydroxyproline from the bones during the culture period. The tumor cells probably resorb bone by releasing lysosomal enzymes and collagenase to break down the bone matrix.[139] It was found that

the release of proteolytic lysosomal enzymes by cultured human breast cancer cells was associated directly with an increase in cAMP and inhibition of microtubular assembly by agents such as colchicine.[139,140] When tumor cells were exposed to agents such as colchicine and vinblastine, which inhibit microtubular assembly, these enzymes were released and the devitalized bones were resorbed.

As further evidence that resorption of bones could occur independently of osteoclast activity, live fetal rat long bone cultures were used, in which osteoclasts were inhibited by treatment with agents such as phosphate and cortisol (Figure 5.18).[139] When these agents were added to the bones together with culture media harvested from human breast cancer cells, the bones were still resorbed, indicating that humoral stimulation of osteoclast activity was not responsible. Histologic sections of these bones revealed no increase in osteoclasts.

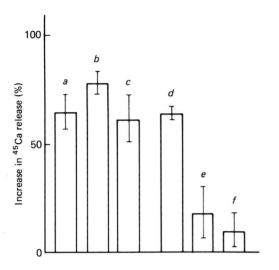

Figure 5.18

Evidence that cultured human breast cancer cells stimulate bone resorption in vitro, and this bone resorption is not affected by drugs which inhibit osteoclast activity such as phosphate and cortisol. a = Breast cancer cells alone; b = breast cancer cells plus phosphate; c = breast cancer cells plus cortisol; d = PGE$_2$; e = PGE$_2$ plus phosphate; f = PGE$_2$ plus cortisol. (Reproduced with permission, copyright 1978 by Macmillan Journals Limited.)[139]

Evidence for osteoclast activation

At present it appears unlikely that the direct cellular effect is more than a minor component of the bone destruction which occurs in metastatic cancer. Scanning electron microscopic studies have shown that bone surfaces occurring adjacent to tumor cells show evidence of extensive osteoclast activity.[141] These osteoclasts are often missed in light-microscopic sections. In a study specifically designed to determine whether there was any evidence of tumor cell bone resorption morphologically, post-mortem vertebral spongy bone samples were examined using scanning electron microscopy. The rat Leydig cell tumor model of the hypercalcemia of malignancy was also examined. Although there was convincing evidence of numerous Howship's lacunae, which in the case of the metastatic tumor lay immediately adjacent to the tumor cells, no evidence could be found of bone resorption which was in any way different from that of the usual osteoclastic resorption. There is other evidence that suggests that osteoclast activation may be the major mechanism. Hypercalcemia of breast cancer is usually responsive to drugs which decrease osteoclast activity, such as plicamycin or the combination of calcitonin and corticosteroids, suggesting that inhibition of osteoclast activity reduces bone resorption in this disease and lowers the serum calcium. Although this is indirect evidence, there is no reason to believe that these drugs could inhibit bone resorption or lower the serum calcium except by inhibiting osteoclasts.

Potential osteoclast activators released in patients with solid tumors

If osteoclast activity is the major cellular mechanism for bone destruction and metastatic bone disease, then it still remains to be determined what the factors are which are responsible for causing this increase in osteoclast activity. There are a number of possibilities. First, factors released by the tumor cells themselves could be responsible. Breast cancer cells have been found to release TGFα and TGFβ, and these growth factors can cause osteoclastic bone resorption. Transforming growth factor α activity has recently been reported in conditioned media from a variety of human breast cancer cells.[142–144] The amount of TGFα activity is strikingly higher in estrogen receptor (ER) negative, hormone-independent cell lines. In ER positive cell lines such as MCF-7, the amount of TGFα activity in the conditioned media is increased when the cells are exposed to estradiol and decreased when they are exposed to anti-estrogens.[143–145] Since TGFα stimulates osteoclastic bone resorption through the epidermal growth factor receptor,[79] this

TGF α-like activity produced by human breast cancer cells is clearly a potential mediator of osteoclastic bone resorption in this situation. Moreover, expression of TGF α mRNA has been a consistent finding in the limited number of human breast cancer cell lines and human breast cancer tissues so far examined.[146,147]

Transforming growth factor β is also frequently expressed by human breast cancer cells. In fact, TGF β is released from breast cancer cells in a biologically active form, which is different from what happens in most tissues where it is released in an inactive form bound to a masking protein. It has been shown to stimulate bone resorption through prostaglandin synthesis.[77] Knabbe et al[148] have recently shown that TGF β released by the ER positive cell line MCF-7 was enhanced by anti-estrogens but decreased by estrogen. Its production by human breast cancer cells is also a potential mechanism for altering rates of osteoclastic bone resorption in this disease.

Prostaglandins released by the breast cancer cells could also be involved in stimulating osteoclastic bone resorption. There have been a number of reports indicating that breast tissue explants excised from patients with both benign and malignant breast disease caused bone resorption when co-cultured with live bone cultures.[149–152] Bone resorption in these co-cultures is inhibited by adding aspirin or indomethacin, agents which inhibit prostaglandin synthesis. Moreover, these tumor extracts contain prostaglandins by assay. However, the significance of these observations remains unclear. Many cells release prostaglandins when exposed to mechanical manipulation.[153] Cells may be stimulated to release prostaglandins because of gentle detachment from the tissue culture flasks or even dispersal in cell culture flasks.[153] Moreover, the source of prostaglandins in these tissues remains unclear. The prostaglandin content of the breast tissue has not been positively correlated with bone lesions or hypercalcemia in patients. It has been found that established human breast cancer cell lines in culture do not release prostaglandins constitutively or cause bone resorption in a similar bioassay.[139] Moreover, more recent studies have suggested that indomethacin or aspirin therapy do little to influence the course of metastatic bone disease in patients with breast cancer.[154] However, in one study explants of human breast cancer tissue co-cultivated with bone cultures caused resorption which could be prevented by indomethacin.[149] In this group's hands, patients who had no bone-resorbing activity in vitro did not have bone metastases 3 years after surgery, whereas a significant number of those whose breast tissue did stimulate bone resorption in vitro had developed bone metastases.[150] However, hormonally manipulated human breast cancer cells have been shown to release prostaglandins (see below and Valentin et al[155]).

There is another factor which may be involved in the osteoclastic bone resorption associated with metastatic breast cancer: breast cancer cells have recently been shown to produce the precursor of the housekeeping protease, cathepsin D.[156] Overexpression of procathepsin D by breast cancer cells may lead to its secretion from the cells. Procathepsin D expression by breast cancer cells may be enhanced by estrogen, and estrogen receptor-positive breast cancers are associated with more prominent bone resorption[167] as well as exacerbations of hypercalcemia following estrogen or anti-estrogen therapy.[162] Following the observations that procathepsin D may be an autocrine mitogen for breast cancer cells at neutral pH, our group has examined the effects of purified procathepsin D on isolated osteoclasts. We were surprised to find that procathepsin D is a powerful stimulant of mature osteoclasts.[157] Procathepsin D acts directly on isolated osteoclasts, stimulating their tartrate-resistant acid phosphatase content and enhancing their capacity to stimulate bone resorption. In addition, it stimulates osteoclastic bone resorption in organ cultures of fetal rat long bones and neonatal mouse calvariae. Whether or not this will turn out to be an important mechanism for the increased bone resorption which occurs in metastatic breast cancer remains to be determined.

It is also possible that other factors not produced by the tumor cells could be involved in causing bone resorption at the metastatic site. Breast cancer cells could stimulate the production of factors such as lymphotoxin, tumor necrosis factor, or interleukin-1 as part of the cell-mediated immune-response mechanism to the presence of the metastatic tumor. These factors are all powerful bone-resorbing factors, could be produced in the bone marrow micro-environment by normal (but activated) marrow cells, and could be responsible for stimulating osteoclastic bone resorption.

Increased renal tubular calcium reabsorption

Most breast cancers are not associated with increased nephrogenous cAMP. However, Percival et al[136] have shown that patients with breast cancer may have increased renal tubular calcium reabsorption. This could be an important component of the hypercalcemia. J A Kanis (pers. commun.) has estimated that renal tubular calcium reabsorption is responsible for approximately 30 per cent of the increase in the serum calcium in some patients with breast cancer-associated hypercalcemia. Earlier, Ralston et al[158] had emphasized that systemic humoral mechanisms acting on the kidney may have been overlooked in the pathophysiology of hypercalcemia in breast cancer, and Isales et al[159] reviewed seventeen cases retrospectively which had been seen over 7 years, and reported that six had increases in nephrogenous cAMP. However, most of these patients had widespread metastatic disease, and coexistent primary hyperparathyroidism was not definitively excluded.

Hormone-mediated hypercalcemia

For many years it has been known that some women with metastatic breast cancer treated with estrogens or anti-estrogens develop acute and severe exacerbations of hypercalcemia which may be fatal.[160–165] Occasionally, hypercalcemia occurs despite a tumor regressing in response to hormones. Approximately 15 per cent of women with ER positive tumors treated with estrogens develop hypercalcemia, whereas approximately 2 per cent of women with metastatic breast cancer develop hypercalcemia as a consequence of treatment with anti-estrogens such as nafoxidine or tamoxifen (see Valentin et al[155] for relevant citations). The mechanism for this sex steroid-provoked and hormone-related hypercalcemia is unknown. In an attempt to unravel this mechanism, the response of cultured human breast cancer cells to incubation with estrogens and anti-estrogens was evaluated in vitro.[155] It was found that in one human breast cancer cell line, the MCF-7 cell line which has estrogen receptors, the cultured tumor cells released prostaglandins of the E-series when incubated with estrogens or anti-estrogens. Production of the prostaglandins was inhibited by indomethacin. Cultured breast cancer cells also released bone-resorbing activity. The production of this bone-resorbing activity was inhibited by co-culture with indomethacin, suggesting that at least part of the bone-resorbing activity was caused by prostaglandins of the E-series. This response was found both in fetal rat long bones and in neonatal mouse calvariae. When a breast cancer cell line which does not have estrogen receptors, the MDA-231 cell line, was incubated with estrogens and anti-estrogens, no bone-resorbing activity was released. The production of bone-resorbing activity by the MCF-7 cell line was also inhibited by treatment with flufenamic acid, which is a compound structurally unrelated to indomethacin and which inhibits prostaglandin synthesis. These results suggest that inhibition of prostaglandin synthesis inhibits the production of bone-resorbing activity on incubation with estrogens or anti-estrogens, and that the increase in bone resorption under these circumstances is related to production of prostaglandins by the cultured breast cancer cells.

This is not the only circumstance in which estrogen receptors are related to increased bone-resorbing activity. Clinical observations suggest that the patterns of metastasis of ER positive and ER negative breast tumors differ.[166–169] ER positive tumors are much more likely to metastasize to bone earlier in the course of the disease, whereas ER negative tumors metastasize to viscera earlier in the course of the disease. It is not yet clear how this relates to osteotropic factors released by tumor cells, although current evidence suggests that ER positive breast cancer cells produce both TGFα and TGFβ, and that TGFβ production is enhanced by incubating the cells with anti-estrogens.[148]

Mechanisms of breast cancer metastasis—the chemoattraction of tumor cells to bone (Figure 5.5)

The cellular and molecular mechanisms by which breast cancer cells metastasize to endosteal bone surfaces are not known. However, clearly these are important in understanding the mechanisms responsible for hypercalcemia in breast cancer, if the notion that hypercalcemia occurs as a consequence of localized bone destruction is valid. When tumor cells spread from the primary site and metastasize to a distant organ, a complex series of events occurs which involves interactions between host cells of various kinds and the tumor cells.

Metastasis is a stepwise selective process which has been reviewed in detail recently.[170] Similar cellular events are probably involved in metastasis of any tumor to any organ. The first event is local invasion by the tumor cells, which involves tumor angiogenesis. Local invasion of vascular channels by tumor cells is an early step in the metastatic process. Clearly, tumor cells have to traverse the walls of vascular channels or lymphatic channels to enter the circulation. Tumor cells must be capable of avoiding host immune-defense mechanisms. It is no surprise that most tumor cells which attempt to metastasize do not survive. However, so many cells are shed from a large primary tumor that only a small proportion needs to be viable to initiate the metastatic process.

The first events which occur are at the primary site. Here, local invasion will be aided by proteolytic enzyme released by the tumor cells,[171–173] and possibly by specific paths of weakness which are present in the adjacent tissues. Cells located in the tumor periphery can attach and enter the lymphatic bloodstream or body cavities.[174–178] Although many cells die at this stage, sufficient numbers may remain viable eventually to form a metastasis. Once tumor cells enter the circulation they come into contact with leukocytes, platelets, and coagulation factors, as well as with other tumor cells.[179–181] This results in the formation of tumor emboli which may be important in the ultimate development of a metastatic deposit.[182] The viability of the tumor cells in these emboli may depend at least partly on the host immune defenses to the presence of the tumor cells.[183] The adhesiveness of the tumor cell emboli to vascular endothelium in tissue sites may be an important factor in determining the actual size and organ specificity of metastases.[174,184,185] Vascularization of these new deposits at metastatic sites will be critical, and it appears to be related to local production of a tumor angiogenesis factor.[186–188]

Why is it, then, that some breast cancer cells should selectively metastasize to bone? Almost 100 years ago, Paget[189] suggested that tumor metastasis depended on a fertile environment which he suggested was the soil in which compatible tumor cells, which he likened to the seed, could flourish and proliferate. It is possible that the host organ could contribute to the development of metastases by providing a local environment in which tumor cell proliferation is enhanced.[190] The possibility that increased rates of bone turnover at local sites could favor tumor metastasis at that site has been suggested by recent studies,[191,192] suggesting that metastatic disease may be localized to an active site of Paget's disease (where rates of bone remodeling are high). In the opinion of the author, it is likely that tumor cells are attracted to bone surfaces selectively by chemotactic signals which are released from bone during the process of remodeling or resorption. Tumor cells released from the primary site will enter the vascular marrow cavity and pass through the wide-channelled marrow sinusoids. Here the tumor cells will be accessible to chemotactic factors which may be produced in high concentrations adjacent to bone surfaces. Several years ago, in vitro methods were developed to study this process.[193,194] The Boyden chamber chemotaxis system was used, and it was found that bone cultures when stimulated to resorb produce a factor or factors which was chemotactic for some tumor cells.[195,196] The conditioned medium harvested from rat long bone cultures was used. This conditioned medium contains the products of the bone-resorption process. The tumor cells examined were the Walker 256 carcinosarcoma cell, which has been shown to cause bone destruction in vitro and which has been used as a model of breast cancer, the EL4 murine lymphoma cell line, and HeLa cells. The chemotactic factor was released from bone independently of the stimulus to bone resorption. When bones were stimulated with PTH, prostaglandins or with the active vitamin D metabolite, similar chemotactic activity was generated. The chemotactic factor was macromolecular, between 3500 and 10 000 daltons, and was stable to heating at 56°C for 60 min. In all of these experiments there was a strong positive correlation between the extent of bone resorption and the chemotactic activity generated for the tumor cells. In other words, the more the bones were stimulated to resorb, the more chemotactic activ-

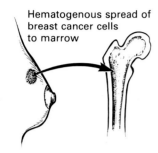

Hematogenous spread of breast cancer cells to marrow

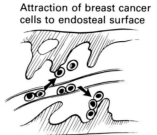

Attraction of breast cancer cells to endosteal surface

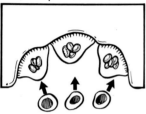

Osteoclast activation by tumor products

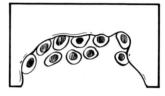

Direct bone resorption by tumor cells

Figure 5.19

Proposed mechanism for the metastasis of breast cancer cells to bone. The initial event is shedding of the tumor cells from the primary site, followed by chemoattraction to bone surfaces, possibly mediated by chemotactic signals such as Type 1 collagen released by remodeling or resorbing bone.[2]

ity was produced by the resorbing bone. More recently, it was found that collagen and collagen fragments were similarly chemotactic for human breast cancer cells.[196] The chemotaxis of the tumor cells is a microtubule- and microfilament-dependent process, and it can be blocked by drugs such as colchicine (which inhibits microtubular function) and cytochalasin B (which inhibits microfilament function).[197]

Recently, it was found that conditioned media harvested from resorbing fetal bone cultures stimulated the growth of Walker 256 rat carcinosarcoma cells.[198] As indicated above, this is a mammary tumor which is associated with osteo-

lytic bone lesions and hypercalcemia. The growth factor activity had a molecular weight of approximately 70 000–80 000. Similar responses were found with other tumor cell lines, including MB-MDA 231 human breast cancer cell lines, TE-85 osteosarcoma cells, a murine fibrosarcoma, and rat embryonic fibroblasts.

There is also other in vivo evidence that bone-derived chemotactic factors exist. Metastatic tumor deposits appear to occur preferentially at sites of bone trauma in rats,[199] as well as at sites of active Paget's disease (cited above). Fisher et al also found that when Walker 256 tumor cells were injected into the aortas of rats whose hind limbs had been previously traumatized by crushing, there was more prominent development of the tumors in the traumatized than in non-traumatized limbs. Presumably, trauma would lead to repair of the bone lesion and subsequent bone remodeling, which could generate chemotactic factors for these tumor cells.

It appears likely that the chemotactic peptide or peptides released by remodeling bone is a fragment of collagen. Fragments of Type 1 collagen, which is the major form of collagen in bone, are chemotactic for the human and animal breast tissue cells[91] (Figure 5.19).

Others have used the Boyden chamber technique to search for tumor cell chemotactic factors. Hayashi et al,[200] Ozaki et al[201] and Ushijima et al[202] have studied the unilateral migration of tumor cells in response to chemotactic factors. These cellular events are complex and possibly multifactorial, but it is clearly important to understand them in order to develop rational therapies to prevent or treat osteolytic metastases in breast cancer. Osteolytic metastases are responsible not only for hypercalcemia, but also for severe pathologic fractures.

The clinical features and treatment of hypercalcemia associated with breast cancer are considered in Chapter 7.

CHAPTER 6

Hypercalcemia associated with hematologic malignancies

Hypercalcemia in myeloma

Myeloma is one of the malignancies most frequently associated with hypercalcemia. Approximately 20–40 per cent of patients with myeloma develop hypercalcemia at some time during the course of their disease.[1] The pathophysiology of hypercalcemia is different in patients with myeloma from that in patients with solid tumors and patients with other types of hematologic malignancies. Hypercalcemia in myeloma always occurs in the presence of extensive bone destruction occurring adjacent to collections of myeloma cells. However, impaired glomerular filtration is also often present in those patients who develop hypercalcemia.

Pathophysiology of hypercalcemia in myeloma

It is obvious that bone resorption and skeletal loss of calcium into the extracellular fluid play an important role in the pathophysiology of hypercalcemia in myeloma (Figure 5.1). However, it is unlikely that hypercalcemia in patients with myeloma is due simply to an increase in bone resorption. Although essentially all patients with myeloma have extensive bone destruction, either in the form of discrete osteolytic lesions or diffuse

osteopenia, only some develop hypercalcemia. Moreover, there is no positive direct correlation between the amount of bone-resorbing activity produced by cultured myeloma cells in vitro and the serum calcium concentration.[2] Since calcium balance studies, when performed, have shown that calcium absorption from the gut is decreased[3] it appears likely that impaired renal calcium excretion, in conjunction with the increased resorption of bone, are responsible for the hypercalcemia.

Glomerular filtration is often impaired in hypercalcemic patients with myeloma, and this impairment may be unresponsive to intravenous fluids. It therefore appears likely that hypercalcemia in myeloma is due to a combination of increased bone resorption associated with irreversibly impaired glomerular filtration. A likely sequence of events is the following. Myeloma cells stimulate increased osteoclastic bone resorption by the mechanisms described below. The extent of bone resorption depends not only on the capacity of the individual tumor cells to activate osteoclasts, but also on the total tumor burden. Those patients who have markedly increased bone resorption which occurs in association with impaired glomerular filtration are those most likely to develop hypercalcemia. In support of this scenario is the observation that measurements of the total body myeloma cell burden and the production of bone-resorbing activity by cultured bone marrow myeloma cells show poor correlations with the level of serum calcium in the patients from whom the marrow cells were derived,[2] unless the state of renal function is also taken into account. Although not closely correlated with the increase in the serum calcium, the amount of bone-resorbing activity released by marrow myeloma cells in vitro is closely correlated with the extent of bone destruction as assessed by clinical parameters. Limitation of the capacity of the kidney to clear an increased load of extracellular fluid calcium because glomerular filtration is impaired will enhance or potentiate hypercalcemia. It appears likely, then, that hypercalcemia in myeloma is due to increased bone resorption occurring in a situation where the capacity of the kidney to clear calcium is impaired, so that the homeostatic mechanism provided by the kidney for clearance of increased loads of calcium entering the circulation is overwhelmed.

Pathophysiology of the bone lesions in myeloma

Almost all patients with myeloma have extensive destructive bone lesions. In fact, this is the most common mode of presentation, since 80 per cent of patients present with disabling bone pain as the major symptom.[1] Destructive bone lesions are often responsible for features other than intractable bone pain, including spontaneous fractures or fractures following trivial injury. The bone lesions have several distinctive patterns. Sometimes patients develop discrete osteolytic lesions which are caused by plasmacytomas. Other patients develop diffuse osteopenia. This occurs when myeloma cells are spread diffusely throughout the axial skeleton and marrow cavity, and it resembles the radiologic appearance seen in postmenopausal osteoporosis. However, in most patients there are discrete osteolytic lesions which are caused by deposits or nests of myeloma cells in the axial skeleton and particularly in the skull, vertebrae, and pelvis.

The destructive bone lesions in myeloma are due to increases in osteoclast activity occurring adjacent to the collections of myeloma cells. This was first shown in a series of thirty-seven patients with myeloma in whom bone was obtained at autopsy and from biopsy samples.[4] The increase in osteoclast activity was found predominantly in

bone sections which contained more than 20 per cent myeloma cells (Figure 6.1). In sites away from the myeloma cell deposits, little or no osteoclast activity was seen. This relationship between the presence of myeloma cells and adjacent osteoclast activity on bone-resorbing surfaces was later confirmed in a larger study with the technique of quantitative bone histomorphometry. In this study, nondecalcified transiliac bone biopsies from 118 patients were examined.[5] There was a clear increase both in osteoclasts and in bone surfaces showing evidence of previous osteoclast activity. However, there was surprisingly no decrease observed in trabecular bone volume. The number of bone-forming surfaces was increased both locally and away from the sites of myeloma cell deposits. The osteoid seams covering these surfaces were reduced in thickness and had a decreased calcification rate. The explanation suggested by the authors for these findings was that the bone-forming activity of individual osteoblasts may be reduced in patients with myeloma.[5]

Clinical data support this conclusion.[6] Isotopic bone scans use labeled bisphosphonates, which are taken up at sites of active mineral deposition and reflect increased activity of osteoblasts. In myeloma bone scans usually show no abnormality, suggesting that osteoblast activity is not enhanced.[7–13] Radiologically, it is rare to see evidence of osteosclerosis adjacent to the discrete lytic lesions. This is in contrast with what is seen in other tumors such as prostate or breast cancers, where osteoblast activity is frequently increased, radiography showing evidence of osteosclerosis (although usually of less-pronounced degree than the osteolytic lesions), and bone scans show 'hot spots' of increased activity or uptake of the labeled bisphosphonate. Serum alkaline phosphatase is a marker enzyme of the mature osteoblast, and serum concentrations of alkaline phosphatase correspond to increased osteoblast activity. Serum alkaline phosphatase is not increased in patients with myeloma unless, for some other reason, osteoblast activity is increased. This may occur, for example, in patients with healing fractures or where there is another tissue source of serum alkaline phosphatase, such as the liver in hepatic disease.

Additional evidence which suggests that the mechanism for bone destruction in myeloma is

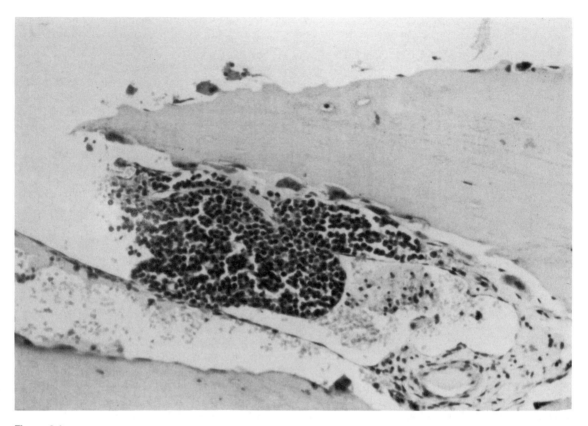

Figure 6.1

The cellular mechanism of bone resorption in myeloma. In this histologic section of bone from a patient with myeloma, myeloma cells can be seen in the marrow cavity adjacent to active bone-resorbing osteoclasts. (Reproduced from *New Engl J Med* with permission.)[4]

primarily osteoclastic is that drugs which inhibit osteoclast activity, such as the bisphosphonates, plicamycin (mithramycin), and calcitonin, are usually very effective in lowering the serum calcium in these patients. This suggests that the hypercalcemia is directly related to the activity of the osteoclasts.

Not all workers have been impressed with the increase in osteoclast activity seen in bone biopsies taken from patients with myeloma.[14] However, it must be remembered that osteoclasts may be difficult to find in routine sections, and

careful searches must be made using sensitive techniques, such as quantitative bone histomorphometry or scanning electron microscopy, before their active involvement in the pathophysiology of the bone lesion can be excluded. Using an in vitro approach, McDonald et al[15] showed that myeloma cells caused release of ^{45}Ca from devitalized bone cultures. This in vitro phenomenon was mediated directly by the myeloma cells independent of osteoclasts, and it had previously been shown for human breast cancer cells.[16] However, for the reasons indicated above, the author thinks it is unlikely that this mechanism plays an important role in the bone destruction which occurs in patients with myeloma.

A few patients with myeloma develop striking osteosclerosis rather than the osteolytic bone lesions.[17] This occurs rarely, possibly in less than 1 per cent of all patients with myeloma. There are other clinical features which tend to distinguish these patients from most other patients with myeloma. They often have the multisystem features which have been referred to as the POEMS syndrome (severe progressive sensory and motor polyneuropathy, organomegaly, endocrine dysfunction including diabetes mellitus, gynecomastia and impotence in males or amenorrhea in females, hyperpigmentation, and papilledema with increased cerebrospinal fluid protein).[18] Peripheral neuropathy is a major component of this syndrome. Hypercalcemia has not been reported in patients with myeloma associated with osteosclerosis or the POEMS syndrome. In the bone biopsies from the 118 patients with myeloma in which histomorphometric data were reported in the study referred to earlier,[5] three patients had marked increases in trabecular bone volume. These patients may have had the osteosclerotic variant of myeloma, although further information was not provided. It is not known why some patients with myeloma should develop osteosclerosis. It appears likely that in these cases the myeloma cells may be making an osteoblast stimulating factor which is responsible. One candidate for this activity may be β_2-microglobulin. It has recently been shown to stimulate bone collagen synthesis in vitro.[19] β_2-microglobulin is apparently produced relatively frequently by myeloma cells although osteosclerosis is relatively rare.[20–22]

Pathogenetic factors

The factors responsible for the increased bone resorption in myeloma are the focus of much investigative attention. Evidence suggests that the effect is mediated by a local factor or factors produced by the myeloma cells, which in turn increases osteoclast activity.[4] However, it is also possible that myeloma cells could stimulate other normal cells around areas of bone resorption to produce factors which activate osteoclasts. In

vitro studies from lymphoid cell lines derived from patients with myeloma, marrow myeloma cells, and explants of bone which contain myeloma cells have shown release by these preparations of a factor or factors which stimulates osteoclast activity in organ fetal rat long bones and neonatal mouse calvariae cultures.[4,23–25] This was first observed in lymphoid cell lines derived from patients with myeloma as well as from patients with other forms of hematologic malignancy.[23] Bone-resorbing activity with similar chemical and biological characteristics to the bone-resorbing activity present in antigen- or phytohemagglutinin-activated peripheral blood leukocyte cultures was present in the conditioned media harvested from these lymphoid cell lines. This bone-resorbing activity is known as osteoclast activating factor, or OAF. One of the lymphoid cell lines in this study was almost certainly derived from the myeloma cells of the patient, since the lambda light-chains released by the myeloma cells had the same antigenic determinants as the Bence Jones protein in the urine of the patient.[23] Bone-resorbing activity present in the culture media was distinguished from the other bone-resorbing factors then known, including PTH, prostaglandins, or the vitamin D metabolites, by dose–response curves, gel-filtration chromatography, heat lability, extraction in lipid solvents, and immunoassay. In a follow-up study, short-term cultures of marrow cells aspirated from patients with myeloma were also evaluated for the production of bone-resorbing activity. This was found in six of the seven marrow cell cultures.[4] The bone-resorbing activity was similar to that found in the culture media harvested from lymphoid cell lines and the bone-resorbing activity present in normal activated leukocyte culture supernatants.[26,27] Again, it was distinguished from PTH, prostaglandins, and vitamin D metabolites. This bone-resorbing activity was not detectable in media harvested from tumor cell cultures derived from patients with other types of hematologic malignancies which are usually not associated with bone lesions or hypercalcemia, such as chronic granulocytic leukemia, chronic lymphocytic leukemia, Waldenstrom's macroglobulinemia, and other lymphoproliferative disorders. Josse et al[25] confirmed these findings, and found the bone-resorbing activity produced by cultured marrow myeloma cells did not affect cAMP accumulation in bone cultures, although its effects on bone were inhibited by indomethacin, suggesting that a prostaglandin-mediated mechanism was involved in its secretion. There was no effect of indomethacin on the resorption of the bone cultures. The OAF produced by activated peripheral blood leukocytes behaved similarly.[28,29] Thus, these studies show that macromolecular bone-resorbing activity is produced by myeloma cells in culture which increases osteoclast activity, but they did not identify this bone-resorbing factor.

Other hematologic neoplasms produce similar bone-resorbing activity. Macromolecular bone-resorbing activity similar to the bone-resorbing activity present in the lymphoid cell lines derived from patients with myeloma, the culture media harvested from activated normal peripheral blood leukocyte cultures and myeloma cell cultures, has been found in various forms of lymphoma. In a patient with lymphosarcoma cell leukemia, the bone-resorbing activity was studied in more detail.[30] This patient was a 75-year-old man who had a neoplasm which behaved in many ways like those associated with T-cell lymphomas caused by the human T-cell lymphoma (HTLV) Type I retrovirus. The bone-resorbing activity produced by the cultured lymphoma cells in vitro eluted from gel-filtration columns in the same fractions as the bone-resorbing activity released by normal activated leukocytes, and was similarly inhibited by cortisol and calcitonin. Some patients with this type of lymphoma also produce excess 1,25-dihydroxyvitamin D. 1,25-Dihydroxyvitamin D appears likely to be synthesized in the neoplastic cells[31] (see below).

The author has been interested in identifying the osteotropic cytokine or cytokines responsible for osteoclastic bone resorption in myeloma. Likely candidates would be interleukin-1,[32–34] lymphotoxin and tumor necrosis factor.[35] Recently, evidence has been gathered which suggests that the bone-resorbing cytokine lymphotoxin plays a major role in the bone destruction which occurs in patients with myeloma.[36] Cultured cells derived from a number of patients with myeloma were studied for evidence of production of bone-resorbing cytokines (Table 6.1). One of the myeloma cell lines was freshly established from a patient who had myeloma with osteolytic bone lesions and hypercalcemia. The author believes it likely that these cells were behaving in culture similarly to the way in which

Table 6.1 Bone-resorbing activity and cytotoxic activity in conditioned medium harvested from cultured human myeloma cells.[36]

Cell source	Bone-resorbing activity*	Cytotoxic activity		
		Without antibody	With antibody to LT	With antibody to TNF
Patient	2.01 ± 0.09	24	<8	19
MC/CAR	1.44 ± 0.10	21	<8	19
L363	2.21 ± 0.19	40	<8	55
1311	1.75 ± 0.10	87	<8	93
1312	1.89 ± 0.10	70	<8	79

*Activity is expressed as the ratio of ^{45}Ca released from cultures of bones of fetal rats whose mothers had received an injection of ^{45}Ca, to the ^{45}Ca released from control bones (no injection); activity was assessed at a 1/40 dilution with fresh BGJ medium. Values represent means $\pm$SEM for four pairs of bone cultures, and all values are significantly greater than 1.0 ($p < 0.05$).

they had behaved in the patient, since the myeloma cells had an identical karyotype in culture as the cells derived from the patient, and the cells in culture secreted free lambda light-chains similar to the monoclonal protein in the serum and urine of the patient. Each of these myeloma cell lines expressed lymphotoxin mRNA. These cell lines also expressed tumor necrosis factor mRNA (Figure 6.2). However, when the conditioned media bathing the cultured cells were examined for biological activity using sensitive cytotoxic assays, it was possible to suppress all of the cytotoxic activity with neutralizing antibodies to lymphotoxin, but there was no effect with neutralizing antibodies to tumor necrosis factor (Figure 6.3, Table 6.2). Evidence of interleukin-1 production by the cells was sought using both bioassays for biologic activity in the conditioned media, and evidence of expression of interleukin-1 α and interleukin-1 β mRNA. No evidence of interleukin-1 production by the myeloma cells was found. In all, five established cell lines were evaluated and a similar pattern of lymphotoxin expression was found in each.

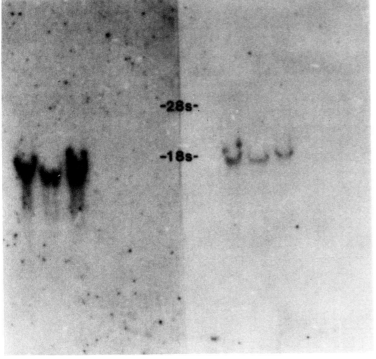

Figure 6.2

Expression of lymphotoxin mRNA in human myeloma cells (lanes a,b,c) shown
by Northern blot analysis (*left*). These tumors also express tumor necrosis factor
mRNA (*right*), but tumor necrosis factor does not appear to be secreted into the
cell culture media. Lanes d and e represent rat myelomas. (Reproduced from *New
Engl J Med* with permission.)[36]

In the same study the effects of recombinant
human lymphotoxin on the serum calcium were
examined in normal mice (Table 6.3).[36] Lymphotoxin was infused subcutaneously via Alzet osmotic minipump for 72 h. A small but significant
increase in the serum calcium was found. The
increase in serum calcium was less than that seen
in most patients with myeloma, although in
patients with myeloma decreased glomerular
filtration and subsequent impairment of renal
calcium clearance would enhance the hypercalcemic response to the production of bone-resorbing
factors.

However, it is unlikely that lymphotoxin is the
sole mediator of osteoclastic bone resorption
produced by human myeloma cells. It was not
possible always to suppress completely the bone-
resorbing activity with neutralizing antibodies to
lymphotoxin. It is therefore possible that there is
an additional bone-resorbing cytokine not yet
identified, which is produced by the myeloma
cells. It is also possible that additional cytokines
will be produced by normal immune cells in vivo
in response to the presence of the malignant
myeloma cells. It is already known that factors
such as tumor necrosis factor will induce production of interleukin-1 by normal immune cells, and
it is possible that the myeloma cell products

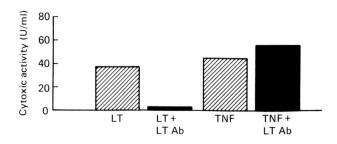

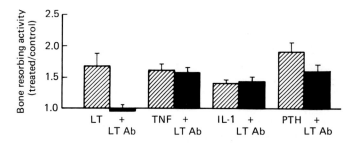

Figure 6.3

Inhibition of bone-resorbing activity and cytotoxic activity with neutralizing antibodies to lymphotoxin. (Reproduced from *New Engl J Med* with permission.)[36]

Table 6.2   Bone-resorbing activity in conditioned medium from myeloma-cell cultures, with and without neutralizing antibody to lymphotoxin.[36]

Sample	Dilution	Bone-resorbing activity*	
		Without antibody	With antibody†
Patient	1/10	2.78 ± 0.15	1.98 ± 0.09
	1/50	2.47 ± 0.25	1.80 ± 0.20‡
	1/100	2.20 ± 0.10	1.50 ± 0.06‡
MC/CAR	1/10	3.36 ± 0.17	2.34 ± 0.29‡
	1/100	1.80 ± 0.34	1.11 ± 0.06‡
L363	1/10	1.62 ± 0.06	1.03 ± 0.02‡
Lymphotoxin	10^{-9} M	1.68 ± 0.20	0.95 ± 0.03‡
Interleukin-1	10^{-10} M	1.61 ± 0.09	1.58 ± 0.07
Tumor necrosis factor	10^{-9} M	1.41 ± 0.04	1.43 ± 0.07

*For explanation of activity, see footnote to Table 6.1.
†Before they were added to the bone cultures, the antibodies were incubated with the factors overnight.
‡Significantly less than in corresponding sample without antibody (p < 0.05).

Table 6.3   Plasma calcium after infusion of recombinant human lymphotoxin or parathyroid hormone in mice.[36]

Infusion	No. of mice	Plasma calcium		
		Day 0	Day 3	Increase
		mg/dl* (means ±SEM)		
Control	24	9.01 ± 0.08	9.45 ± 0.05	0.44 ± 0.09
Parathyroid hormone				
2.5 μg/day	8	9.28 ± 0.15	9.96 ± 0.18	0.68 ± 0.26
5.0 μg/day	12	9.07 ± 0.11	11.09 ± 0.53	2.02 ± 0.52†
Lymphotoxin				
5.0 μg/day	4	9.08 ± 0.12	9.85 ± 0.14	0.77 ± 0.11
10.0 μg/day	8	8.89 ± 0.11	10.03 ± 0.16	1.14 ± 0.23†
20.0 μg/day	8	9.36 ± 0.13	10.36 ± 0.13	1.00 ± 0.16†

*To convert to millimoles per liter, multiply by 0·2495.
†Significantly greater than increase in mice receiving the control infusion (p < 0.05, according to analysis of variance using the Student–Newman–Keuls test).

would also cause normal immune cells to produce bone-resorbing cytokines. Such activities would not be detectable in the in vitro systems employed, which utilize cultured myeloma cell lines.

It is certainly possible that other bone resorbing cytokines may be involved in osteoclastic bone resorption in myeloma. Interleukin-6 is a growth factor for myeloma cells which is produced in excessive amounts in patients with severe myeloma.[37] Its properties have been described in Chapter 3. It is still not clear whether it is a stimulator of bone resorption in vivo. It appears to act as a growth factor on osteoclast precursors and may work in conjunction with other factors such as interleukin-1 or tumor necrosis factor in stimulating osteoclastic bone resorption.[38] There have been conflicting reports on the effects of interleukin-6 on bone resorption in organ culture. In some reports no effect is seen and in others it was found to stimulate bone resorption. Most workers agree that it has no effect on organ cultures of fetal rat long bones or neonatal mouse calvariae. A likely possibility is that it works in conjunction with other cytokines in myeloma to stimulate osteoclast formation and osteoclast activity.

There have also been some reports that interleukin-1 is involved in the bone destruction which occurs in myeloma.[39,40] In the human tumors that the author has examined, evidence of interleukin-1 production has been undetectable.[36] However, this is not to say that it is not produced in vivo in excessive amounts, since it may be produced by host cells in response to the tumor or tumor products. It also remains possible that different myeloma cells will produce different bone resorbing cytokines. Unraveling the various cytokines involved in bone destruction in myeloma will require further investigation.

The diagnosis of hypercalcemia in myeloma

The clinician must take care in making the diagnosis of hypercalcemia in patients with myeloma. There are now a number of well-reported cases of myeloma as well as other gammopathies where the abnormal plasma proteins bind to calcium and lead to an increase in the total serum calcium concentration. Of course, this does not necessarily mean that the ionized calcium concentration is increased. A patient with a total serum calcium which is increased above the normal range may, in fact, have a normal ionized calcium if an abnormal calcium-binding protein is present in the serum. Under these circumstances the usual correction formulae for estimation of ionized serum calcium which employ the serum albumin concentration will give misleading results. The most commonly used correction formula is to allow 0.8 mg calcium for every gram the serum albumin is less than 4 g/dl. However, binding of calcium to a monoclonal protein is not a common occurrence, and only a few well-documented cases have been reported.[41] Misinterpretation of the total serum calcium can also be readily avoided by measuring ionized calcium directly when it is suspected, if techniques for making this measurement are available.

The serum alkaline phosphatase is often not increased in myeloma, unlike other types of malignancy associated with osteolytic bone disease. In patients with myeloma there is usually little evidence for increased osteoblast activity, and the serum alkaline phosphatase is a parameter of the activity of the mature osteoblast. An exception may occur in patients with healing fractures. However, this is also uncommon, since most patients with myeloma are elderly, and older people (unlike younger adults and children with bones which are turning over more rapidly) may not show an increase in serum alkaline phosphatase response during fracture healing.

Renal function is almost always impaired in patients with myeloma and hypercalcemia at the time of presentation. In fact, as a general rule, if a new patient presents with hypercalcemia and it appears that they have malignant disease, then myeloma is clearly one of the first conditions considered. This is particularly true if the impairment of glomerular filtration is resistant to intravenous fluid therapy.

Unlike patients with primary hyperparathyroidism and most patients with other forms of malignant disease and hypercalcemia, the serum phosphorus is usually increased in patients with myeloma. The elevation in the serum phosphorus usually reflects the fixed impairment of glomerular filtration. Thus, in many ways the electrolyte pattern for the patient with myeloma and hypercalcemia at the time of presentation may resemble vitamin D intoxication or sarcoidosis, where both serum calcium and phosphorus are increased, and there is impaired renal function.

Hypercalcemia is a poor prognostic indicator in patients with myeloma. In the author's experience it usually means that the patient is in the last few months of life. However, patients should not die from hypercalcemia with the modern medical therapies which are available for lowering the serum calcium in this condition. When patients develop hypercalcemia they usually have widespread bone destruction, an extensive tumor burden, and impaired glomerular filtration. Of course, all of these features are powerful negative prognostic indicators.

Treatment of hypercalcemia and bone lesions in myeloma

The treatment of hypercalcemia and the osteolytic bone lesions in myeloma is similar in principle to the treatment of hypercalcemia and bone metastases in patients with solid tumors. The principles are, where possible, to treat the underlying disease and to employ medical treatment to reverse the effects of hypercalcemic factors produced by the tumor cells on target organs. However, there are some special aspects to the treatment of myeloma which should be considered. When myeloma is complicated by hypercalcemia, the patients almost always have impairment of renal function. This may be due to Bence Jones proteinuria, hyperuricemia, urinary tract infection, and, in occasional patients, amyloidosis. Sometimes hypercalcemia itself is responsible for the impairment of renal function. However, often the impairment of renal function is fixed and is not reversible by fluid administration. For this reason vigorous rehydration with isotonic saline should be modified in patients with myeloma, and care should be taken during rehydration to avoid left ventricular failure and subsequent pulmonary edema. Fixed impairment

of renal function also limits the choice of drugs available for the medical treatment of hypercalcemia. Plicamycin (mithramycin) and phosphate therapy are hazardous in this situation and should be avoided where possible. The author has found particularly beneficial responses to the combination of corticosteroids and calcitonin in this type of hypercalcemia. In myeloma, hypercalcemia almost always responds to this combination with a fall in the serum calcium into the normal or near-normal range. Many patients will respond to corticosteroids alone, which will often be used as part of the treatment of the malignancy itself, but there are some patients in whom corticosteroids are not effective in controlling osteoclastic bone resorption. In these, calcitonin in doses of 200–400 MRC units subcutaneously each 12 h is almost invariably effective in the author's experience.[42,43] These patients should be kept on corticosteroid therapy. Some patients will respond to calcitonin transiently for a period of 4 or 5 days, and in these calcitonin can be withdrawn for several days and then reinstituted. The advantages of the combination of calcitonin and corticosteroids are that they have no deleterious effects on renal function.

Plicamycin (mithramycin) is an effective inhibitor of bone resorption, and will work in patients with myeloma in standard doses of 15–25μg/kg body weight, but it is directly nephrotoxic and the side-effects are not only more frequent, but also cumulative in patients with impaired renal function. For this reason its use should be avoided, except in those patients in whom corticosteroids and calcitonin are ineffective.

The newer bisphosphonates amino-hydroxypropylidene bisphosphonate (APD) and dichloromethylene bisphosphate (Cl_2MDP) are also very effective in lowering the serum calcium and relieving bone pain in patients with myeloma.[44,45] These two agents are now widely used in Europe, and they can be used either intravenously or orally. They are not currently available for use in the USA. Recently, intravenous etidronate has been used in the treatment of patients with hypercalcemia of malignancy, including several patients with myeloma, and the patients showed evidence of both clinical and biochemical improvement without side-effects. These patients received 7.5 mg/kg body weight per day for 7 days.[46,47] Intravenous etidronate is available for use in the USA.

There have been conflicting reports on the use of fluoride in myeloma for over 20 years.[48–52] About 10 years ago it was suggested that fluoride therapy could lead to positive calcium balance and improve the radiologic and morphologic appearances of the bone lesions in patients with myeloma.[52] However, fluoride does not inhibit bone resorption and a number of studies have suggested that there are no beneficial effects on the symptomatology or on the clinical course of the disease. The author cannot see the rationale for the use of fluoride in the treatment of patients with myeloma, and does not think that it is warranted at present.

More patients with myeloma have painful bone lesions than have hypercalcemia. For this reason there is a great need for an agent which would relieve bone pain and bone destruction in these patients. Several groups have shown that the newer bisphosphonates APD and Cl$_2$MDP do lead to symptomatic improvement in bone pain, and have found evidence that these agents cause inhibition of bone resorption in patients as assessed by decreases in the urinary excretion of calcium and hydroxyproline.[44,45] Although neither of these agents is available at present in the USA, one of this bisphosphonate family of drugs will almost certainly become available in the next few years, and should achieve a useful place specifically for the treatment of osteolytic bone disease in myeloma.

Hypercalcemia in hematologic malignancies other than myeloma

Hypercalcemia is far more common in myeloma than in any other type of hematologic malignancy. However, it also occasionally occurs in patients with malignant lymphomas, and even sometimes in patients with acute or chronic forms of leukemia. There is good reason to believe that the mechanisms are different in many of these cases from those which are responsible in myeloma (Figure 5.1).

Hypercalcemia in malignant lymphomas

Other than for myeloma, hypercalcemia occurs more commonly in the malignant lymphomas than it does in any other form of hematologic malignancy. It occurs most frequently in T-cell lymphomas, but also occasionally in B-cell lymphomas and in histiocytic lymphomas. In these diseases, bone lesions are far more common than hypercalcemia. Bone lesions also appear more frequently than hypercalcemia in children with these disorders. If a child presents with osteoporosis or osteopenia of unknown cause, then a malignant lymphoma or even an acute leukemia should be an early consideration. In an adult who presents with osteopenia and spinal cord compression, malignant lymphoma should be a major diagnostic consideration.

Another type of lymphoma in which bone lesions are frequent is Burkitt's lymphoma. This disorder, which occurs most frequently in the jaws of African children, is almost always associated with local osteoclastic bone resorption.[53] Hypercalcemia has been reported,[54] but is rare. This is the one case of hypercalcemia in Burkitt's lymphoma of which the author is aware. In this case, manipulation of serum calcium by corticosteroid therapy revealed quite clearly that PTH was not the mediator of the hypercalcemia, since, when the patient was hypercalcemic, serum PTH concentrations were suppressed. As the serum calcium was lowered with therapy, serum PTH concentrations began to rise.

There has been much recent interest in the possibility that 1,25-dihydroxyvitamin D is the mediator of hypercalcemia in some patients with T-cell lymphomas (see below) and B-cell lymphomas.[55] Hypercalcemia has been reported in patients with adult T-cell lymphoma (ATCL) associated with the HTLV. The HTLV is a type C retrovirus which has been associated with a form of T-cell lymphoma described in patients in the USA, Japan, and the West Indies.[56–59] This tumor is characterized by lymphocytosis, lymphadenopathy, and an aggressive course with the rapid appearance of disseminated skin lesions. These patients also present with a paraneoplastic syndrome of increased bone turnover manifested by abnormal bone scans and hypercalcemia. Whereas hypercalcemia is an infrequent complication of most malignancies including lymph-omas, the incidence of hypercalcemia in patients with ATCL is greater than 50 per cent.[60,61] Indeed, in one recent study, five of eleven patients with ATCL presented with clinical manifestations of hypercalcemia, and all eleven patients developed hypercalcemia at some time during the course of their disease.[58] Although some patients have had lytic bone lesions due to tumor invasion, biopsy of lytic lesions usually fails to reveal tumor involvement and shows only increased osteoclas-tic activity.[62] There have also been several well-documented cases in which hypercalcemia occur-red, but no lytic lesions could be detected.[61,63] Moreover, tumor extracts obtained from one patient have been shown to cause bone resorp-tion in mouse calvariae in vitro and induce hypercalcemia in mice in vivo.[63] These findings are consistent with the hypothesis that the hyper-calcemia associated with this form of T-cell lym-phoma is mediated by a humoral mechanism. Although histological examination of bone from patients with ATCL and hypercalcemia has con-sistently revealed increased osteoclast resorp-tion, hormones known to mediate bone resorp-tion by increased osteoclast activity have not been elevated in these patients. Specifically, PTH levels are typically suppressed and PTH has not been detected in tumors obtained from these patients. Similarly, increased levels of prostaglan-din E in blood or tumor tissue have not been found, and the hypercalcemia does not respond to the administration of indomethacin. Macro-molecular bone resorbing activity was found in HTLV-infected cell culture supernatants.[64] Another stimulator of bone resorption and hyper-calcemia is 1,25-dihydroxyvitamin D. In this regard it is of particular interest that several recent reports have demonstrated increased con-centrations of 1,25-dihydroxyvitamin D which correlated with the presence of hypercalcemia in four patients with ATCL.[65-68] The finding of increased serum 1,25-dihydroxyvitamin D levels in these patients in the presence of PTH sup-pression and significant renal impairment sug-gests that lymphoma cells were the site of the production of increased concentrations of 1,25-dihydroxyvitamin D which resulted in the syn-drome of increased bone turnover and hypercal-cemia. However, although 1,25-dihydroxyvitamin D may be involved in the hypercalcemia that occurs in some of these cases, in the majority it

appears that serum 1,25-dihydroxyvitamin D con-centrations are not increased (T. Matsumoto, pers. commun.).

Recently the capacity of HTLV-transformed cord blood lymphocytes to metabolize 25-hydroxyvitamin D has been examined in vitro.[31] It was found that these lymphocytes, which incorporate the HTLV virus, metabolize 25-hydroxyvitamin D to a metabolite which co-elutes with authentic 1,25-dihydroxyvitamin D by high-performance liquid chromatography (HPLC) over three different HPLC systems. In addition, the metabolite eluting from these columns in the same position as 1,25-dihydroxyvitamin D resorbs bone in vitro (Table 6.4), in a similar manner to that of 1,25-dihydroxyvitamin D (Figures 6.4–6.6). This metabolite also competes with labeled 1,25-dihydroxyvitamin D for binding to receptors on bone cells (Figure 6.7). Definitive identification of this metabolite was confirmed by determination of its molecular structure by mass spectroscopy. Analysis of the trimethylsilyl-derivative of the HTLV metabolite by mass spec-troscopy indicated that it had the same fragmen-tation pattern as that produced by 1,25-dihydroxyvitamin D (Figure 6.8).

Table 6.4 Bone-resorbing activity of the vitamin D metabolite produced by incubation of HTLV-transformed cord blood lymphocytes with 25-hydroxyvitamin D for 24 h. Bone resorption is expressed as the percentage ^{45}Ca released and as the treated/control ratio, with any ratio greater than 1.0 indicating stimulation of bone resorption.[31]

Agent	Percentage release of ^{45}Ca	Bone-resorbing activity (treated/control ratios of ^{45}Ca release)
Control	16.1 ± 0.7	—
Metabolite	29.3 ± 3.5	1.82 ± 0.22*
1,25(OH)$_2$D (10^{-8} m)	30.2 ± 3.2	1.88 ± 0.20*
(10^{-7} m)	42.5 ± 4.6	2.64 ± 0.29*

*Significantly different from control at $P < 0.01$.

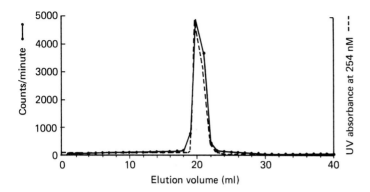

Figure 6.4

Elution pattern of ³H-vitamin D metabolites produced by incubating ³H-25-hydroxyvitamin D with HTLV-transformed cord blood lymphocytes for 24 h.[31] The vitamin D metabolites were extracted, and chromatographed over a methylene chloride column. They elute in the same position as authentic 1,25-dihydroxyvitamin D.

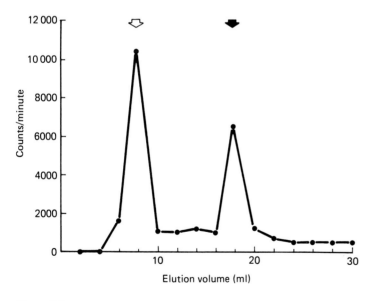

Figure 6.5

Elution pattern of the ³H-vitamin D metabolites produced by incubating ³H-25(OH)D with HTLV-transformed cord blood lymphocytes for 24 h. The elution volumes of 25(OH)D and 1,25(OH)$_2$D are indicated by the open and closed arrows respectively.[31]

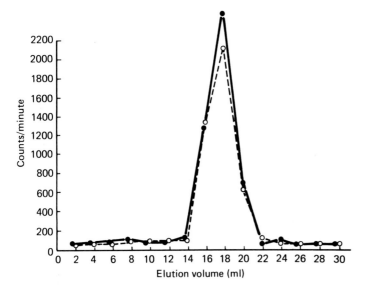

Figure 6.6

Elution pattern of the ^{3}H-vitamin D metabolite produced by incubating ^{3}H-25(OH)D with HTLV-transformed cord blood lymphocytes (solid circles) compared to that of authentic ^{3}H-1,25(OH)$_2$D (open circles).[31]

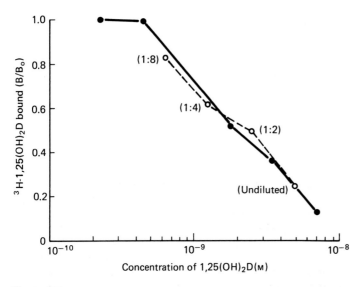

Figure 6.7

Displacement of ^{3}H-1,25(OH)$_2$D from intracellular 1,25(OH)$_2$D receptors prepared from rat osteosarcoma cells by 1,25(OH)$_2$D (closed circles) or the HTLV-metabolites produced by incubating 25(OH)D with HTLV-transformed cord blood lymphocytes (open circles).[31]

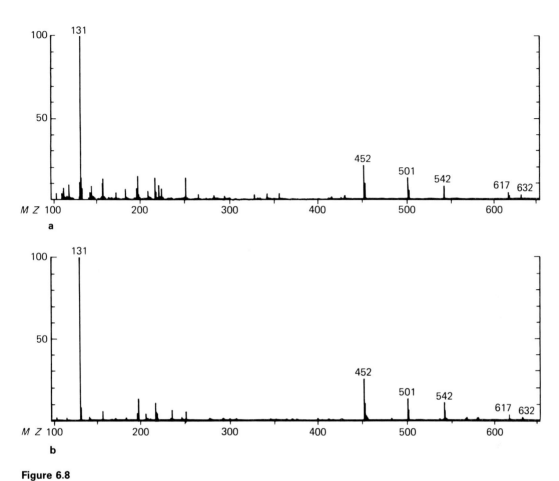

Figure 6.8

Positive-ion electron impact (70 eV) mass spectrum of (a) the trimethylsilyl ether derivative of 1,25-dihydroxyvitamin D and (b) the vitamin D metabolite produced by incubating 25-hydroxyvitamin D with HTLV-transformed cord blood lymphocytes.[31]

It is also likely that lymphokines are produced in malignant lymphomas associated with hypercalcemia. Particularly in the case of the T-cell lymphomas, a whole range of cytokines are produced by these cells, and it has been found in preliminary studies that there is macromolecular bone-resorbing activity produced by the tumor cells.[64] Thus, in many of these cases there may be a dual mechanism including both production of 1,25-dihydroxyvitamin D by the malignant cells as well as the production of lymphokines which resorb bone.

There is some controversy over the relative frequency of abnormalities in 1,25-dihydroxyvitamin D production in patients with lymphoproliferative disorders associated with hypercalcemia. In the hands of some investigators, this has been a relatively common phenomenon, whereas in other studies, it has been rare.[69,70] Hypercalcemia is extremely common in patients with adult T-cell leukemia (approximately 50–100 per cent of patients develop hypercalcemia during the course of the disease). Hypercalcemia is also relatively more common in patients with lymphoproliferative disease than it is in patients with solid tumors. In the largest study of this type, over 4000 patients admitted to the National Institutes of Health with malignant disease were studied.[71] It was found that 10.4 per cent of patients with lymphoproliferative disease had a total plasma calcium greater than 11 mg/dl (2.75 mmol/l), whereas in patients with solid tumors, the frequency of total plasma calcium greater than 11 mg/dl was only 5.6 per cent. Adams et al[72] studied 16 patients with hypercalcemia and lymphoproliferative disorders prospectively. Of their 16 patients, 7 had HTLV-I associated malignancies (2 with adult T-cell lymphoma, 5 with HTLV-I-related B-cell lymphoma or Hodgkin's disease). Eight of the 16 patients had an increase in plasma 1,25-dihydroxyvitamin D. Thus in the experience of this group, increased serum 1,25-dihydroxyvitamin D is quite common. In contrast, Motokura et al[73] studied a large group of patients in Japan with HTLV-I-associated lymphoma and hypercalcemia and found no evidence for increased serum 1,25-dihydroxyvitamin D, but rather evidence for increased production of PTH-related protein.[70] There have been no reports in any patients as yet of an increase both in PTH-related protein and 1,25-dihydroxyvitamin D.

One key difference between myeloma and the lymphomas is that although both the malignant cells are in the marrow cavity immediately adjacent to endosteal osteoclasts, in myeloma there is frequently impairment of renal function caused primarily by the production of Bence Jones proteins by the myeloma cells.

A number of years ago the author had the opportunity to study in detail a patient with a lymphoma which was then called lymphosarcoma cell leukemia.[30] This patient was a 71-year-old man who presented with anemia and was found on physical examination to have peripheral lymphadenopathy and hepatosplenomegaly. At the time of presentation, he had hypercalcemia (serum calcium 13.6 mg/dl) and a form of leukemia in which the circulating abnormal cells were large lymphocytes with cleaved nuclei. Similar cells were present in the lymph nodes. Neoplastic cells from this patient were established in culture and found to produce a bone-resorbing factor which was partially purified by gel-filtration chromatography and found to be a macromolecular compound of similar size to the cytokines which are now known to resorb bone, such as lymphoto-xin, tumor necrosis factor, and interleukin-1. The effects of this factor on bone were inhibited by cortisol, and the patient's serum calcium fell when he was treated with corticosteroid therapy. Nothing is known of the vitamin D status of this patient. It appears likely that his hypercalcemia was due, at least in part, to production of this macromolecular cytokine by the tumor cells.

Hypercalcemia in Hodgkin's disease

Hypercalcemia does not occur commonly in Hodgkin's disease. Only one episode of hypercalcemia was noted by Canellos[60] during the course of 190 cases of advanced Hodgkin's disease. Walker[74] also noted only one single episode among sixty-one patients over a period of 1 year. At the time of presentation hypercalcemia appears to be even rarer, with only about ten case reports being noted in the literature.[60,74–77] Vitamin D therapy has been associated with hypercalcemia and Hodgkin's disease by Kabakow et al.[75] Their patient was a 48-year-old woman who developed hypercalcemia and lymphadenopathy while taking low doses of vitamin D. The serum calcium seemed to be vitamin D-dependent, since it abated when vitamin D was withdrawn and calcium was reduced in the diet. However, it recurred later in association with advanced Hodgkin's disease. More recently Breslau et al[65] noted hypercalcemia associated with increased serum 1,25-dihydroxyvitamin D occurring in non-Hodgkin's lymphoma, and there have been several further reports of a similar syndrome associated with Hodgkin's disease.[66–68,78,79] The serum calcium in these cases correlates well with serum 1,25-dihydroxyvitamin D levels.

Hypercalcemia is fairly unusual in Hodgkin's disease, but it can occur in patients with advanced disease, particularly in males with bulky intraabdominal disease localized to the abdominal cavity. The mechanism for the hypercalcemia appears to be at least in part related to 1,25-dihydroxyvitamin D, but may also be related to other factors. Davies et al[68] produced convincing evidence that 1,25-dihydroxyvitamin D was the predominant factor in the pathophysiology of hypercalcemia in a newly diagnosed patient with Hodgkin's disease whose hypercalcemia was worse on exposure to intense sunlight. This patient improved with successful chemotherapy for the Hodgkin's disease, and there was an increase in serum 25-hydroxyvitamin D but not 1,25-dihydroxyvitamin D following rechallenge with ultraviolet irradiation. It is likely that Hodgkin's tissue may possess the 1α-hydroxylase enzyme similar to that found in renal tissue. However, it may be premature to conclude that increased serum 1,25-dihydroxyvitamin D is the complete explanation for the hypercalcemia. Lemann and colleagues administered 1,25-dihydroxyvitamin D to normal volunteers, but failed to elevate the serum calcium levels above normal limits, despite the occurrence of blood levels of 1,25-dihydroxyvitamin D comparable with those of hypercalcemic Hodgkin's patients.[80,81] However, these individuals do develop hypercalciuria, and Moses and Spencer[82] showed that in autopsies in eleven patients with Hodgkin's disease but with normocalcemia, five had nephrocalcinosis. Thus, it is likely that there are two mediators of hypercalcemia in Hodgkin's disease. It appears that when hypercalcemia does occur in Hodgkin's disease, it is most likely in the patients with extensive tumors with very large tumor bulk.

Hypercalcemia in chronic lymphocytic leukemia (CLL)

Hypercalcemia is rarely associated with the most common form of chronic lymphocyte leukemia, where the circulating cells appear to be morphologically normal B-lymphocytes. However, a few cases have been reported.[60,83] When a patient does present with hypercalcemia in CLL, it is important for the physician to look carefully for other causes of the hypercalcemia, such as primary hyperparathyroidism or another malignancy. In those cases where the hypercalcemia is directly related to CLL, the factors which are responsible for causing hypercalcemia are unknown. One patient seen by the author[84] was a 73-year-old woman who had multiple bone lesions including pathologic fractures of the right humerus and left hip without a history of trauma. Bone sections contained typical CLL cells which were characterized by surface marker studies showing evidence that they were monoclonal B-cells. Areas of bone adjacent to these cells

showed large numbers of osteoclasts, suggesting the neoplastic cells were producing a local humoral mediator of osteoclast activity.

Hypercalcemia in animal lymphomas

Spontaneous lymphosarcomas occur commonly in aging domestic dogs, and these lymphosarcomas are frequently associated with hypercalcemia. A number of animals with this condition were studied[85] and it was shown that immunoreactive PTH concentrations were undetectable in the serum, but would increase into the normal range when the serum calcium was lowered by the administration of corticosteroid therapy. Aspirin and indomethacin had no effect on the hypercalcemia, suggesting that prostaglandin synthesis was not related to the hypercalcemia. Bone-resorbing activity was detected in cultures of the abnormal lymphoid cells, and it was suggested that hypercalcemia in these animals was due to production of a macromolecular bone-resorbing factor by the tumor cells.

Hypercalcemia in acute leukemia

Hypercalcemia has occasionally been described in acute leukemia, although it is also unusual in this condition, and is probably more frequent in children than it is in adults. As with the lymphomas, bone lesions occur more frequently than florid hypercalcemia. In fact, children presenting with acute leukemia not infrequently have bone lesions, and occasionally diffuse osteopenia or osteoporosis may be the means of presentation. Bone lesions (and hypercalcemia) can occur in both acute granulocytic leukemia and acute lymphocytic leukemia. The author has seen one patient, an adult in his forties, with hypercalcemia complicating acute eosinophilic leukemia. Hypercalcemia has also been described during the phase of acute transformation of chronic myelogenous leukemia. Recently there have been several cases of hypercalcemia reported in patients with myelofibrosis. In none of these cases has the mechanism for the bone lesions or hypercalcemia been determined.

Treatment of hypercalcemia due to malignancy

The serum calcium can be adequately lowered in the great majority of patients with hypercalcemia due to malignancy or to any other cause. However, many treatment modalities are available, an indication of the fact that there is no single form of treatment which is either universally safe or effective. The therapy for hypercalcemia should be individualized so that each hypercalcemic patient is dealt with on a case-by-case basis after careful consideration of a number of issues. In selecting treatment for any patient, these issues include the cause of the hypercalcemia, the rate of rise of the serum calcium, the specific contraindications to particular forms of therapy (in particular renal failure), and the underlying objective of therapy in the patient under consideration. One potentially important issue is the pathogenetic mechanism responsible for the hypercalcemia. Where hypercalcemia is due predominantly to increased absorption of calcium from the gut, or to increased resorption of bone, ideal therapies would be those which inhibit these processes specifically. Unfortunately such specific therapies are still not available. In this chapter the medical treatment of hypercalcemia due to malignancy is reviewed. In most of these patients hypercalcemia is due to a combination of increased bone resorption and decreased renal calcium excretion. The medical managment of hypercalcemia due to the other major causes is discussed in other chapters.

In hypercalcemia due to malignancy the best treatment is clearly that which inhibits or ablates the neoplasm. Unfortunately, this is often not possible, and this chapter focuses on the medical managment of the increase in serum calcium. In patients with malignancy-associated hypercalcemia the pathophysiologic mechanism responsible for hypercalcemia varies from tumor to tumor. In most patients the serum calcium rises progressively, and often fairly rapidly, so that the disorder in calcium homeostasis is unstable (Figure 7.1). Parfitt[1] has referred to this type of hypercalcemia as 'disequilibrium hypercalcemia'. These are important points to bear in mind when consideration of choice of therapy is being made.

The agents available to the clinician for the medical treatment of hypercalcemia are numerous. They are listed in Table 7.1.

Indications for treatment

There is no widespread agreement on the indications for active treatment for hypercalcemia. It is the author's belief that all patients who are hypercalcemic should be treated, because the serum calcium tends to rise in a steadily progressive manner (Figure 7.1). Moreover, symptoms due to hypercalcemia (for example, nausea,

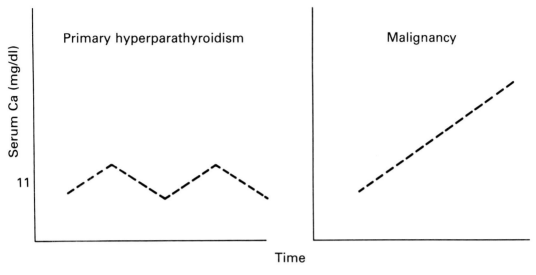

Figure 7.1

Differences between hypercalcemia in patients with malignancy and in patients with primary hyperparathyroidism. In malignancy serum calcium tends to rise progressively (unstable or 'disequilibrium' hypercalcemia). In primary hyperparathyroidism the serum calcium may be maintained relatively constant for prolonged periods ('equilibrium' hypercalcemia).

Table 7.1   Therapy available for hypercalcemia of malignancy.

1   Ablation of tumor

2   Intravenous fluids

3   Furosemide

4   Bisphosphonates

5   Calcitonin/glucocorticoids

6   Plicamycin

7   Phosphate

8   Indomethacin

9   Investigational
   gallium nitrate
   cisplatinum
   WR-2721 (Chapter 5)

vomiting, and confusion) can be readily confused with the symptoms of terminal malignant disease or some of the other therapies given for malignant disease, such as radiation therapy or chemotherapy, and are very distressing for the patient who is suffering from them. For this reason hypercalcemic symptoms should be prevented by the active treatment of early or mild hypercalcemia. However, the choice of therapy can vary. Patients with mild hypercalcemia who are asymptomatic (serum calcium usually less than 12.5–13.0 mg/dl (3.1–3.25 mmol/l)) can be treated with non-urgent forms of therapy, whereas patients who are symptomatic or have a serum calcium greater than 13 mg/dl (3.25 mmol/l) require more immediate therapy.

Maintenance therapy for hypercalcemia

Maintenance or non-urgent therapy for hypercalcemia is discussed before emergency therapy

Table 7.2   Maintenance treatment of hypercalcemia.

1   Oral bisphosphonate

2   Oral phosphate

3   Glucocorticoid (± calcitonin)

4   Plicamycin

5   Indomethacin

since most patients present with a serum calcium level below 13.0 mg/dl (3.25 mmol/l). Maintenance therapy is indicated in those patients who are ambulant, asymptomatic, and can tolerate oral forms of therapy. There are many choices, and these are considered in turn. They are listed in Table 7.2.

Bisphosphonates

The bisphosphonates are a family of synthetic compounds which are stable analogues of pyrophosphate. The basic structural difference between pyrophosphate and the bisphosphonates is the substitution of the P–O–P bond of pyrophosphate with a P–C–P bond of the bisphosphonates (Figure 7.2). Consequently, the bisphosphonates are resistant to enzymatic hydrolysis by pyrophosphatases which are present in the intestine and kidney, so they have a prolonged half-life. Since they were found originally to be very effective in preventing the build-up of sludge on pipes, they were developed as potential dental agents to prevent calcification and plaque formation. This effect is related to their capacity to impair crystal growth by binding to hydroxy-apatite, thereby inhibiting mineralization.[2,3] During these earlier studies on their mode of action on bone, it was found that they were also effective inhibitors of osteoclastic bone resorption.[4] This is the basis for their usefulness in the treatment of bone diseases, and appears to be a universal property of all the bisphosphonates which have

so far been developed. These effects on bone occur at doses which seem to have few significant effects in other tissues. However, since their first use as therapeutic agents in the late 1960s, they have had an unfortunate history. The first agent to be marketed was etidronate. This agent was an effective inhibitor of bone resorption and was useful in the treatment of Paget's disease, myositis ossificans, heterotropic calcification, osteoporosis associated with paraplegia, and tumoral calcinosis (for a review, see Russell and Smith[5],) but resulted in impaired mineralization and caused osteomalacia. Newer-generation agents were developed which were powerful inhibitors of bone resorption but had lesser effects on mineralization. These included clodronate (Cl_2MBP) and APD.[6,7] Clodronate was tested extensively in the late 1970s and early 1980s, and

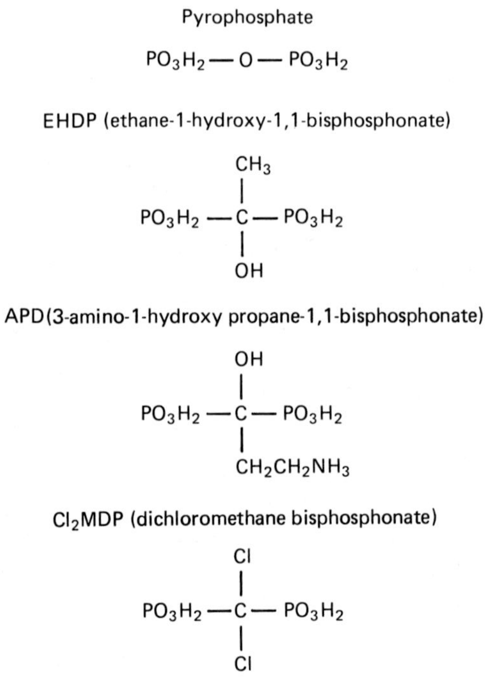

Pyrophosphate

$$PO_3H_2 — O — PO_3H_2$$

EHDP (ethane-1-hydroxy-1,1-bisphosphonate)

$$PO_3H_2 — \overset{\overset{\textstyle CH_3}{|}}{\underset{\underset{\textstyle OH}{|}}{C}} — PO_3H_2$$

APD (3-amino-1-hydroxy propane-1,1-bisphosphonate)

$$PO_3H_2 — \overset{\overset{\textstyle OH}{|}}{\underset{\underset{\textstyle CH_2CH_2NH_3}{|}}{C}} — PO_3H_2$$

Cl_2MDP (dichloromethane bisphosphonate)

$$PO_3H_2 — \overset{\overset{\textstyle Cl}{|}}{\underset{\underset{\textstyle Cl}{|}}{C}} — PO_3H_2$$

Figure 7.2

Structures of the bisphosphonate compounds compared with pyrophosphate.

was found to be effective not only in Paget's disease but also in the treatment of hypercalcemia and osteolytic bone disease. However, three patients (of a total of more than 600) on long-term therapy in investigative studies developed acute leukemia and, although a direct connection between the clodronate therapy and the later development of acute leukemia was never shown, the drug was withdrawn from further testing in the USA. Clodronate is still used extensively in Europe. The newer drug was APD, which is also very effective in Paget's disease and hypercalcemia of malignancy. It has not yet been tested extensively in the USA. Both APD and clodronate are undergoing clinical studies in the USA, and it seems likely that either these or newer-generation bisphosphonates will be the agents of choice in the treatment of hypercalcemia or bone destruction associated with malignancy because of their powerful effects as inhibitors of osteoclastic bone resorption.

Mode of action

In the treatment of hypercalcemia, the bisphosphonates cause a lowering of the serum calcium by decreasing osteoclastic bone resorption. The mechanism by which they inhibit osteoclastic activity is not known. It has been suggested that they may interfere with mineralized bone surfaces to prevent mature osteoclasts from resorbing bone. However, in vitro studies now suggest that they have a direct cellular effect. Long-term cultures of human bone marrow show that bisphosphonates inhibit the formation of cells with the osteoclast phenotype.[8] In this culture system, bisphosphonates exhibit similar potencies to those they possess in vivo. APD and Cl_2MBP do not inhibit the proliferation of osteoclast precursors. However, these compounds decrease the formation of cells with the osteoclast phenotype in these cultures, as determined by cross-reactivity with a monoclonal antibody which preferentially recognises osteoclasts. Some of the bisphosphonates clearly have effects on immune cells which could influence osteoclast activity by a secondary effect.[9] They are all active in bone organ culture systems in vitro, although it has not been possible to correlate their relative potencies in vitro in these systems with their in vivo activity.

Bijvoet and his colleagues[10] have argued that the bisphosphonate APD may interfere with the migration of osteoclast precursors to bone surfaces, or the subsequent differentiation of the precursors to the mature cells. They did not find similar effects with etidronate or Cl_2MBP. There has been one suggestion that bisphosphonates may be toxic to osteoclasts. Using isolated rat osteoclasts incubated with Cl_2MBP on bovine bone slices,[11] there was no apparent effect of Cl_2MBP on isolated osteoclasts separated from bone surfaces. The authors concluded that when Cl_2MBP becomes adsorbed onto bone surfaces, osteoclast injury occurs when osteoclasts subsequently resorb bone.

The bisphosphonates have many effects on intermediary metabolism of bone cells. Whether these are important in their effects on osteoclastic bone resorption is not known. As mentioned above, they clearly have a role in inhibiting mineralization by binding to hydroxyapatite to prevent crystal growth and dissolution. These effects are related more to their capacity to inhibit mineralization rather than to their effects to decrease osteoclastic bone resorption.

Mode of administration

Etidronate

Etidronate is available in the USA for use both as an oral and as an intravenous agent. It is used in doses of 5–10 mg/kg body weight per day orally. At higher doses the drug is very likely to cause impaired mineralization of newly formed bone. Absorption from the gut is variable, but like most of these compounds is probably less than 10 per cent.

Etidronate is now also being used as an intravenous agent. It has recently been released for the treatment of hypercalcemia of malignancy by the Food and Drug Administration (FDA). It is effective in causing a fall in the serum calcium in most patients when used intravenously, although this fall may be slower than that seen with other agents. Etidronate is not effective in treating hypercalcemia when used orally without preceding intravenous therapy.[12] It may be given intravenously in doses of 7.5 mg/kg body weight per day for 7 days, followed by oral therapy, and is effective in lowering the serum calcium in at least 75 per cent of patients.[13]

Clodronate

Clodronate can be used either orally or intravenously. Some patients with hypercalcemia tolerate oral clodronate poorly and for these reasons it may be preferable to use intravenous clodronate. One suitable regimen is to use 300 mg per day for 5 days. The drug should be added to 0.5 liters of normal saline and infused over a minimum period of 2 hours. Clodronate has been used in a range of doses from 100–300 mg/day for periods of between 3 and 10 days. It is likely that weekly administration of a single intravenous dose of clodronate would also be an attractive method for controlling hypercalcemia.

Clodronate can be given orally to patients with hypercalcemia of malignancy in doses of 0.8–3.2 g as a single dose or two doses per day without significant known toxicity.

Pamidronate

Amino-hydroxypropylidene bisphosphonate (also called amino-hydroxypropane bisphosphonate and widely known as APD) has the generic name of pamidronate. It can be used either orally or intravenously. The intravenous route is preferred in patients with hypercalcemia because the orally administered drug is absorbed unreliably and associated with side-effects such as mouth ulceration. There have now been many studies on the use of intravenously administered APD. Although initially APD was given in doses of approximately 15 mg/dl for 6–9 days or until serum calcium had fallen,[14] more recently it has proved effective when used in one or just a few intravenous infusions.[14–20] The response is determined by the amount administered. Most patients will respond to 30–60 mg given as one or two intravenous infusions. Patients with severe hypercalcemia (>4 mmol/l or 16 mg/dl) may require 90 mg. If 60–90 mg are given, the chances of transient hypocalcemia are greater, although this is not of clinical significance. More than 90 per cent of patients should have a serum calcium which is rendered normal by this therapy. APD should be given intravenously over a period of 2–24 hours in 0.9 per cent saline. The infusion rate should not exceed 7.5 mg–15 mg/hr, and the concentration in the infusion fluid should not exceed 15 mg/125 ml.

Efficacy

Etidronate

Etidronate is not effective when used orally in the treatment of hypercalcemia or in primary hyperparathyroidism.[12,21] Since it is useful when used orally in Paget's disease, the reason that oral etidronate is so ineffective in the treatment of hypercalcemia is not known. It may be more effective when used orally after a previous response to an intravenous infusion (Figure 7.3). Possibly part of the reason for its lack of efficacy when used orally is related to its poor absorption from the gut. However, since it is effective when used orally in Paget's disease, this explanation is not entirely satisfactory.

Clodronate

Clodronate is very effective when used either orally or intravenously. There have been numerous studies showing how effective it is both in hypercalcemia of malignancy and in osteolytic bone disease, as well as in primary hyperparathyroidism and parathyroid carcinoma.[22–25]

Pamidronate

Amino-hydroxypropylidene bisphosphate is also extremely effective (Figure 7.4(a)) when used either intravenously or orally for patients with hypercalcemia of malignancy or osteolytic bone disease.[12,26] Sleeboom and Bijvoet[27] claim that it is effective in 95 per cent of patients with hypercalcemia, a much better response than that reported with etidronate or clodronate. Other investigators who have used this drug as a single or multiple intravenous infusions report similar response rates when adequate doses are given. Its effects are probably more rapid than those of etidronate or clodronate, and in a recent study of 48 patients randomly allocated to either APD,

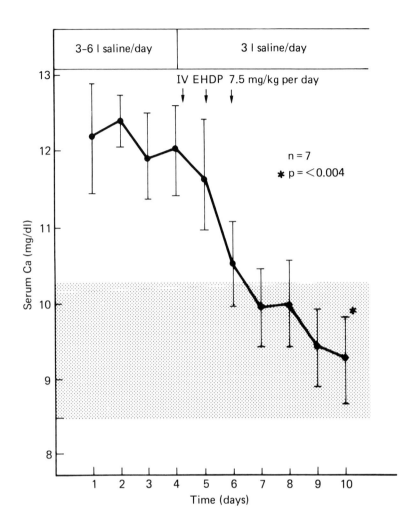

Figure 7.3

Effects of intravenous etidronate on serum calcium in patients with
hypercalcemia of malignancy.[35] (Conversion of traditional units to SI units:
serum calcium 1 mmol/l = 4 mg/dl.)

clodronate or etidronate, APD appeared to normalize serum calcium more effectively and its
effects were more prolonged[28] (Figure 7.4(b)).

Following a single intravenous infusion of APD,
the serum calcium should return to the normal
range between 3–7 days (mean 4
days).[14,15,17,18,29–32] The duration of action of APD

is variable. In most patients, the serum calcium stays within the normal range for 3
weeks.[15,17,30,32,33] This can be anticipated if total
doses of APD > 30 mg are administered.
Thiebaud et al[17] showed that recurrence was
much less likely when patients were treated with
larger doses of APD. Although comparable data

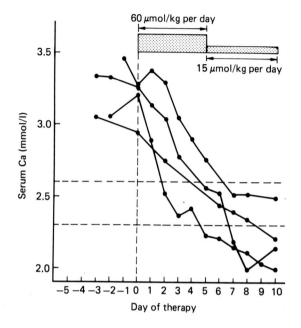

Figure 7.4(a)

Effects of APD on serum calcium in patients with hypercalcemia of malignancy.[12]

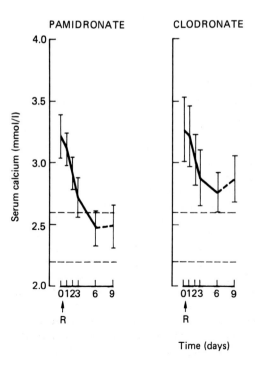

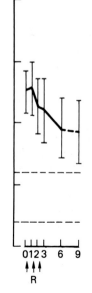

Time (days)

Figure 7.4(b)

Effects of different bisphosphonates on serum calcium in patients with hypercalcemia of malignancy. Patients treated with pamidronate received a single intravenous infusion of 30 mg in 0.9 per cent saline over 4 hours. Patients treated with clodronate received a single intravenous infusion of 600 mg in 500 ml 0.9 per cent saline over 6 hours. Patients treated with etidronate received 7.5 mg/ kg body weight in 500 ml 0.9 per cent saline over 2 hours for 3 consecutive days (redrawn from Ralston et al[28]).

are not available for the other bisphosphonates, this appears to be a very favorable response. Most patients will respond to retreatment just as well as they responded to the initial treatment, although the occasional patient may show a less satisfactory response with repeated treatments. When APD is available, it should be given as a single intravenous infusion of 15–90 mg, depending on the severity of the hypercalcemia, over 2–24 hours in 0.9 per cent saline. In patients in whom a rapid response is required, calcitonin should be given concomitantly for the first 72 hours.[34] APD can be used again by intravenous infusion as required.

Toxicity

Etidronate is associated with two forms of toxicity, impaired mineralization of bone and increases in serum phosphate. The impairment of mineralization is due to its effects to bind to hydroxyapatite, and results in increases in osteoid tissue. This effect is dose related, being more likely to occur when used orally in doses greater than 10 mg/kg body weight per day for more than 3 months, and may precipitate bone pain and pathologic fractures. The increase in serum phosphate is due to an increase in tubular phosphate reabsorption, and probably has no clinical significance. It seems to occur only in man, and has not been found in studies in experimental animals. Etidronate also has the capacity to limit the hydroxylation of 25-hydroxyvitamin D to 1,25-dihydroxyvitamin D, possibly by a direct effect on the kidney.

Neither clodronate nor APD have any appreciable effect to impair mineralization at the doses used for the treatment of hypercalcemia. However, clodronate has been associated with several cases of acute leukemia occurring in patients taking long-term therapy (see above). Although it has never been proven that there is a direct relationship between the clodronate and the acute leukemia, and in at least several of the patients another potential etiologic factor for acute leukemia was present, the drug was withdrawn in the early 1980s from further clinical testing in the USA. However, it is still used extensively and successfully in Europe. Amino-hydroxy-propylidene bisphosphate has been associated with some relatively minor toxicity.

These include the development of mouth ulcers in patients given earlier preparations of the oral form of APD which was unformulated and administered in water, fever in some patients (which appears to be transient), and changes in the peripheral blood leukocytes (in particular, lymphopenia). The mechanism for these effects is unknown. Some patients treated with large doses of APD develop transient hypocalcemia. This is rarely symptomatic and the total serum calcium does not drop below 8 mg/dl (2 mmol/l).

Treatment with APD (or any other bisphosphonate) can be monitored by following fasting urinary calcium/creatinine ratios as well as serum calcium. Fasting urinary calcium/creatinine ratios reflect bone resorption. In patients who respond to APD, urinary calcium/creatinine will decrease as bone resorption is inhibited. However, in those few patients who respond poorly to APD, if urinary calcium/creatinine remains increased then bone resorption is the underlying pathogenetic mechanism for hypercalcemia, and more APD or another inhibitor of bone resorption is indicated. In contrast, if patients respond poorly to APD and urinary calcium/creatinine is low, then it is likely that increased reabsorption of calcium from the renal tubules is the major pathogenetic mechanism responsible for hypercalcemia, and more APD is unlikely to help.

Place in managment

At present in the USA the only useful form of bisphosphonate therapy available for the treatment of hypercalcemia is intravenous etidronate. Etidronate is ineffective when used orally without preceding intravenous therapy.[7] Either APD or a new-generation bisphosphonate which has none of the side-effects of earlier agents and greater efficacy should become available for widespread use very soon. In the meantime intravenous etidronate 7.5 mg/kg body weight per day for 3–7 days, followed by oral therapy in low doses (5–10 mg/kg body weight per day) appears to be a useful and relatively safe therapeutic approach. Etidronate appears to be most effective when it is used concomitantly with approximately 3 l/day intravenous saline.[35] When used in this way, the serum calcium of most patients will be normalized within 4 or 5 days.

There is such widespread interest in the bis-phosphonates for the treatment of hypercalcemia of malignancy that several recent publications have been devoted entirely to this topic. The interested reader is referred to them for further information. These symposia have focused independently on APD (pamidronate), Cl$_2$MBP (clodronate) and EHDP (etidronate).[36–39]

Plicamycin (also called mithramycin)

Plicamycin was developed as a cytotoxic drug during the 1960s and was found to be effective in certain embryonal type tumors.[40,41] It was found incidentally that many patients treated with plicamycin also suffered a profound fall in the serum calcium. Because of this effect, it was tried as an antihypercalcemic drug empirically, and was found to be very effective.[12,42–46] Since the early 1970s it has remained a major agent in the drug treatment of hypercalcemia of malignancy.

Mechanism of action

Plicamycin is an agent which inhibits DNA-dependent RNA synthesis.[47–49] This is probably the mechanism by which it affects tumor cells. Whether this effect has anything to do with its mode of action on bone-resorbing cells is unknown. As an antiproliferative factor it could impair the generation of new osteoclasts from replicating precursors. Its mode of action on bone resorption is relatively slow, so this may be the mechanism by which it impairs the resorption of bone by osteoclasts. It has been shown to inhibit bone resorption in organ culture, and when bone resorption is inhibited it is usually irreversible.[50,51] However, this does not necessarily imply toxicity. Although a toxic effect of plicamycin on bone cells could explain the apparently irreversible impairment of bone resorption, it is also possible that an inhibitory effect on the formation of new osteoclasts by impairment of replication could deplete the pool of cells available to respond to osteoclast-stimulating agents such as PTH. This could lead to the apparent irreversibility of its inhibitory effects on osteoclastic bone resorption in an organ culture system.

Mode of administration

Plicamycin is administered intravenously either as an infusion or as a bolus. The package insert recommends that it should be administered as an

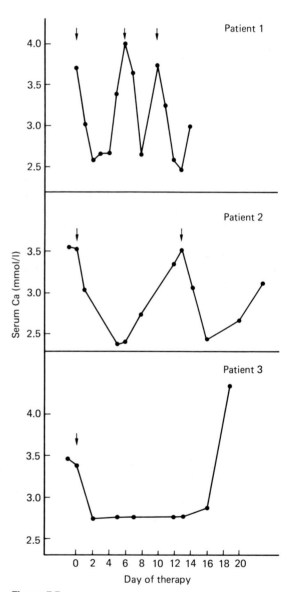

Figure 7.5

Effects of mithramycin (25 μg/kg body weight per day) on serum calcium in three patients with the hypercalcemia of malignancy. Note the different patterns of response. Time of mithramycin infusion is indicated by the arrows. Note the rebound effect in patient 3 (see text).[12]

infusion over at least 10 min. This is probably wise. In most situations it is not warranted to administer it as a bolus, since it does not lower the serum calcium rapidly. The author recommends giving it in a dose of 15–25 μg/kg body weight, administered over 4 h, and then not repeating the infusion until the serum calcium rises again. However, there is no general agreement on how often plicamycin should be administered. There are a number of different regimens. Some clinicians give it daily for five successive days and then await a relapse before treating again. Others give the initial infusion and then wait to see whether there is a fall in the serum calcium. If there is, they then wait until the serum calcium has increased before re-treating.

There is a possible rebound effect from the effects of plicamycin,[12] but how this can be anticipated or avoided is not known. The difficulty is anticipating how long a patient will remain normocalcemic after plicamycin administration. Some patients remain normocalcemic for a few days, some for several weeks. Hypercalcemia may develop very rapidly in the patient who relapses after plicamycin therapy. One patient seen by the author, a patient with breast cancer, had a very satisfactory response to plicamycin, after which she remained normocalcemic for 14 days. However, she progressed from her baseline serum calcium of 11.2 mg/dl (2.8 mmol/l) to a hypercalcemic crisis with a serum calcium of 17.5 mg/dl (4.4 mmol/l) within 6 h (Figure 7.5).[12]

Efficacy

Plicamycin is very efficacious in causing a fall in the serum calcium. It has been found to be effective in approximately 80 per cent of patients with the hypercalcemia of malignancy.[12] Its maximum effect occurs between 24 and 36 h after administration. Of course, it is more effective in patients with increased bone resorption as a primary cause of the hypercalcemia. In the author's experience, when plicamycin is ineffective, so too are most other antihypercalcemic drugs.

Toxicity

Plicamycin has considerable toxicity (summarized by Mundy and Martin).[52] It causes liver damage,

leading to an increase in hepatocellular enzymes. It also causes impairment of renal function, and it causes bleeding which may or may not be related to thrombocytopenia. Its nephrotoxic effect is a serious disadvantage to its use in the treatment of hypercalcemia. Many patients with hypercalcemia have impaired renal function, and plicamycin will impair renal function further in these patients. Its other toxic effects are cumulative and, since its major rate of elimination is via the kidney, these are particularly likely to occur in patients with poor renal function.

Place in therapy

Plicamycin remains a major drug in the treatment of hypercalcemia of malignancy. One of the major drawbacks to its use (apart from its toxicity) is the inconvenience of its administration, which is by intravenous infusion only. It should be used only in those patients with normal or near-normal renal function. It should be used carefully, with close monitoring of renal function, liver function tests, and peripheral blood leukocyte and platelet counts.

Calcitonin

Calcitonin is a natural product of the parafollicular cells (also called the C-cells) of the thyroid gland. It was recognized in 1962 by Copp et al as a serum-calcium lowering agent.[53] Its synthesis, secretion and actions are considered in some detail in Chapter 3. It was purified during the late 1960s and the gene was cloned 10 years later. It has an interesting gene structure, since the calcitonin gene also encodes a calcitonin gene-related peptide (CGRP) which seems to have primary vascular and neuromuscular effects. There is differential expression of calcitonin and CGRP in different parts of the body. Release of calcitonin from the thyroid gland is regulated by the serum calcium concentration, as well as by gastrointestinal peptides such as gastrin and glucagon. It has an important, although limited, place in the medical therapy of hypercalcemia.[54–56]

Mode of action

Calcitonin inhibits osteoclastic bone resorption (Chapter 3). However, its effects on osteoclastic bone resorption are transient and reversible. Despite the continued presence of calcitonin, bone resorption is not permanently inhibited. This resistance to the continued presence of calcitonin, which is seen both in vitro and in vivo, is known as 'escape' (described in more detail in Chapter 3). Calcitonin causes decreased formation of osteoclasts, and probably dissolution of preformed osteoclasts into mononuclear cells. It also causes contraction of the cytoplasmic membranes of osteoclasts.

Calcitonin also has effects on the gut and the kidney. It leads to the excretion of calcium and phosphate both in the urine and in the feces. It is unlikely that the effects on gut and kidney are of major importance in the maintenance of calcium homeostasis. However, it is possible that the initial calcium-lowering effect of calcitonin administration could be due to inhibition of renal tubular calcium reabsorption,[56,57] working in combination with inhibition of bone resorption.

There is no evidence that calcitonin has any major effect on osteoblasts or bone formation.

Mode of administration

The only form of calcitonin which has been used widely in the USA is parenteral salmon calcitonin. This synthetic peptide is adequately potent and has a prolonged half-life in vivo, which lessens the need for frequent administration. It can be used either intramuscularly or subcutaneously, and it is equally effective by either route. It should be used in doses of 50–200 MRC (Medical Research Council) units each 12 h. Human calcitonin is now also available. Human calcitonin can be used intramuscularly, subcutaneously, or intravenously. It has similar biological potency to salmon calcitonin.

Recently attempts have been made to circumvent the problems associated with the necessity for intramuscular or subcutaneous administration of calcitonin by the use of intranasal calcitonin. There is little information available yet for the efficacy of this preparation in the hypercalcemia of malignancy. However, it is possible that this form of administration will eventually turn out to be useful in the long-term management of hypercalcemic patients who are ambulant and outside the hospital. Recently, salmon calcitonin has been administered to patients as rectal suppositories containing 300 MRC units three times daily for 7 days with good results, similar to those which might be expected with parenteral administration.[58] If patients will tolerate this form of therapy, then it is potentially very useful in the maintenance treatment of ambulant patients with hypercalcemia of malignancy.

Toxicity

Calcitonin is a naturally occurring compound, and there is little significant toxicity associated with its use. In large doses it is occasionally associated with mild nausea and vomiting. Since synthetic salmon calcitonin is antigenic, it occasionally causes skin rashes and allergic reactions. For this reason it has been recommended that patients should be skin tested before beginning treatment.

Results of treatment in patients with hypercalcemia

Calcitonin has been shown to be effective in a number of therapeutic studies in patients with hypercalcemia (Figures 7.6 and 7.7). However, the beneficial effects are usually very short-lived when calcitonin is used alone, not usually persisting for more than 24–48 h. When calcitonin is used with corticosteroids, the beneficial effects are more prolonged.[54,55] The response is often not complete, and in the experience of the author about 20 per cent of patients show no response. Unfortunately, there do not appear to be any careful dosing studies of the use of calcitonin in hypercalcemia, although the timing of administration and the spacing of doses may be critical for seeing maximal benefits.[59] Calcitonin may also be effective in other forms of hypercalcemia, including that due to immobilization.

Place in therapy

Calcitonin is a nontoxic agent which usually lowers the serum calcium and can be used in the presence of impaired renal function. For this

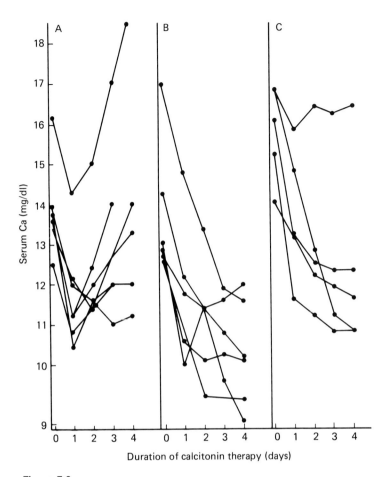

Figure 7.6

Effects of calcitonin alone and the combination of calcitonin and glucocorticoids in patients with hypercalcemia of malignancy. A, patients treated with calcitonin alone. B, patients with lymphoproliferative disorders treated with the combination of calcitonin and glucocorticoids. C, patients with solid tumors treated with a combination of calcitonin and glucocorticoids. (Reproduced with permission from *Ann Intern Med.*)[54] (Conversion of traditional units to SI units: serum calcium 4 mg/dl = 1 mmol/l.)

reason there is no apparent reason why the agent should not be used when patients with urgent hypercalcemia first present. The beneficial effects are of only short duration, but can be prolonged when corticosteroids are administered concomitantly. The institution of calcitonin treatment together with isotonic saline and corticosteroids is recommended at the time that treatment is begun. Alternatively, calcitonin can be used in conjunction with bisphosphonates, since the bis-

phosphonates, although very effective, are usually slow acting.

Calcitonin may also have a place in the long-term management of hypercalcemia, although its place here is less secure. When used for long-term management it must be given with corticosteroids. It is rather inconvenient to use in this situation, because the currently available preparations require parenteral administration. It is also likely that it will need to be withdrawn

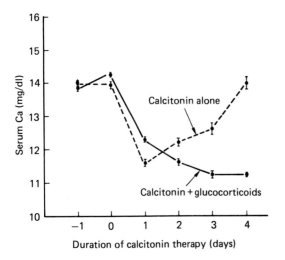

Figure 7.7

Effects of calcitonin and glucocorticoids compared with calcitonin alone in the treatment of patients with the hypercalcemia of malignancy. The combination produced a prolonged beneficial effect where calcitonin alone produced only a transient effect followed by rapid relapse to the pretreatment level of serum calcium. These patients had not responded to glucocorticoids alone.[52] (Conversion of traditional units to SI units: 4 mg/dl = 1 mmol/l.)

periodically for continued effectiveness. A regimen of treatment with calcitonin for five days out of seven is recommended, with withdrawal over the weekends. An example of the effectiveness of this approach is described by Mundy and Martin and shown in Figure 7.8.[54] With the recent introduction of intranasal preparations of calcitonin, the problems associated with parenteral administration to the ambulant patient may be avoided.

Phosphate

Phosphate therapy has been used for many years in the treatment of hypercalcemia. The major mechanism of action probably differs according to whether phosphate is given orally or intravenously. Whereas the use of phosphate as an oral agent is to be recommended and is relatively safe, intravenous phosphate is associated with so much toxicity that, in the opinion of the author, it should be avoided.

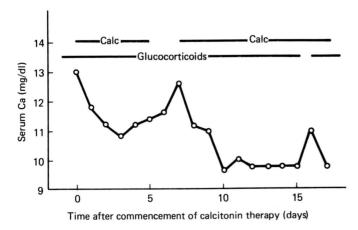

Figure 7.8

Effects of calcitonin and glucocorticoids in the longterm management of a patient with renal failure and the hypercalcemia of malignancy due to myeloma. Calcitonin was withdrawn for two days out of every seven, and the serum calcium was maintained in an acceptable range which did not cause symptoms.[52] (Conversion of traditional units to SI units: 4 mg/dl = 1 mmol/l.)

Mode of action

Oral phosphate probably works by impairing calcium absorption from the gut, and by inhibition of bone resorption.[52] Increases in phosphate concentration inhibit osteoclastic bone resorption in vitro.[60] The mechanism is not entirely clear, but may involve a physicochemical mechanism.[61]

Intravenous phosphate probably works predominantly by causing the precipitation of calcium in the extracellular fluid with phosphate. This is responsible for the immediate calcium-lowering effect of the intravenous preparations. The toxicity associated with this form of therapy is high (see below).

Mode of administration

Oral phosphate is given in divided doses of 2–3 g/day as neutral sodium phosphate. It can be given as tablets, as an elixir, or as suppositories. Intravenous phosphate is given in concentrations of 50–75 mmol/24 h. In the USA it has to be prepared specially for each administration by a pharmacist.

Toxicity

The major complication of oral phosphate therapy is diarrhea. The mechanism for diarrhea is unknown, but is dose-related. Oral phosphate therapy can also lead to renal stones, to muscle cramps and pain, and to soft-tissue calcification.[62,63] This is most likely to occur in patients with impaired renal function or with preexisting high serum phosphorus levels. The major side-effect of intravenous phosphate is life-threatening soft-tissue calcification. This can lead to precipitation of calcium in the kidney, lungs, heart, and other organs. This is a serious complication which has led to the death of a number of patients treated with this form of therapy.[64]

Efficacy

Oral phosphate is effective in lowering the serum calcium in those patients who have a serum phosphorus which is not increased at the beginning of therapy. In patients with a serum phosphorus in the high normal range (>4 mg/dl), oral phosphate is not effective and is potentially hazardous, as its administration increases the

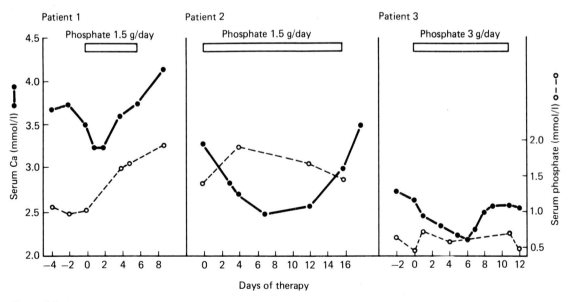

Figure 7.9

Effects of oral phosphate therapy on the serum calcium and serum phosphate in three patients with hypercalcemia of malignancy.[12]

chances of soft-tissue calcification. A few patients will respond initially, but the favorable response cannot be maintained (Figure 7.9).[12]

Place in the treatment of hypercalcemia

Oral phosphate is a useful form of treatment in those patients who have normal renal function and whose serum phosphorus concentration is low (<3.8 mg/dl).[12] The author uses it in all patients who can tolerate it if they have a serum phosphorus of <3 mg/dl, even when using other antihypercalcemic agents. Intravenous phosphate should not be used except in those rare situations in which no other form of therapy is available or the patient may die from severe hypercalcemia.

patients with widespread malignancy, who have a limited prognosis. Glucocorticoids have multiple effects on calcium homeostasis, including inhibition of bone resorption in some circumstances, and impaired absorption of calcium from the gastrointestinal tract. They are usually ineffective against hypercalcemia associated with PTH excess.

The mode of action of glucocorticoid therapy was studied in some detail in a patient with lymphosarcoma cell leukemia and hypercalcemia, and it was found that the glucocorticoids were effective in lowering the serum calcium when used in the patient, and they inhibited production of the bone-resorbing activity by the tumor cells in vitro.[66] In patients with lymphomas where 1,25-dihydroxyvitamin D was an inhibitor they may be exerting their major effect by inhibiting gut absorption of calcium.

Glucocorticoids

About 30 per cent of all patients treated with glucocorticoids for the hypercalcemia associated with nonparathyroid malignancy respond with a fall in the serum calcium.[12] However, in some of these patients the response is not complete and the serum calcium does not return to the normal range. Moreover, responsiveness to glucocorticoids is unpredictable. Glucocorticoids are most likely to be effective in patients with hematologic malignancies, particularly those with myeloma. However, they are not effective in all patients with myeloma.[54] They are less likely to be effective in patients with hypercalcemia due to breast cancer or lung cancer, although there are occasional patients who respond very well. They are almost always ineffective in patients with primary hyperparathyroidism who do not have bone disease.[65] In those patients who do respond, glucocorticoids are very useful agents, because they can be used orally and the side-effects of glucocorticoid therapy are not important in patients with hypercalcemia of malignancy, since most patients do not survive for more than 3–6 months. Most of the side-effects associated with glucocorticoid therapy occur only when the drug is administered for prolonged periods (usually more than 3 months). This is not an important consideration in many

Indomethacin and aspirin

The agents which inhibit prostaglandin synthesis have been used in the treatment of hypercalcemia of malignancy in those circumstances where prostaglandins are thought to be playing a pathogenetic role. However, most workers who have used these agents have been disappointed by their lack of efficacy. It is hard to know how many patients are likely to respond to indomethacin, although the experience of most people would suggest it is less than 20 per cent, and probably less than 10 per cent. An unusual example of a good response to indomethacin is shown in Figure 7.10. The failure of indomethacin in most patients is disappointing because in the mid-1970s there were a number of case reports suggesting that patients with hypercalcemia were responding to indomethacin or aspirin,[67] and then a larger study by Seyberth et al[68] suggested that aspirin or indomethacin inhibited prostaglandin synthesis in these patients, and lowered the serum calcium. However, the evidence in this report that the serum calcium was responding was less convincing than that showing decreased synthesis of prostaglandins.

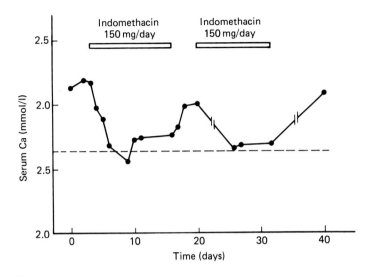

Figure 7.10

Beneficial effects of indomethacin on serum calcium in a patient with hypercalcemia of malignancy due to squamous cell carcinoma of the lung. This good response is unusual with indomethacin therapy.

Gallium nitrate, cisplatinum and WR-2721

Cytotoxic drugs have been found to cause a decrease in the serum calcium in some circumstances. The most prominent example has been plicamycin. However, recently other cytotoxic drugs which are used in the treatment of various forms of cancer have also been found to lower the serum calcium, and these are being investigated as potential agents in the treatment of the hypercalcemia of malignancy. Two such agents are gallium nitrate and cisplatinum. Gallium nitrate was used originally as a diagnostic agent for scanning.[69] Because it also had anti-tumor effects, it was then developed as a cytotoxic agent. In early human studies, transient hypocalcemia was occasionally found.[70] This led to it being developed as a potential anti-hypercalcemic agent. In this sense, its history is similar to that of plicamycin, which is also a cytotoxic drug found serendipitously to have a calcium-lowering effect. It appears to be more effective than calcitonin in the treatment of hypercalcemia.[71] In preliminary reports, it has been suggested that it is several orders of magnitude less potent on a molar basis than APD at inhibiting bone resorption in rats.[72] Cisplatinum has been used effectively in an animal model of the human hypercalcemia of malignancy.[73] It has also been shown to normalize the serum calcium in a series of nine patients with hypercalcemia of malignancy.[74] In these patients there was no detectable reduction in tumor size, and the average duration of response was 38 days. About two-thirds of the patients who were treated showed a good response. The dose used was 100 mg/m². These patients also received concomitant vigorous hydration with isotonic saline, no significant toxicity was found, and the drug was well toler-ated. Cisplatinum causes a decrease in serum magnesium, which could be responsible for the effect of lowering the serum calcium, although studies in vitro suggest that its calcium-lowering effects are unrelated to magnesium metabolism. However, the limiting factor with the widespread use of both of these drugs is their potential toxicity. One of the major problems may be nephrotoxicity. Since many patients with hyper-calcemia, particularly with the hypercalcemia of malignancy, present with impaired renal function, this is likely to be a serious limitation to their usefulness. However, it is also possible that future careful studies will reveal a dose of these drugs at which they are useful in the treatment of hypercalcemia without deleterious side-effects. Both drugs are promising and worthy of further investigation.

WR-2721 is an interesting organic triophos-phate compound which was developed originally for use as an anti-cancer agent. As with gallium nitrate and plicamycin, WR-2721 was found to cause hypocalcemia during early clinical studies.[75] WR-2721 clearly has multiple modes of action. It inhibits PTH secretion both in vitro and in vivo[76,77] and also inhibits renal tubular calcium reabsorption.[78] WR-2721 has an effect on renal tubular calcium reabsorption which is indepen-dent of PTH.[78] It also inhibits osteoclastic bone resorption in organ cultures.[79] WR-2721 has been effective in lowering the serum calcium in several patients with parathyroid carcinoma.[80,81]

WR-2721 has been used as a calcium-lowering agent in an animal model of the humoral hyper-calcemia of malignancy.[82] In this model, WR-2721 lowered the plasma calcium and increased urin-ary calcium excretion. This effect was very rapid. This result suggests that in this model renal tubular calcium reabsorption plays an important pathogenetic role,[78] and that this agent has potential as an anti-hypercalcemic drug. It is a relatively non-toxic agent: it causes nausea and vomiting in small numbers of patients,[83,84] and is also associated with somnolence and sneezing.[84,85]

The choice of therapy for nonurgent hypercal-cemia associated with malignancy is wide. Figure 7.11 summarizes a personal experience in ap-proximately fifty hypercalcemic patients treated

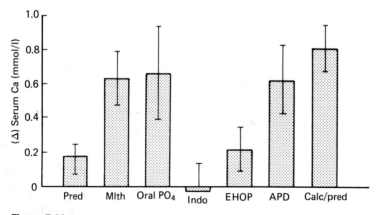

Figure 7.11

Effects of a series of antihypercalcemic agents used in a therapeutic study of the hypercalcemia of malignancy in fifty patients. Note that most patients with hypercalcemia respond well to therapy. The choice of therapy should be individualized to take into account the potential side-effects of each of the agents and tolerance to their administration.[12]

with the agents available in 1979. Most drugs are effective, and the choice is best tailored for the individual patient, particularly taking into account potential toxicity. In most patients the author believes a bisphosphonate to be the preferred choice at present.

Urgent therapy of the hypercalcemia of malignancy

Urgent therapy for hypercalcemia is indicated in patients who are symptomatic from hypercalcemia or who have a serum calcium which is >13 mg/dl. In these patients this level of serum calcium may be potentially life-threatening. At this level the serum calcium is likely to be unstable and can rise very rapidly to higher levels which can threaten the patient's life. These patients need to be treated actively and urgently. Agents which are likely to precipitate the hypercalcemia, such as recent treatment with estrogens or anti-estrogens if the patient has breast cancer, or therapy with thiazide diuretics which impair renal calcium clearance, should be discontinued. The choices are summarized in Table 7.3.

Fluids

Patients with severe hypercalcemia are often profoundly dehydrated. These patients may be depleted of 5–10 l extracellular fluid volume,

Table 7.3 Emergency treatment of hypercalcemia.

Intravenous saline and
1 Intravenous bisphosphonate
2 Calcitonin/glucocorticoids
3 Plicamycin

similar to patients with diabetic ketoacidosis or the diabetic hyperosmolar nonketotic syndrome. There are a number of similarities between severe hypercalcemia and the diabetic hyperosmolar syndrome. In both there may be a profound diuresis. In the case of the diabetic this is due to an osmotic diuresis, whereas in patients with hypercalcemia there is failure of urine concentration due to the ineffectiveness of antidiuretic hormone (ADH) to promote free water reabsorption in the collecting tubules. Volume depletion leads to enhanced sodium reabsorption in the proximal tubules. Since the reabsorption of calcium and sodium are closely linked in the proximal tubules, volume depletion also causes calcium reabsorption, which increases the serum calcium further. In patients with hypercalcemia sodium repletion breaks this vicious cycle, impairs renal tubular calcium reabsorption, and leads to a calcium diuresis. Therefore, the most appropriate fluid for rehydration in these hypercalcemic patients is intravenous isotonic saline, since if a sodium diuresis can be provoked, then calcium will also be excreted in the urine together with sodium. Many of these patients can be treated in a general hospital ward, although some will be so dehydrated that they are best treated with central venous pressure monitoring and with the facilities of an intensive care unit. However, it is only a minority of patients with hypercalcemia who have life-threatening elevations of serum calcium (the author's estimate is less than 10 per cent of all patients who present with hypercalcemia). In the majority of patients vigorous rehydration will lower the serum calcium, in some down to the normal range.[86] However, this fall is only transient, and since hypercalcemia is in most cases associated with increased bone resorption, serum calcium will again begin to rise. Fluid replacement has no beneficial effect on the increase in bone resorption, although a saline diuresis will promote a calcium diuresis. A common error in this situation is to rehydrate patients alone, without additional therapy directed at the increase in bone resorption. However, fluid repletion and volume expansion will lower the serum calcium only transiently if given alone. These patients also need therapy which inhibits bone resorption.

Patients with severe hypercalcemia frequently have other electrolyte problems. They often have a mild impairment of glomerular filtration, which is reversed by rehydration. If renal failure is fixed, then the most likely associated malignancies are myeloma or an obstructive uropathy associated with lymphoma. Occasionally the serum phosphorous is increased, and as a consequence of hypercalcemia and hyperphosphatemia the patient may develop nephrocalcinosis, which can also cause renal failure. Some patients with cancer and hypercalcemia have hypokalemia, which may be related to many factors including vomiting, cytotoxic drug therapy, or injudicious or overly vigorous use of loop diuretics. Serum phosphorous may be decreased both in patients with hypercalcemia of malignancy and in those with primary hyperparathyroidism due to renal phosphate wasting.

On a number of occasions, the author has noted hypernatremia in patients with hypercalcemia. This has almost invariably occurred during the resuscitation phase, and nearly all of the patients have been elderly. The clinical setting is almost always the same. Hypernatremia occurs in a patient with hypercalcemia of any cause— primary hyperparathyroidism, in hematologic malignancies, and in solid tumors associated with hypercalcemia. The hypercalcemia is almost always severe, and the patient is moribund and unable to drink or indicate thirst. The patient has been resuscitated with large volumes of saline given intravenously, which is the recommended therapy in this situation. Serum sodium following institution of therapy has risen to the high 150s or 160 mmol/l. When isotonic saline is replaced with hypotonic fluids, the serum sodium has fallen satisfactorily.

In the author's opinion, the pathogenesis of this syndrome of hypernatremia occurring in association with hypercalcemia may be related to the limited capacity of the kidney to handle a sodium load under circumstances where the patient is severely hypercalcemic. It is well known that hypercalcemia interferes with the actions of ADH on the collecting tubules of the kidney to promote water reabsorption (Chapter 4). In the presence of impaired ADH action the patient cannot reabsorb water adequately, and will pass a copious dilute urine. Under these circumstances, the patient cannot be expected to clear a large sodium load (in this case given to promote a calcium diuresis). Of course, hypernatremia will be enhanced if the patient is moribund, the thirst mechanism is impaired, and the patient cannot drink hypotonic

fluids. The situation is similar to that which occurs in the moribund patient with central diabetes insipidus, who also rapidly becomes hypernatremic without access to hypotonic fluids.

Clearly this is an important situation to recognize. The clinical features of hypernatremia are similar to those of hypercalcemia. They include impairment of conscious state with confusion, stupor, and coma. Milder cases are associated with changes in personality and irritation. This complication of therapy is potentially fatal, but is easily prevented or reversed. Other causes of hypernatremia should be considered and excluded. These are extrarenal free-water loss from the skin or lungs, renal free-water loss from central diabetes insipidus or an osmotic diuresis associated with diabetes mellitus, or primary excess of sodium due to excess salt administration without access to water.

Bisphosphonates

In the author's experience, the most effective drugs currently available for the urgent treatment of hypercalcemia of malignancy are the bisphosphonates. These drugs belong to a class of compounds with metal-complexing properties that are powerful inhibitors of bone resorption and in addition retard the growth and dissolution of crystals of poorly soluble calcium salts.[4] They are discussed in more detail earlier in this chapter. Their effects on bone occur at doses that usually have no serious effects on other biologic systems. All of the bisphosphonates that have been used in this situation (etidronate, clodronate, and APD) are effective, although etidronate seems to be effective only if it is first given intravenously. Any one of these three drugs may be used in the doses indicated earlier. These drugs lower the serum calcium in most patients after 3–5 days of therapy, although APD may be effective more often than etidronate. It remains unknown whether clodronate will be released in the USA, although it is now being used widely in Europe. Etidronate has now been released in an intravenous preparation for use in this country for this cause. It is best used as an intravenous infusion of 7.5 mg/kg body weight per day, accompanied by 3 l intravenous isotonic saline. This combination is effective in most patients, causing a return of the serum calcium to the normal range within 5 days.[35] The manufacturers warn against its use in renal failure, and this is a potential limitation to its widespread application.

The major toxicity associated with the etidronate is impairment of bone mineralization. This is not so much of a cause for concern in the emergency treatment of hypercalcemia as it is in the long-term maintenance therapy of diseases such as Paget's disease. The author has used APD without seeing any evidence of severe toxicity. Fever develops in some patients, and when it is used orally some patients may develop mouth ulceration. If administered rapidly the bisphosphonates have the potential for worsening renal function.[87,88] However, when patients are well hydrated and the agents are given slowly, the risk is probably small if the serum creatinine is <2 mg/dl (226 μ mol/l).

Plicamycin

Plicamycin is an efficacious agent in the treatment of severe hypercalcemia, and it causes a decrease in the serum calcium concentration to the normal range within 24–48 h of administration in most patients to whom it is given. It is discussed in more detail above. The major problem with its use in the urgent treatment of hypercalcemia is that many of these patients have impaired renal function. Since the side-effects of plicamycin are cumulative and its major route of clearance is by the kidneys, this is a serious disadvantage to its use. The author has found another problem with the use of plicamycin in the emergency situation—it is impossible to predict how long a response to plicamycin will be sustained. Some patients may respond for weeks, whereas for some patients the response will last only for several days, and the rebound of serum calcium following successful treatment may be very rapid, occurring over 12 h or so.[12]

Calcitonin and glucocorticoids

Calcitonin alone and calcitonin in combination with glucocorticoids rapidly lower the serum calcium in many patients with hypercalcemia due to malignancy.[54] This is a safe and effective form of therapy which can be widely used, without fear of toxicity, in patients with renal failure or with fluid accumulation. These drugs in combination are effective inhibitors of bone resorption both in vitro and in vivo[54,66] and may also promote a calcium diuresis.[56] They can be administered before hydration is complete. Before the availability of the parenteral bisphosphonates the author believed they were the most effective and safest forms of urgent treatment of the hypercalcemia. In the author's experience they work in 70–80 per cent of patients. Although they do not always lower the serum calcium to the normal range, they usually lower it to a level which is tolerable. The reason for using the combination is that glucocorticoids seem to prevent the escape phenomenon which occurs when calcitonin is used alone. Moreover, this combination is very rapidly effective, usually working within 4–6 h of the first dose. The best responses are found in patients who have hematologic neoplasms such as myeloma.[54] In fact, the author has not yet seen a patient with myeloma who did not respond to this combination. Responses occur in patients who were previously resistant to glucocorticoids alone, suggesting that the combination is more effective than either agent alone. Calcitonin can be used in a dose of 200 MRC units each 12 h intramuscularly or subcutaneously. Glucocorticoids should be given in maximal doses during the initial stages of therapy. For example, 100 mg hydrocortisone hemisuccinate can be given intravenously by infusion or by intravenous bolus every 6 h. After the patient responds, the glucocorticoid dose can be carefully tapered.

Loop diuretics

The powerful loop diuretic furosemide has been a very fashionable form of therapy for the hypercalcemia of malignancy in the USA. In fact, the evidence that this diuretic given in large amounts is useful in the treatment of hypercalcemia of malignancy is rather unimpressive. The rationale for its use is that loop diuretics promote sodium excretion, which in turn causes a calcium diuresis. This is certainly true in the normal individual, but the doses required to produce any effect in the patient with such an unstable situation as hypercalcemia is less clear. The evidence that this therapy is beneficial comes from a study in sixteen patients with hypercalcemia where the patients were given very large doses of furosemide (100 mg every 2 h) and there was a considerable, albeit transient, fall in the serum calcium.[89] All of the patients received fluid in addition to furosemide, and it was impossible to unravel the effect of the fluids from the effects of furosemide, since fluids alone are effective in lowering the serum calcium.[86] The use of such large doses of intravenous furosemide in patients with hypercalcemia is not a convenient form of therapy. It certainly requires the facilities of an intensive care unit, and careful monitoring of other fluid and electrolyte losses. The author has found that in clinical practice most physicians give hypercalcemic patients much smaller doses of furosemide than those recommended by Suki et al.[89] The problem is that, in a patient who is not fully rehydrated, furosemide is likely to worsen the hypercalcemia rather than relieve it. For example, a patient with hypercalcemia who has depleted extracellular fluid volume already has increased calcium reabsorption from the proximal convoluted tubule, and the effects of a potent loop diuretic which will deplete the extracellular fluid volume further will cause even more calcium reabsorption. The major place for loop diuretics in the treatment of hypercalcemia is in the patient who requires fluid therapy but is overhydrated or has cardiac failure. There is no documented evidence that oral loop diuretics will lower the serum calcium in the patient with hypercalcemia.

Intravenous phosphate

Albright first suggested intravenous phosphate for the treatment of hypercalcemia over 50 years ago,[90] but it did not become a popular form of therapy until Goldsmith and Ingbar[91] demonstrated its efficacy in a number of patients. This was further supported by a study by Fulmer et al[92] which showed that intravenous phosphate was clearly superior to cortisone or sodium sulfate in lowering the serum calcium in hypercalcemic patients. They suggested that a dose could be selected which was both safe and effective. However, since then there have been numerous reports indicating that intravenous phosphate is an extremely dangerous form of therapy. A number of patients have died following intravenous phosphate therapy, due to soft-tissue calcium deposition occurring in the lungs, kidney, and in other organs such as the heart and brain.[93–96] Because it is such a dangerous form of therapy, its routine use cannot be recommended. However, in a life-threatening situation where no other form of therapy is available and the patient is likely to die otherwise, a single infusion of up to 75 mmol may be used.

The mechanism by which intravenous phosphate causes a fall in the serum calcium is probably by causing soft-tissue calcium deposition. However, phosphate may have other effects on calcium homeostasis. It could cause an impairment of bone resorption and inhibit calcium absorption from the gut. It is much more likely that these effects are more relevant when phosphate is given orally rather than intravenously.

Dialysis

Dialysis is occasionally used in patients with hypercalcemia due to malignancy when the patients are in renal failure and other therapies are contraindicated or ineffective.[97] The choice of drug therapy for hypercalcemia in this situation is extremely limited. Renal failure is most likely to occur in patients with hypercalcemia of malignancy if they have myeloma, since this malignancy is frequently associated with impaired glomerular filtration. Of course, dialysis is not an effective longterm form of treatment for hypercalcemia, because hypercalcemia is usually due to increased bone resorption. Its use should therefore be limited to the short-term situation, particularly where the patient has a reversible form of renal failure. Either peritoneal dialysis or hemodialysis may be effective in lowering the serum calcium in this situation.

CHAPTER 8

Primary hyperparathyroidism

Primary hyperparathyroidism is due to excess and inappropriate production of PTH by the parathyroid glands, and the hallmark of the disease is hypercalcemia. It is now an extremely common endocrine disease, particularly in the aging female population, and has a markedly variable clinical presentation which frequently confuses the clinician. The manifestations of the disease are due not just to the effects of PTH on the skeleton and kidneys, but also to 1,25-dihydroxyvitamin D, which is produced by the kidney in response to excess circulating concentrations of PTH. Most of the clinical features occur as a consequence of chronic hypercalcemia.

History

The description of the features of the clinical syndrome caused by hyperfunction of one or more parathyroid glands has evolved considerably over the past 50 years. The parathyroid glands were first described by Sandstrom.[1] The bone disease of primary hyperparathyroidism is called osteitis fibrosa cystica. This was the descriptive term given by Von Recklinghausen and, as Albright and Reifenstein[2] point out in their monograph (pages 54 and 55), it is entirely misleading. However, since it has been in use for almost 100 years, it is not realistic to change it now. This particular form of metabolic bone disease was described by Von Recklinghausen[3] in three skeletons, at least one of which in retrospect did not have osteitis fibrosa cystica, but rather polyostotic fibrous dysplasia, with the characteristic shepherd's crook deformity of the femur. Von Recklinghausen described the bone disease, but he did not recognize that the parathyroid glands were the cause. Askanazy[4] noted the association of osteitis fibrosa cystica with a tumor in the neck, and Erdheim[5] reported that the parathyroid glands may be enlarged in some patients with metabolic bone diseases. However, it was a further 20 years before the parathyroid glands were shown to be the cause of this particular type of metabolic bone disease.

The first description of primary hyperparathyroidism as a clinical entity was made by the surgeon Mandl in Europe in 1926.[6] The patient, whose name was Albert, was a 39-year-old Viennese streetcar conductor who presented with a deformity of the left side of the pelvis, the left femur and foot, and the right knee. The softening of his bones was interpreted to be due to an inadequate parathyroid gland response, so he was treated with parathyroid extracts and parathyroid tissue was transplanted into him. Because he failed to improve, Mandl searched for a parathyroid gland tumor. An apparent parathyroid adenoma was found, and the patient improved dramatically after its removal. However, the disease recurred and, looking back over this case many years later, Cope et al[7,8]

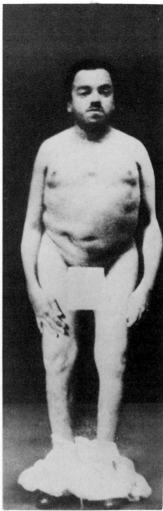

Figure 8.1

Captain Charles Martell, the initial patient with documented primary hyperparathyroidism studied in the USA by Bauer and co-workers. A brief history of this patient is presented in the text.

suggested that sections from the original tumor indicate that the diagnosis was more likely chief cell hyperplasia, and ascribed the recurrence of hypercalcemia to this cause. It should be remembered that chief cell hyperplasia was not recognized as an entity until 1958. (Interestingly,

Albert's sister also had a form of bone disease and may also have had hyperparathyroidism.)

The most widely known and dramatic case in the USA was that of Charles Martell, a seaman, who at the end of the First World War began to suffer a progressive crippling disease associated with recurrent renal stones (Figure 8.1). His history has been documented in a number of publications,[9–12] and described in more detail by Albright and Reifenstein (pages 56–7). In 1918 he was a master mariner in the merchant marines in good physical condition, an active athlete, 6′1″ tall. At the time of presentation to Bellevue Hospital in January 1926 he was a cripple who had suffered multiple fractures and had lost 7″ in height. His bone lesions showed a presence of multiple cysts, and generalized osteopenia, and he was hypercalcemic. The disease was also characterized by recurrent renal calculi. He was followed for several years, and careful metabolic balance studies were performed. Bauer et al[10] noted that the disease in this patient seemed to produce identical metabolic effects to those seen in laboratory animals treated for prolonged periods with parathyroid extract, and they postulated that excess secretion of PTH from a hyperfunctioning parathyroid gland was responsible. The patient underwent a series of unsuccessful neck explorations until finally a parathyroid gland adenoma was found in the mediastinum. However, by this stage the unfortunate man had progressive renal failure, and he died in 1932.

This experience so impressed Albright that he decided to study patients with this disease in detail. He had told his students that if they were interested in studying patients with primary hyperparathyroidism, they should advertise for patients with stones and then investigate those with hypercalcemia. He himself established a bone and stone disease center at the Massachusetts General Hospital to gather patients with this particular clinical syndrome. Not surprisingly, and due in part to Albright's dominance of clinical endocrinology at that time, primary hyperparathyroidism became well recognized as a disease characterized by bones, stones, and abdominal groans over the next 30 years.

In the past 25 years several changes in current medical practice have altered our concepts of primary hyperparathyroidism dramatically. First, it is now realized that when this condition is sought, it is quite often found. Boonstra and Jackson[13] found it to be present in one out of every 1000 people in the adult population. Now that routine measurement of serum calcium is performed on serum samples, the annual incidence rate may be as high as 250 new patients

per million population per year.[14,15] This has changed our notions about many features of the disease dramatically, including its mode of presentation, its age- and sex-distribution, and its natural history.

There have been other recent landmarks which have also changed our concepts of this disease. The introduction of the PTH radioimmunoassay and the subsequent improvements in the assay have improved the differential diagnosis of hypercalcemia considerably, and have also aided in localization of abnormal tissue in those patients requiring recurrent surgery. The recognition of related diseases such as the multiple endocrine neoplasia syndromes and familial hypocalciuric hypercalcemia have also broadened our concepts and increased our ability to recognize this disease when it presents with associated conditions.

The epidemiology of primary hyperparathyroidism

The understanding of the clinical presentation of primary hyperparathyroidism has changed remarkably in the past decade. The epidemiology, mode of presentation, and frequency of this disease are now realized to be entirely different from what was previously thought until about 10–15 years ago. This change in knowledge is due to the widespread use of the autoanalyzer and routine determination of the serum calcium on blood specimens sent to clinical laboratories for routine electrolyte determinations. Since the routine measurement of serum calcium has become common practice, primary hyperparathyroidism is clearly a common condition in the asymptomatic patient, almost as common as Graves' disease. The disease was formerly thought of as a rare disease of bones, stones, and abdominal groans and psychic moans, which occurred predominantly in patients between the ages of 20 and 50 years, and mainly in females. It is now realized that most patients are diagnosed by coincidence at a time when they have no symptoms which can be linked to primary hyperparathyroidism. The typical patient is an elderly female in whom the disease was not suspected before the serum calcium was determined. These observations have come from two epidemiological studies, one performed in the USA and one in

Table 8.1 Age- and sex-distribution of primary hyperparathyroidism at time of presentation.

Age (years)	Hellstrom and Ivemark[32] (Sweden) 138 patients		Muller[135] (Netherlands) 208 patients		Mallette et al[31] (USA) 57 patients		Mundy et al[14] 111 patients	
	Males	Females	Males	Females	Males	Females	Males	Females
0–19	1	1	2	1	—	2	—	—
20–29	1	4	8	5	—	—	—	2
30–39	8	9	7	8	9	4	—	—
40–49	8	21	10	13	14	18	2	5
50–59	7	23	8	19	9	23	5	11
60–69	3	12	4	14	9	9	6	6
70–79	—	2	1	2	2	4	8	38
80+	—	—	—	1	—	—	—	17
	28	72	39	64	42	58	21	79

The numbers given are percentages of the total numbers of patients in that particular study. The percentages are rounded to the nearest whole number.

Table 8.2 Mode of presentation of primary hyperparathyroidism.

Presentation	Helstrom and Ivemark[32] (Sweden) 138 patients	McGeown[136] (UK) 177 patients	Watson[137] (UK) 100 patients	Wang[111] (US) 431 patients	Mundy et al[14] 111 patients
Accidental (no symptoms)	—	3	14	1	57
Renal (stones or nephrocalcinosis)	79	72	47	55	7
Bone disease	20	17	13	21	—
Psychiatric disorder	—	—	2	1	5
Acute hypercalcemic syndrome	—	2	—	2	14
Gastrointestinal symptoms	—	4	12	12	4
Symptoms of hypercalcemia (lethargy, polyuria, etc.)	—	2	8	4	8
Hypertension	1	—	4	1	5

The numbers given are percentages of the total numbers of patients in that particular study. The percentages are rounded to the nearest whole number.

England.[14,15] The changing patterns of mode of presentation and clinical features of primary hyperparathyroidism at the time of presentation are outlined in more detail under the section on clinical features, and the data are provided in Tables 8.1 and 8.2. At present patients rarely have overt bone disease at the time of diagnosis, and only infrequently do they have a history of recurrent renal calculi. The annual incidence rate is approximately 250 new patients per million population per year.[14,15] The prevalence, which reflects the number of patients in a population with the disease at any one time, is probably about 1 : 1000, based on a study in which 26 000 consecutive serum calcium measurements were made.[13] Until 10 years ago most information on the mode of presentation and epidemiology of primary hyperparathyroidism came from referral centers. As these were mostly surgical referrals, there was almost certainly undue emphasis on younger patients with prominent clinical features such as bone or stone disease. Thus, until the mid-1970s, information on the mode of presentation, epidemiology, and frequency of this disease was totally biased by patient selection.

and 5.0 g in weight, although occasional adenomas may weigh up to 10–20 g. A normal parathyroid gland weighs 25–50 mg. In general the larger the adenoma, the more prominent the clinical features. The normal glandular tissue may be compressed by the adenoma and appear as a rim of normal tissue forming a pseudo-capsule around the adenoma. The adenomas consist of a homogeneous collection of cells (in most cases chief cells), and are devoid of fat cells or other cells of the parathyroid gland, such as the oxyphil cells. This is one of the distinguishing features which may aid the pathologist in distinguishing an adenoma from normal tissue. Most parathyroid adenomas involve the inferior glands. In about 5–10 per cent of patients the gland is in an abnormal location, such as in the thymus, within the thyroid, or in the mediastinum adjacent to the carotid sheath, the pericardium or in the retro-esophageal space. Occasionally, a parathyroid cyst occurs within a hyperfunctioning adenoma. It is probable that most parathyroid cysts occur within parathyroid adenomas, and thus may represent areas of previous hemorrhage or necrosis. Nonfunctioning adenomas of the parathyroid gland probably occur, but their frequency is entirely unknown.

Pathology

Primary hyperparathyroidism may be due to a solitary parathyroid gland adenoma, hyperplasia of four parathyroid glands, or carcinoma of a parathyroid gland. A solitary parathyroid adenoma is by far the most common cause, being responsible for the clinical syndrome in approximately 85 per cent of patients. Hyperplasia is responsible in the majority of the remaining patients. Carcinoma of the parathyroid gland is fairly rare, there being less than 200 cases so far reported in the world literature.

Adenoma

Most adenomas are composed of the chief cells of the parathyroid, and a few of transitional cells between chief cells and oxyphil cells. Adenomas are rarely composed only of oxyphil cells. Parathyroid gland adenomas are usually between 0.5

Hyperplasia

Most investigators find that about 15 per cent of patients with primary hyperparathyroidism have hyperplasia of all four glands. The hyperplasia usually involves the chief cells, although occasionally it involves the water-clear cells (in about 1 per cent of all cases). Very rarely, hyperplastic glands from the same patient may show a different pattern of cellular involvement. For example, one gland may show chief cell hyperplasia, but another gland may show involvement of the oxyphil cells. Hyperplasia is almost always present in patients who have primary hyperparathyroidism associated with multiple endocrine neoplasia. It is not possible to diagnose hyperplasia on clinical grounds before surgery, since the clinical features are indistinguishable from those of patients with primary hyperparathyroidism due to a parathyroid adenoma.

The frequency of parathyroid gland hyperplasia as a cause of primary hyperparathyroidism has been the subject of some controversy, some groups believing that hyperplasia is much more common than is generally cited.[16] However, if hyperplasia were more common, then it would be expected that surgically treated patients would get recurrences of hypercalcemia more frequently. In reality, recurrence is fairly uncommon, occurring in less than 10 per cent of all surgically treated cases. Part of the problem in distinguishing adenoma from hyperplasia is in the difficulty in the histologic distinction between normal parathyroid tissue and abnormal parathyroid tissue, and particularly between adenomatous and hyperplastic tissue. Moreover, hyperplastic glands may vary considerably in size, and this can also lead to confusion in distinguishing hyperplasia from single or multiple adenomas. Most investigators believe that the presence of an adenoma in more than one gland is very rare (less than 2 per cent of all cases).

Carcinoma

Carcinoma of the parathyroid gland is, fortunately, rare. Estimates have been made in the literature that approximately 2 per cent of all patients with primary hyperparathyroidism have a carcinoma of one of the parathyroid glands responsible, but this is clearly an overestimate. The disease is clearly relatively much less frequent as a cause of primary hyperparathyroidism, particularly now that mild or asymptomatic benign hyperparathyroidism is being recognized so frequently (see Table 8.2). In a recent review, Shane and Bilezikian[17] noted that only a little more than 100 patients had been described with this condition since 1938. The condition can occur at any age and, although there is a slight female preponderance, it is relatively more common in males than benign primary hyperparathyroidism. Hypercalcemia is usually much more severe (>14 mg/dl in 70 per cent) and these patients frequently have bone and stone disease (nephrocalcinosis as well as nephrolithiasis). The clinical course is usually prolonged. About 50 per cent of patients survive for 5 years and one-third for 10 years. Occasionally these tumors metastasize,

although local invasion and recurrence is more common. It may be difficult to distinguish between an adenoma and a carcinoma at the histologic level, and the diagnosis may be made on invasive or metastatic behavior of the tumor rather than on the morphologic characteristics of the cells. Histologic features which are characteristic of a carcinoma include adherence to surrounding structures, capsular invasion by the parathyroid cells, and the presence of many mitotic figures which are not seen frequently in nonmalignant adenomas. Carcinomas also frequently contain dense connective tissue strands or septra which run through the tumor to disrupt the gland architecture. The tumor may be locally invasive, involving blood vessels. Spread is usually by lymphatics, but it may also involve liver, lungs, or bone by hematogenous spread. Patients die usually because of the effects of PTH excess, so surgery to remove the primary and the secondaries is warranted. Medical treatment with bisphosphonates, mithramycin or calcitonin and steroids may relieve symptoms by lowering the serum calcium. Radiation therapy has not been successful.

Multiple endocrine neoplasia Types I and II

Primary hyperparathyroidism may occur in association with multiple endocrine neoplasia (MEN) Types I and II. Type I is associated with pituitary adenomas and pancreatic adenomas. The pituitary adenomas may be growth-hormone secreting and associated with acromegaly or prolactinomas, causing the galactorrhea–amenorrhea syndrome. In MEN Type II, primary hyperparathyroidism is associated with pheochromocytoma, medullary carcinoma of the thyroid, and, in some cases, mucosal neuromas. It is a fairly uncommon association of MEN II. It occurs more frequently in MEN I, but usually becomes apparent after the age of 30 years, unlike the situation in familial hypocalciuric hypercalcemia, where hypercalcemia may be manifest from birth. These syndromes are discussed in more detail below.

Etiology

The cause of primary hyperparathyroidism is unknown. As noted above, most patients with primary hyperparathyroidism have either an adenoma or carcinoma of one single parathyroid gland, or hyperplasia of all four glands. However, it is often not easy to distinguish between an adenoma and hyperplasia, and as indicated above, some investigators have considered that adenomas and four-gland hyperplasia in fact represent variants of the same underlying disorder. This view was strengthened by an earlier study suggesting that parathyroid adenomas in fact represented a polyclonal disorder consistent with a multicellular origin.[18]

However, it seems that single parathyroid gland adenomas are more likely to represent a monoclonal disorder arising from an abnormality in a single cell, whereas four-gland hyperplasia is more likely to represent a multicellular process and to be due to an external stimulus to parathyroid cell growth and secretion. Recent studies using techniques of molecular biology have suggested that this concept is likely to be correct. Arnold and Kronenberg[19,20] applied two independent molecular genetic tests which suggested that a monoclonal abnormality is present in most parathyroid adenomas. First, they found the parathyroid gene had undergone a clonal rearrangement in several examples and second, using X-chromosome inactivation analysis of clonality, they found that the DNA hybridization patterns of six adenomas from women with the disease indicated monoclonality compared with results from five hyperplastic glands.

Recently a growth factor which stimulates the replication of parathyroid gland cells freshly dispersed in culture has been identified in the serum of patients with MEN Type I.[21] Although this factor has been linked to parathyroid gland hyperplasia in the MEN I syndrome, it is apparently not present in most other patients with parathyroid gland hyperplasia. This factor may be related to the expanding fibroblast growth factor (FGF) family. The target cell appears to be the endothelial cell of the parathyroid gland vasculature. The parathyroid gland does not appear to be the source of this factor, since total parathyroid gland ablation does not remove it from the serum. This potential relationship of this parathyroid stimulant to the FGF family is interesting

from two points of view. First, it stimulates parathyroid endothelial cells and the FGFs are powerful endothelial cell stimulants. Second, it suggests a possible relationship to the INT-2 oncogene, since the predicted product of the INT-2 oncogene is homologous to basic FGF and recent findings suggest that the mutant gene in MEN I is localized to the pericentric region of chromosome 11, a region that is near INT-2.[22]

Several other external stimuli have been linked to parathyroid gland hyperplasia. One possible example is prolonged hypocalcemia which causes secondary hyperparathyroidism and associated parathyroid gland hyperplasia. It has been widely accepted for many years that hyperplastic glands associated with a history of prolonged hypocalcemia could become adenomatous (so-called 'tertiary' hyperparathyroidism), although this has never been proved definitively. This syndrome has been described in patients with secondary hyperparathyroidism due to chronic renal failure, and it is probably a manifestation of persistent parathyroid cell secretory mass rather than a change in the pathology of the tissue from hyperplasia to an adenoma. After renal transplantation, patients may develop persistent hypercalcemia, which lasts for some months until the parathyroid glands slowly involute. Hypercalcemia in these circumstances does not necessarily mean a change in the underlying parathyroid gland histopathology.

Other factors have been implicated in the pathophysiology of primary hyperparathyroidism. The history of previous irradiation is present in some patients. Several centers have suggested that up to one-third of patients have a previous history of neck irradiation,[23] and that a significant number (up to 10 per cent) of these patients are at risk from the development of primary hyperparathyroidism.[23] However, irradiation of the neck was frequently practised in children after the Second World War, usually because of tonsillar problems. Thus, since primary hyperparathyroidism is such a common disease, it is possible that this association may be coincidental. Many secretagogues for PTH other than calcium have been described both in vitro and in vivo, but there is no current evidence that any of these factors are responsible for causing parathyroid adenomas to develop.

In both primary hyperparathyroidism due to parathyroid gland adenomas and four-gland hyperplasia, serum concentrations of PTH and calcium are increased. There must therefore be some defect in the abnormal parathyroid cell whereby it does not recognize the extracellular calcium concentration as a normal cell does. The normal parathyroid gland is almost unique among secretory cells in that it is negatively regulated by extracellular calcium. In the normal parathyroid cell, there is a very sensitive negative-feedback relationship between serum calcium and the release of PTH from the parathyroid gland. Investigators have coined the term 'set-point' to describe the relationship between PTH release and extracellular fluid calcium. The 'set-point' is defined as the extracellular fluid calcium concentration at which half-maximal inhibition of PTH release occurs. In dispersed cells from neoplastic parathyroid gland tissue studied in vitro to determine the relationship between extracellular fluid calcium and PTH release, wide variations have been found in the 'set-point' in parathyroid tissue both in vitro and in vivo.[24,25] The site of the defect in the parathyroid cell between sensing the extracellular fluid calcium signal and responding to it by releasing PTH is unknown.

It is clear that the increase in extracellular fluid calcium in primary hyperparathyroidism fails to suppress both parathyroid cell replication and parathyroid hormone secretion. Since extracellular fluid calcium appears to control parathyroid function through a receptor which is coupled to intracellular effector systems via one or more G proteins, parathyroid cell proliferation and PTH secretion could be due to a number of possible changes. These possibilities include alterations in the sensitivity of the putative receptor to calcium, the presence of a putative antibody acting as an agonist to the calcium receptor, alterations in the sensitivity of the receptor to extracellular fluid calcium, alterations in the coupling of the receptor to its effector systems, or distal defects in the response to these effectors.[26] Although some of these abnormalities have been observed experimentally,[27] and could be responsible in some cases of familial hyperparathyroidism,[28] whether they occur as a cause or effect of the disease is unknown. However, in some situations, and particularly in patients with parathyroid gland hyperplasia, the set-point for PTH release in response to serum calcium may be normal, but the mass of cells is so great that circulating PTH is increased. It appears that even normal para-thyroid cells cannot be completely suppressed by increasing the calcium bathing them (Chapter 1). Thus, when the total secretory mass of parathyroid tissue is greatly increased, this basal nonsuppressible PTH may be sufficient to cause the clinical features of primary hyperparathyroidism.

Clinical features

The clinical features of primary hyperparathyroidism have been described in detail a number of times over the past 50 years, and the reader is referred to several extensive reviews[2,29–31] (Tables 8.1 and 8.2). To a large extent, each author's description has been determined by his particular experience of the disease, and this has changed considerably as the means of making the diagnosis has improved. Older accounts emphasize the striking and devastating effects of the disease on the skeleton, the kidneys, the central nervous system, and the gastrointestinal tract. More recently the disease is far more commonly recognized in patients who are asymptomatic and who have none of these features.

The clinical features of primary hyperparathyroidism are protean. It has long been debated which of these symptoms are due to PTH directly and which are due to chronic elevations in the serum calcium. This has never been adequately resolved. It appears likely that the majority are related directly to the serum calcium, although some changes (for example, renal tubular acidosis and renal phosphate wasting) are clearly due to the direct effects of PTH on target organs. The common clinical features at the time of presentation are shown in Table 8.2.

Hypercalcemia

Hypercalcemia is almost certainly present in all patients with primary hyperparathyroidism at some time during the course of the disease. The hypercalcemia may remain relatively constant for many years, with only slight oscillations in the serum calcium about an increased mean (Chapter 1). This is in contrast with patients with the

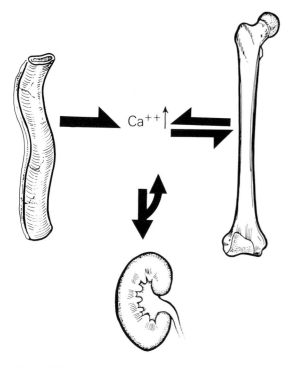

Figure 8.2

Abnormalities in calcium homeostasis in patients with primary hyperparathyroidism. These patients show increased bone turnover, increased renal tubular calcium reabsorption, and increased absorption of calcium from the gut. The increase in the absorption of calcium from the gut is mediated by 1,25-dihydroxyvitamin D, which is produced in the renal tubules in response to PTH.

acting together on the formation and activation of the osteoclast. In the kidneys 1,25-dihydroxyvitamin D and PTH work together to increase renal tubular calcium reabsorption, which causes a decrease in the fractional excretion of calcium. The effects of these combined actions of PTH and 1,25-dihydroxyvitamin D working in concert result in hypercalcemia. However, despite the increased circulating levels of PTH and 1,25-dihydroxyvitamin D in the majority of patients, the increase in serum calcium remains steady and fixed.

The primary target organ for PTH and the pathophysiology of hypercalcemia in this disease has been argued for more than 50 years. The problem in part is that it is difficult to study the effects of PTH on any isolated organ and the subsequent effects on extracellular fluid calcium in isolation. Bone, kidney, and gut are all targets (either directly or indirectly) for PTH action. However, in many patients with stable increases in the serum calcium, renal calcium excretion is in the normal range (and presumably equal to net bone resorption and net gut absorption of calcium), and so renal tubular calcium reabsorption must be the major mechanism for the increase in serum calcium in these patients.

Renal

Primary hyperparathyroidism causes a number of changes in both structure and function in the kidney. The major anatomical consequences are nephrolithiasis and nephrocalcinosis. Nephrolithiasis and nephrocalcinosis usually do not occur together.[2,32] Nephrocalcinosis is rare, and usually occurs in those patients with severe longstanding disease and impaired renal function. It is due to precipitation of calcium and phosphate in the renal tubular epithelium and in the renal interstitial tissue. It is seen more frequently in patients with bone disease, and in patients with parathyroid carcinoma. In individual cases it is difficult to determine whether renal failure is a result or a cause of nephrocalcinosis. Certainly the increase in serum phosphorus which occurs as a consequence of impaired glomerular filtration is required for precipitation of calcium–phosphate salts in the kidneys.

hypercalcemia of malignancy where the rise in serum calcium is steadily progressive if the patient is not successfully treated. The hypercalcemia of primary hyperparathyroidism is due to the combined effects of PTH and 1,25-dihydroxyvitamin D on calcium transport across the gut, kidney, and bone (Figure 8.2). The increase in calcium absorption from the gut is due to 1,25-dihydroxyvitamin D, which is produced by the proximal renal tubule cells in response to increased circulating PTH concentrations. The increase in bone resorption is due to the combined effects of PTH and 1,25-dihydroxyvitamin D

However, this precipitation can further impair renal function. In most cases it appears likely that nephrocalcinosis may occur as a consequence of decreased renal function, but when it occurs it leads to a further decline in renal function.

Nephrolithiasis or the formation of calcium-containing renal stones is also a frequent finding in patients with primary hyperparathyroidism. This was a much more frequent finding in patients with primary hyperparathyroidism before the advent of the autoanalyzer became the mechanism by which this disorder was commonly recognized. Now, about 5–10 per cent of patients with recurrent calcium-containing renal stones can be expected to have primary hyperparathyroidism. These patients usually do not have bone disease. The stones may be mixed stones or may consist of calcium oxalate and calcium phosphate.[33] Stones may cause urinary-tract infections and chronic interstitial nephritis, or may lead to hydronephrosis. The benefits of surgery and correction of primary hyperparathyroidism on the course of renal-stone disease have not yet been well documented.

In addition to these anatomic or morphologic changes, primary hyperparathyroidism causes a number of functional changes in the kidneys. These functional changes may occur in the absence of nephrocalcinosis or nephrolithiasis. The most common change is renal tubular acidosis. Parathyroid hormone acts on the proximal tubule to inhibit bicarbonate reabsorption, and to cause an increase in bicarbonate wasting and a consequent hyperchloremic acidosis. In addition to renal tubular acidosis there may also be phosphaturia, aminoaciduria, and glycosuria. The decrease in phosphate reabsorption is the predominant effect of PTH on the proximal tubule. In addition to these effects there is a decrease in renal concentrating ability, probably due to an impairment in the capacity of ADH to cause water reabsorption in the collecting ducts of the kidney. The pathogenesis of the changes in renal tubular function is complex. The possibilities are that they could be due to a direct effect of PTH or due to hypercalcemia, or to a combination of both (which seems most likely). It is unlikely they are solely due to hypercalcemia, since hypercalcemia from other causes is not usually associated with these particular changes in renal tubular function (except for the change in concentrating capacity). It is also conceivable that they could be related to the phosphate depletion which occurs as a result of the action of PTH on the proximal tubule to impair phosphate reabsorption.

Skeletal

Although most patients with primary hyperparathyroidism have histologic abnormalities in bone, in only a relatively small number are these clinically apparent. Skeletal abnormalities are varied. Rarely they may be severe, with bone pain, loss of height, deformity, bone cysts, and susceptibility to fracture. This is the classic bone disease of primary hyperparathyroidism, which was called osteitis fibrosa cystica when it was first described 100 years ago. Now it is rarely seen, and it is a very unusual finding in patients with primary hyperparathyroidism at the time of presentation. Bone disease that is severe enough to cause symptoms is present in less than 15 per cent of patients at the time of diagnosis, and radiologic osteitis fibrosa cystica is now present in less than 5 per cent of all patients at the time of diagnosis. The reason that osteitis fibrosa cystica is unusual at the time of presentation is that primary hyperparathyroidism is diagnosed by routine determinations of the serum calcium at an earlier stage of the disease process. However, many patients with primary hyperparathyroidism may have diffuse or vertebral osteopenia at presentation, and this is a much more common manifestation of skeletal abnormalities in this disease than osteitis fibrosa, according to a large series of 319 patients from the Mayo Clinic[34] and in a study in San Francisco of eighty-seven patients with primary hyperparathyroidism.[35] In the Mayo Clinic study none of the patients had osteitis fibrosa but fourteen had compression fractures of vertebral bodies and diffuse osteopenia of the spine. However, since most patients with primary hyperparathyroidism are postmenopausal females, this is hardly surprising. Photon absorptiometry of the single-beam type is a very sensitive technique for detecting cortical loss in bone mineral, and it may be particularly useful for following the progress of the bone disease in primary hyperparathyroidism.[36]

However, evidence of osteitis fibrosa may be more common at the morphologic level. Jowsey and Riggs[37] noted that each of twenty patients with primary hyperparathyroidism had bone turnover increased by about fivefold, with increased bone resorption and bone formation surfaces, using the technique of microradiography. Only about one-third of these patients had radiological abnormalities. Just how frequently patients with primary hyperparathyroidism have bones which appear completely normal on bone histomorphometry is difficult to say. It is likely that most have some detectable morphologic abnormalities. The presence of PTH excess in vivo causes a number of striking changes in bone morphology (Figure 8.3). There is a marked increase in osteoclastic bone resorption. This is characterized morphologically by increased numbers of osteoclasts and increased percentages of trabecular bone surfaces undergoing resorption or showing evidence of recent resorption. The osteoclast lacunae may be a little more shallow than those caused by osteoclasts stimulated to resorb by other mechanisms. The mode of action of PTH on the

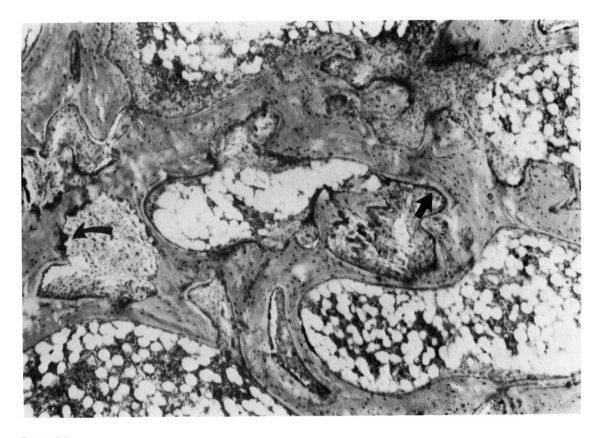

Figure 8.3

Histologic sections of bone in a patient with primary hyperparathyroidism and osteitis fibrosa cystica. There is marked evidence of osteoclastic bone resorption (curved arrow), with marrow fibrosis and increased osteoblast activity (straight arrow).

osteoclast is described in Chapter 2. However, the effects of PTH excess are not limited to the osteoclast. There also are marked changes in osteoblast function, including an increase in osteoblast number. There is also increased osteoid volume, the osteoid which is present is of woven nature, and there are increased numbers of trabecular bone surfaces which are covered by osteoid. In addition there is increased peritrabecular marrow fibrosis. Dynamic histomorphometric parameters show increased mineral accretion rates and increased distances between tetracycline labels, indicating increased rates of bone turnover. However, the ratio between double labels and single labels may be less than is expected, because the woven osteoid causes the appearance of broad single bands occurring because of abnormal tetracycline uptake. Parathyroid hormone excess does not always cause evidence of bone loss. When given in low doses and intermittently, the major effect may be an anabolic response. The rates of bone formation may be also greatly influenced by the state of mobilization. In immobilized patients resorption may be by far the predominant abnormality, and parameters of formation may be suppressed (Chapter 9).

Morphologic bone changes are far more sensitive than radiographs for detecting hyperparathyroid bone disease. The major radiographical abnormalities in hyperparathyroid bone disease occur as a consequence of loss of cortical bone. However, the morphologic changes will reflect cellular abnormalities not only in cortical bone, but also in trabecular bone, and will be present in many more patients than would be suspected by taking X-ray photographs alone. It has been suggested for many years by authors such as Belanger, Meunier, and Bordier that patients with primary hyperparathyroidism may have osteocytic osteolysis.[38–40] Osteocytic osteolysis correlates with changes in PTH and serum calcium.[40] However, the significance of osteocytic osteolysis has been questioned by Jones and Boyde[41] who have suggested that this observation represents an artifact of preparation.

Changes in bone radiology

Changes in skeletal X-ray photographs are seen in less than 20 per cent of all patients with primary

hyperparathyroidism. The changes that may be seen include osteopenia, subperiosteal resorption of bone, osteosclerosis, loss of the lamina dura, and cystic lesions in the skeleton.[42,43]

Osteopenia

Osteopenia is currently the most common skeletal finding in patients with primary hyperparathyroidism at the time of presentation. It may differ from osteoporosis because the loss of bone mineral is not always homogeneous. For example, there may be loss of bone mineral in specific sites such as the skull, phalanges, and clavicles.

Subperiosteal resorption

Subperiosteal bone resorption is seen best in the phalanges (Figure 8.4), but is also seen at the acromioclavicular joint, at the symphysis pubis, and in the sacroiliac joints. It is essentially diagnostic of hyperparathyroidism, being seen in both primary hyperparathyroidism and secondary hyperparathyroidism, particularly that associated with chronic renal disease. An occasional patient with hypercalcemia of malignancy has been found to have subperiosteal resorption, but this is rare. Subperiosteal resorption is seen most frequently and easily along the radial aspect of the middle phalanges in the posterior–anterior view, and may be associated with erosions of the tufts of the terminal phalanges. The earliest lesions are seen near the base of the phalanx and are localized. Similar lesions may occur in the skull to give a mottled or 'salt-and-pepper' appearance (Figure 8.5). Another characteristic location for radiologic changes is in the distal third of the clavicles, which may show a distinctive tapering.

Osteosclerosis

Osteosclerosis is uncommon, but is clearly found in some patients with primary hyperparathyroidism. It may be either diffuse or localized. It is seen much more commonly in secondary hyperparathyroidism associated with chronic renal failure. In this situation it may occur in between 5 and 20 per cent of patients. It is also much more common in children. Osteosclerosis may occur as a result

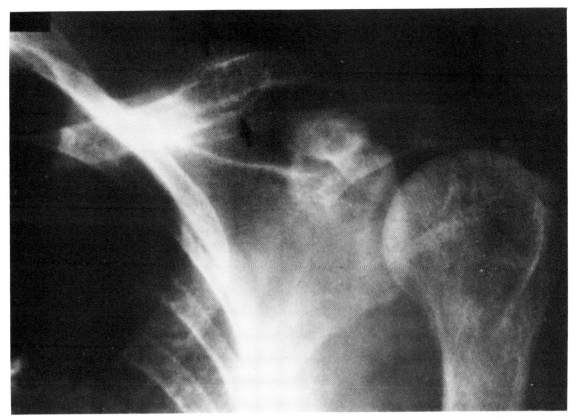

Figure 8.4

Radiograph of the clavicle of a patient with primary hyperparathyroidism showing the characteristic features of subperiosteal resorption and cystic destruction (arrow). These changes are reversed following successful parathyroid surgery.

of the anabolic effects of PTH on the skeleton (Chapter 3).

Loss of the lamina dura

Loss of the lamina dura of the teeth is found in most, but not all, patients who have osteitis fibrosa cystica. The lamina dura is not regained after surgical treatment of the primary hyperparathyroidism. The lamina dura tends to be thinner in patients with primary hyperparathyroidism than the thinning in the lamina dura which occurs following malocclusion of the teeth. This is not a

feature that is looked for often nowadays, and is not present in the majority of asymptomatic patients at the time of presentation.

Cystic lesions

Two types of cystic lesions are seen in patients with primary hyperparathyroidism: true bone cysts and brown tumors. These have similar features on X-ray examination and may be difficult to distinguish. Bone cysts are filled with fibrous tissue and occur in subperiosteal locations. They do not resolve after surgical cure of

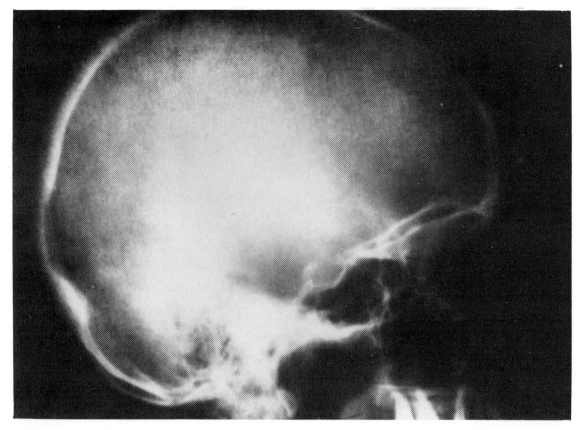

Figure 8.5

Radiograph of the skull of a patient with primary hyperparathyroidism, showing characteristic features of 'salt-and-pepper' skull, with mottling due to increased osteoclastic bone resorption.

primary hyperparathyroidism. In contrast, the brown tumors are filled with osteoclasts, osteoblasts, and woven bone which is poorly mineralized. These are referred to as osteoclastomas. These lesions resolve when primary hyperparathyroidism is treated successfully by surgery.

Neuromuscular and neuropsychiatric

Neuromyopathic disorder

A specific neuromyopathic disorder has been described in patients with primary hyperpara-

thyroidism.[44] It is neuropathic rather than myopathic. It consists of atrophy of the Type II muscle fibers, which are involved much more frequently than Type I muscle fibers. The microscopic findings are those of denervation without accompanying inflammatory changes. The mechanism responsible for these changes is entirely unknown. The motor nerve conduction velocities remain normal but the electromyogram (EMG) shows muscle abnormalities, including polyphasic potentials in involved muscles and high-amplitude/long-duration and high-amplitude/short-duration polyphasic potentials.

The disorder is characterized by weakness and

easy fatiguability associated with atrophy of proximal muscle groups, with the lower extremity being involved more severely than the upper extremity. The psoas, gluteus medius, and hamstrings of the lower extremity, and the deltoids and biceps of the upper extremity are most often affected. This finding may be the presenting feature. There are no or minimal sensory abnormalities. Serum SGOT, aldolase, and CPK are all normal. Improvement in all of these abnormalities occurs with successful treatment of the primary hyperparathyroidism. The frequency of this finding is unknown, but it was estimated to be more than 80 per cent in Patten's series. The frequency is unknown because muscle biopsies are not performed routinely in primary hyperparathyroidism, and patients may not complain of muscle weakness.

Psychiatric changes

A whole range of psychiatric changes may be seen in patients with primary hyperparathyroidism. These include impaired mentation, emotional lability, depression, psychoneurosis, personality changes, psychomotor retardation, memory impairment, and occasional overt psychosis. It has frequently been noted that many patients who feel uneasy and have difficulty in coping with common domestic situations improve dramatically with a much greater sense of well-being following successful surgery. The psychiatric changes are probably due to the increase in serum calcium rather than the elevation in PTH, since they occur in many patients with hypercalcemia of other causes.

Gastrointestinal changes

Several gastrointestinal changes may be seen in patients with primary hyperparathyroidism. These include peptic ulcer disease and pancreatitis.

Relationship with peptic ulcer disease

Whether there is a true association between primary hyperparathyroidism and peptic ulcer disease remains controversial. It is clear that patients with primary hyperparathyroidism may

have the multiple endocrine neoplasia Type I syndrome and have an associated Zollinger–Ellinson syndrome, but this is a rare association. In patients with primary hyperparathyroidism but without MEN I, it is not clear that there is a direct relationship with peptic ulcer disease. Some studies have reported that between 10 and 30 per cent of patients with primary hyperparathyroidism have peptic ulcers, but since up to 10 per cent of the general adult population have peptic ulcer disease, it is not clear that there is a direct relationship. The problem is that more than 40 per cent of patients with primary hyperparathyroidism have epigastric distress (probably related to hypercalcemia itself) and, consequently, the upper gastrointestinal tract is more frequently evaluated in patients with primary hyperparathyroidism than it is in the general population. Thus, it is possible that frequent evaluation of the gastrointestinal tract is leading to the diagnosis of unsuspected but unrelated peptic ulcer disease.

If there is a real relationship between primary hyperparathyroidism and peptic ulcer disease, then it may be related to increased gastrin secretion. It is well documented that hypercalcemia leads to increased gastrin secretion, which in turn leads to an increase in the basal acid output. In hypoparathyroidism, acid secretion is diminished unless the serum calcium is >7 mg/dl.

Pancreatitis

There is an infrequent but clear association between acute pancreatitis and primary hyperparathyroidism, first described in two patients by Cope et al in 1957.[45] It has been reported to occur in 12 per cent of patients, although this is clearly an overestimate. Acute pancreatitis may be the presenting feature in some patients with primary hyperparathyroidism. It is more frequent when the hypercalcemia is severe. Five clinical types have been described: acute, acute postoperative (occurring in patients with unsuspected primary hyperparathyroidism); recurrent; chronic with pain, which is the most common; and chronic without pain. Many patients with this association have pancreatic calcification on abdominal radiography.

The mechanism for the association between acute pancreatitis and primary hyperparathyroidism is completely unknown, but many potential mechanisms have been suggested. It is probable that the primary hyperparathyroidism precedes the pancreatitis. It may be that pancreatitis is related to hypercalcemia rather than to increased PTH concentrations, because the association may be seen in other hypercalcemic disorders, such as FHH. One suggestion has been that hypercalcemia predisposes to calcium deposition and obstruction of the pancreatic ducts. Hypercalcemia could also facilitate the conversion of trypsinogen to trypsin, which may have destructive effects on pancreatic tissue, or there could be a direct effect of PTH itself on the pancreas to cause inflammation. There have also been suggestions that pancreatitis could lead to primary hyperparathyroidism, although this appears frankly unlikely. The mechanisms suggested for pancreatitis leading to primary hyperparathyroidism have been that hypocalcemia could lead to parathyroid hyperplasia, or pancreatitis could lead to increased glucagon secretion, which could in turn lead to a low serum calcium and PTH secretion. As a practical clinical point, it should be remembered that when acute pancreatitis occurs in a patient with primary hyperparathyroidism, the serum calcium may be normal because of the effects of the acute pancreatitis to lower the serum calcium, due to intervening acute renal failure or calcium-soap deposition in pancreatic tissue.

Articular

Five types of joint disease have been described in patients with primary hyperparathyroidism.[46-49] These include chondrocalcinosis, gout, degenerative osteoarthritis, ankylosing spondylitis, and avulsion of tendons.

Chondrocalcinosis

This disease is also called pseudogout. The frequency of this condition in primary hyperparathyroidism has been estimated at between 7.5 and 18 per cent of patients. It is more likely to occur in older patients. The immunoreactive PTH is increased in many patients with chondrocalcinosis, although the significance of this observation is not clear, because earlier PTH assays were used. The episodes may occur immediately after parathyroidectomy. The joint fluid contains thromboid crystals associated with infiltration with polymorphonuclear leukocytes, with the crystals showing weak positive birefrigence. The crystals may be deposited in articular cartilages and in the menisci.

Gout

Hyperuricemia is frequent in patients with primary hyperparathyroidism, and is presumably due to decreased renal clearance of urate. Acute episodes of gout are not uncommon in patients with primary hyperparathyroidism. Thus, gout and chondrocalcinosis may occur in the same patient, and 20 per cent of patients with pseudogout also have hyperuricemia. The distinguishing points between gout and chondrocalcinosis are the types of joints involved, the severity of the pain in gout, the presence of crystals in the joint fluid, and the characteristic response of gouty arthritis to colchicine.

Other articular manifestations

A rare form of degenerative osteoarthritis has been described in patients with primary hyperparathyroidism.[46] This type of joint disease probably occurs as a consequence of the bone disease. The bone disease may cause subchondral fractures near the joints, which lead in turn to a traumatic synovitis and then in turn to effusions, joint instability, and calcific periarthritis involving the synovium. The condition may resemble rheumatoid arthritis radiologically.

An association between ankylosing spondylitis and primary hyperparathyroidism has been reported.[50] It is likely that this merely represents the chance simultaneous occurrence of two unrelated conditions.

Avulsion of tendons may occur spontaneously in patients with primary hyperparathyroidism in association with ectopic calcification.[51,52] This is particularly likely to occur in the quadriceps muscle.

Cardiovascular

Hypertension is frequently present in patients with primary hyperparathyroidism, particularly in the elderly patient.[53–55] However, essential hypertension is so common in the elderly population that it is difficult to know whether it is directly related to primary hyperparathyroidism. Hypertension is often not resolved by successful surgery,[56] and this has led to the conclusion that the association with hyperparathyroidism may be chance and that hypertension should not *per se* be an indication for surgery. The severity of the hypertension is correlated with renal function. However, it has long been known that infusions of calcium can increase blood pressure[57] (Chapter 4), and this remains an open issue.

Other manifestations

Skin necrosis

Skin necrosis can occur in patients with any cause of hypercalcemia.[58] It is rare in primary hyperparathyroidism and is usually preceded by pruritus. It involves precipitation of calcium and phosphate in the subcutaneous and epidermal tissues. Its rarity in primary hyperparathyroidism is presumably due to the low levels of serum phosphorus in most patients with this cause of hypercalcemia.

Band keratopathy

As with skin necrosis, this is not frequent in primary hyperparathyroidism unless the serum phosphorus is increased, and probably represents precipitation of calcium and phosphate crystals in the cornea. In extreme cases, it may be seen microscopically, but is easier to detect under the slit-lamp. It is frequently seen in those situations where both the serum calcium and serum phosphorus are increased, such as vitamin D intoxication and sarcoidosis.

Primary hyperparathyroidism in special situations

Pregnancy Occasional patients with primary hyperparathyroidism become pregnant. There is very little information in the literature on this topic, which is surprising. Less than 100 cases

have been reported. Since primary hyperparathyroidism is such a common disease, it seems likely that patients with primary hyperparathyroidism must have decreased fertility for pregnancy to occur so uncommonly. Either that, or most patients are asymptomatic and go through pregnancy without sign either in the mother or in the fetus of the underlying maternal disease. Most of the cases that have been reported have been recognized retrospectively following severe hypocalcemia occurring in the newborn child.[59] In these patients, there is a high fetal wastage, with increased risk of tetany and death in the neonatal period. Severe neonatal hypocalcemia is probably related to the prolonged suppression of PTH secretion which occurs as a consequence of hypercalcemia in the mother. Anecdotal reports on this topic suggest that primary hyperparathyroidism should be treated on its own merits during pregnancy.[60–67] If a patient has mild and asymptomatic primary hyperparathyroidism, then neck exploration should probably be avoided during pregnancy since it may lead to miscarriage. If more severe or symptomatic hypercalcemia is present, primary hyperparathyroidism should be treated surgically to avoid clinical manifestations of neonatal hypocalcemia in the child.[68] Pregnant patients with primary hyperparathyroidism can be treated surgically at any stage of pregnancy and good results anticipated. The neonatal hypocalcemia may last for 4–5 months, and permanent hypoparathyroidism has even been reported. A neonatologist should be present at the time of delivery in anticipation of severe hypocalcemia.

Normal calcium homeostasis is disordered during pregnancy. The fetus is hypercalcemic relative to the mother. Maternal calcium balance is also altered, presumably to supply the mineral demands of the fetus. Maternal serum 1,25-dihydroxyvitamin D and calcium absorption from the gut are increased during all trimesters, a state which has been called 'absorptive hypercalciuria'.[69] Serum PTH and urinary cyclic AMP secretion are appropriately suppressed. The cause of increased 1,25-dihydroxyvitamin D production is unknown, but may be due to placental production of the vitamin D metabolite.[70] In addition, it has now been demonstrated in sheep that fetal parathyroid glands produce the PTH-related protein (PTH-rP) and this may enhance calcium transport across the placenta.[71,72] PTH-rP appears to mediate these effects through a mechanism which is independent of the PTH receptor.

Childhood hyperparathyroidism[73–77]   Primary hyperparathyroidism may occur in childhood, but is rare under the age of 10 years. There have been just a few cases reported in the literature. When it does occur, bone disease and joint problems appear to be more frequent than renal stones, and the hypercalcemia is frequently severe (often in the region of 15 mg/dl (3.75 mmol/l)). Children with hypercalcemia may present with problems of coping with school life, and particularly with personality problems related to the elevation in the serum calcium. Primary hyperparathyroidism in childhood will probably be recognized more frequently in the future as the serum calcium is determined more routinely. It should be remembered that the multiple endocrine neoplasia syndromes associated with primary hyperparathyroidism do not usually present during the first decade of life, in contrast with FHH, which is almost always expressed before the age of 10 years. This latter condition is asymptomatic and benign, and children with this disease should not have neck exploration.

Occasionally primary hyperparathyroidism occurs in the neonatal period. Fortunately this is rare. It is usually due to four-gland hyperplasia, and may require surgery or medical therapy. Some of these patients may be in kindreds with FHH (Chapter 9). If surgery is required, then the parathyroid tissue should be cryo-preserved for later possible implantation in the forearm.

Parathyroid crisis   Occasionally, patients with primary hyperparathyroidism may have a sudden rise in the serum calcium, and will present in a shock-like state accompanied by coma and fever. The serum calcium is often greater than 18 mg/dl (4.5 mmol/l) and may be as high as 25 mg/dl (6.25 mmol/l). The clinical state is probably related to the rate of rise in the serum calcium. This condition is a medical emergency, and is frequently associated with death. The patients are severely dehydrated and may develop renal vein thrombosis.

Recently, the author has noted a variation on parathyroid crisis occurring in the elderly.[14] This may be a relatively common mode of presentation in elderly patients who present with a story of acute deterioration in conscious state associated with dehydration which may be ascribed to progressive cerebral vascular disease or senile dementia. This syndrome is analogous to that of diabetic nonketotic hyperosmolar coma, which is characterized by acute episodes of dehydration and hyperglycemia leading to coma, and which is commonly seen in elderly mild diabetics. Patients with primary hyperparathyroidism presenting in this manner have probably all had the disease for a number of years, were untreated, and would have been unrecognized if the serum calcium had not been measured. This syndrome of parathyroid precoma can be diagnosed by evaluation of the serum calcium and blood urea nitrogen in any patient who undergoes rapid and unexplained deterioration of conscious state, associated with polyuria and dehydration. The precipitating event may be an acute infection. Five patients were noted who presented with this syndrome, all elderly women of over 70 years.[14] Their condition was characterized by progressive confusion, disorientation, drowsiness, and lethargy. The serum calcium in these elderly patients was only mildly elevated, and in all was less than 13 mg/dl.

MEN (Multiple Endocrine Neoplasia)   Primary hyperparathyroidism occurs in association with other endocrine neoplasms. There are a number of different syndromes, but the most important and the most frequent are referred to as familial MEN Type I and familial MEN Type II. Some believe that familial hypocalciuric hypercalcemia is also a variant of familial hyperparathyroidism. This is discussed in more detail in a later chapter.

Familial MEN Type I is an autosomal dominant condition which is characterized by hyperfunction of the parathyroid glands, the pancreatic islets and hormone-secreting cells in the anterior pituitary. The clinical features may be indistinguishable from those of primary hyperparathyroidism due to a parathyroid gland adenoma, as far as the hyperparathyroidism is concerned. Probably all of the patients with familial MEN Type I have parathyroid gland hyperplasia, although the enlargement of the glands may be asymmetric. The condition usually becomes apparent between the ages of 20 and 40 years, and it is equally common in men and women. It probably accounts for between 1 and 10 per cent of all patients with primary hyperparathyroidism. The cause is not known. It appears likely that there must be some common mechanism which links the abnormalities in the parathyroids with those that occur in the pancreatic islets and the anterior pituitary. Recently, it has been found that the plasma of patients with familial MEN Type I contains a growth factor for cultured bovine parathyroid cells. This growth factor appears to be a protein of approximately 50 000 daltons, and it was distinguished from all of the known parathyroid cell secretagogues.[24] The source of this factor is unknown, although it appears unlikely that it arises from the parathyroid glands, since it persists in the plasma in patients after total removal of the parathyroids. There are also some rare variants of familial MEN Type I, including several kindreds with familial parathyroid adenomas.

Familial MEN Type II is not as common as familial MEN Type I. This syndrome is associated with hyperfunction of the parathyroids, medullary carcinoma of the thyroid, and pheochromocytoma. Primary hyperparathyroidism is always due to parathyroid gland hyperplasia in this syndrome. It is usually found in patients undergoing operation for medullary carcinoma of the thyroid. Most of these patients are normocalcemic. This is also an autosomal dominant condition and there is no reasonable clue as to the etiology. As with familial MEN Type I, there are many rare variants.

Recent studies have clarified the genetic and molecular basis for the parathyroid gland hyperplasia which occurs in both MEN Type I and MEN Type II syndromes. It appears likely that MEN I is associated with deletion of a regulatory gene on chromosome 11. In separate studies, Thakker et al[78] and Friedman et al[79] demonstrated that allelic deletions were present on chromosome 11 in patients with primary hyperparathyroidism associated with the MEN I syndrome. This suggests the possibility of a similar mechanism to the so-called 'two-hit' theory of Knudson[80] proposed for retinoblastoma. This concept suggests that the first hit or mutation is inherited, whereas the second hit occurs in somatic cells. Interestingly, Friedman et al[79] showed a similar abnormality in chromosome 11 in some patients with sporadic parathyroid adenomas. Thakker et al[78] showed linkage between the MEN I gene and the oncogene

INT-2, which is of interest because the product of this oncogene bears some homology to basic fibroblast growth factor, and molecular rearrangements between the PTH gene and the region containing INT-2 have previously been shown by Arnold et al[19,20] in sporadic parathyroid adenomas. One other interesting point arising from these studies is that the apparent hyperplasia which occurs in MEN Type I represents a monoclonal rather than a polyclonal disorder, whereas the latter appears to be present in sporadic parathyroid gland hyperplasia. It is thus unclear why patients with parathyroid gland hyperplasia are the patients who seem to have a circulating parathyroid growth factor present in their circulation, since a parathyroid growth factor would be more likely to produce a polyclonal disorder than a monoclonal disorder.

The mechanism for primary hyperparathyroidism appears to be different in MEN Type II. In this condition, although parathyroid cell hyperplasia is also present, several groups have located the abnormality near the centromere on chromosome 10, but have been unable to show deletion of genetic material in patients with this disorder.[81]

The familial hyperparathyroidism syndromes have been described recently in detail,[23] and the reader is referred to this comprehensive review for further information on this topic.

Elderly In recent years primary hyperparathyroidism has been recognized more frequently in the aging population. In fact, the majority of patients are now females over the age of 60 years, and many are over the age of 70 years. Primary hyperparathyroidism has an age- and sex-distribution similar to that of postmenopausal osteoporosis.[14] Recognition of the disease in this group of people has been aided by the use of multichannel autoanalyzers on which serum calcium is routinely determined. These elderly patients frequently complain of symptoms such as lethargy, weakness, lassitude, weight and appetite loss, constipation, and nocturia, which are frequent complaints in the elderly and would not necessarily lead to suspicion of hypercalcemia. Moreover, many have evidence of changes in mental status or even dementia. As described above, it has been found that some of these elderly patients may, in response to an intercurrent illness, develop an episode of parathyroid crisis associated with severe dehydration and acute change in mental status. The major issue, of course, is how to treat these patients. There are risks of operation in the elderly, even when an experienced and expert parathyroid surgeon is available. Most workers in this area are adopting an aggressive policy towards parathyroid gland surgery in the elderly, since some patients show marked improvement of symptoms which were not previously ascribed to the presence of hypercalcemia. However, asymptomatic patients usually do well with conservative treatment, and some patients will not be suitable for surgery because of the presence of cardiovascular or other serious disease, or because of limited life expectancy. In elderly women an alternative form of treatment is the use of estrogens or progestogens, which have been very successful in lowering the serum calcium in some patients. Some patients with mild hypercalcemia may be followed, but these should be monitored carefully, because any recurrent illness could provoke a parathyroid crisis.

Unusual complications

There are several unusual complications of primary hyperparathyroidism which should be mentioned. The first is spontaneous infarction. This may cause a remission of the primary hyperparathyroidism,[82] but is extremely rare. Infarction of a parathyroid adenoma has been induced by radiologists deliberately as a form of treatment. Sudden hemorrhage into a parathyroid adenoma can cause acute hypercalcemia and pain in the neck.[83] Again, this is an extremely uncommon event.

Diagnosis of primary hyperparathyroidism

The diagnosis of primary hyperparathyroidism is usually straightforward. The primary diagnostic test is the PTH immunoassay, and a clear elevation in serum PTH associated with longstanding hypercalcemia in a patient with normal renal function is diagnostic of the condition. However, there are some patients in whom the diagnosis is suspected but is difficult to confirm. As indicated below, there are still some difficulties with the PTH radioimmunoassay. In some patients evidence of longstanding hypercalcemia will not be available. In some the picture will be confused by the presence of another condition which can coexist with primary hyperparathyroidism, such as cancer. In these patients the diagnosis is usually based on the accumulation of pieces of evidence, none of which is conclusive. Several years ago Watson et al[84] reported the use of discriminate analysis, the discriminate functions being derived from plasma phosphate, alkaline phosphatase, chloride, bicarbonate, urea, and erythrocyte sedimentation rate. They found that these tests were probably more reliable than, or at least as reliable as, the PTH radioimmunoassays then available. An alternative is for the experienced clinician to examine each of these tests individually, consider them together with the history, and then make a decision.

There are several special tests which can be used in unusual situations. These include the steroid suppression test and the thiazide provocative test. The latter is not sufficiently specific to be widely useful. However, the steroid suppression test can be very helpful in some cases. It is particularly useful with a mild elevation in the serum calcium in a patient without skeletal abnormalities. These patients essentially never respond with a fall in the serum calcium if they have primary hyperparathyroidism. In contrast, if sarcoidosis or another granulomatous disease is responsible for the hypercalcemia, then they almost always respond in 10 days and, if cancer is the cause of hypercalcemia, then about 30 per cent will respond. The information that this test provides is that primary hyperparathyroidism can be ruled out in those patients who do respond.

In 1964 Wills and McGowan[85] suggested that the serum chloride may be a useful index of the presence of primary hyperparathyroidism. They suggested that patients with primary hyperparathyroidism have a serum chloride of greater than 102 mmol/l, whereas in patients with hypercalcemia from other causes the serum chloride is low. This is, by and large, an extremely useful test, since about 80 per cent of patients with primary hyperparathyroidism will have a serum chloride which is consistently >102 mmol/l, whereas hypercalcemia due to other causes is almost always accompanied by a serum chloride of <100 mmol/l.

PTH radioimmunoassay

Parathyroid hormone radioimmunoassays are widely used in the differential diagnosis of hypercalcemia, and they allow a diagnosis of primary hyperparathyroidism to be made in the majority of cases. Several different assays are available commercially. Although there have been major recent improvements, there are still some difficult problems with this assay, and the immunoreactive PTH concentration in the serum should still be regarded as an index of parathyroid gland secretion rather than a direct measurement of circulating biologically active PTH.

Clinical expectations of PTH immunoassays

The major usefulness of the PTH assay is in the differential diagnosis of hypercalcemia. The assay should be able to distinguish patients with primary hyperparathyroidism from patients with hypercalcemia due to malignancy. It is not necessary to use the assay to recognize disease—that can be done by measurement of the serum calcium. The crux of the issue in the patient with hypercalcemia is that the assay should distinguish primary hyperparathyroidism from the other causes of hypercalcemia. The PTH assay may also be useful in the differential diagnosis of hypocalcemia. The assay should distinguish patients with primary hypoparathyroidism from those with pseudohypoparathyroidism.

Several facts should be borne in mind when the clinician uses this test in the differential diagnosis of hypercalcemia. First, with single-site radioimmunoassays, 50–95 per cent of patients from different series with primary hyperparathyroidism have an increase in serum immunoreactive PTH. Second, between 0 and 95 per cent from different series of patients with malignancy have detectable or increased immunoreactive PTH (although the more recent assays show no more than minor elevations, and in less than 10 per cent of patients). Third, of patients who have an immunoreactive PTH concentration greater than twice normal (in the absence of renal failure), all have primary hyperparathyroidism. Fourth, no

patients with malignancy (without hyperparathyroidism) have an immunoreactive PTH which is more than twice normal. A few patients with primary hyperparathyroidism have immunoreactive PTH concentrations which are normal or even low. Therefore, a reasonable conclusion is that if the immunoreactive plasma PTH concentration is clearly increased, certainly more than twice normal in a patient who has normal renal function, then the patient has primary hyperparathyroidism. However, if the immunoreactive PTH concentration is less than twice normal, then the patient may have primary hyperparathyroidism or the patient may have malignancy causing the hypercalcemia.

There are considerable variations in individual assay techniques for the measurement of immunoreactive PTH. It is clear that (a) some patients with primary hyperparathyroidism have low or normal immunoreactive PTH in all assays currently available, and (b) some patients with hypoparathyroidism have normal, and in some assays even high, immunoreactive PTH. Results from individual assays still show some variability[86] (see also Chapter 4). Using different assays, patients with the same disease give markedly divergent results. This has been noted in the hypercalcemia of malignancy,[87,88] in patients with renal calculi and in primary hyperparathyroidism, and has now been well documented in several studies both in the USA and in Europe.[86,89,90]

The major reason why the PTH assay has been so difficult to interpret is that PTH circulates as a number of fragments of different biological activity and immunoreactivity. These fragments are generally classified as either C-terminal fragments or N-terminal fragments, depending on which portion of the molecule the antibody is directed towards. The C-terminal fragments are biologically inert and have a relatively long half-life, but on the other hand they are a fair index of parathyroid gland activity. In contrast, the N-terminal fragments are biologically active, have a shorter half-life, and are useful in conjunction with neck catheterization for localization. Biologically inert fragments of PTH continue to be released from the parathyroid glands in patients with nonparathyroid hypercalcemia, and are a cause of misleading immunoassay results. C-terminal assays have been used more frequently in the past than N-terminal assays, because the C-terminal fragment of the molecule seems to be more immunogenic. Current N-terminal or mid-molecule assays are gaining wider acceptance, particularly in patients with impaired renal function or when it is necessary to obtain an index of immediate gland secretion such as in neck catheterization and localization. Mallette[91] uses a mid-region specific assay in which overlap in PTH values between primary hyperparathyroidism and malignancy is only about 5 per cent.

At present C-terminal assays are still useful in many patients as a diagnostic index of primary hyperparathyroidism, and they indicate a diagnosis of primary hyperparathyroidism when they are clearly increased. In patients in whom renal function is impaired, and in some patients with indeterminant C-terminal assay results, N-terminal assays have been advocated, but these have been generally disappointing (see below). The value of using the serum calcium in conjunction with immunoreactive PTH concentrations is dubious, because in most assays it does not improve the distinction between malignancy and primary hyperparathyroidism, and hypercalcemia does not entirely suppress immunoreactive PTH.

The recently developed two-site immunoradiometric assay technique (IRMA) should surpass the previous PTH radioimmunoassays. This technique utilizes two antisera which are directed at the N-terminal and C-terminal ends of the PTH molecule. Thus, only the intact hormone is detected, and problems with cross-reactivity with circulating PTH fragments are avoided. IRMA assays are now commercially available and they are becoming widely used. Although their capacity to discriminate between patients with primary hyperparathyroidism and hypercalcemia of malignancy, or between hypoparathyroidism and normals, has still not been widely assessed, early results suggest that they represent a considerable improvement over currently available standard radioimmunoassays.[92,93]

Recent comparative assessments of commercial PTH assays

The current status of the PTH radioimmunoassays available to clinicians in the USA for the differen-

tial diagnosis of hypercalcemia was addressed in a brief report by Lufkin et al.[86] These workers essentially repeated an earlier study by Raisz et al[89] in which aliquots of serum obtained from patients with the hypercalcemia of malignancy and primary hyperparathyroidism were compared using different assays. The earlier study by Raisz et al showed that there was marked variability in the results obtained from assay to assay. The reason for repeating the study was that in the interim 8 years there had been technological improvements in PTH assays, with increased capacity to detect specific regions of the molecule including the biologically active N-terminal region, as well as mid-region and intact portions of the molecule. Lufkin et al found that the great majority of patients with primary hyperparathyroidism had an increase in circulating immunoreactive PTH concentrations in all assays. However, in most assays there was a small number of patients with proven primary hyperparathyroidism who had immunoreactive PTH concentrations which were in the normal range and overlapped with those of patients with the hypercalcemia of malignancy. This is the crux of the problem as far as the value of the PTH radioimmunoassay in the differential diagnosis of hypercalcemia is concerned. These assays might be expected to show suppressed levels of immunoreactive PTH in patients with the hypercalcemia of malignancy, since PTH expression by the tumors is clearly not the mechanism of hypercalcemia in nonparathyroid malignancies. However, in all assays the majority of patients had circulating PTH concentrations spread throughout the normal range, and in some there was even a slight increase in the PTH. The mechanism for this slight increase, which seems to have been more common with earlier C-terminal assays, is unknown. It is possible that it could be due to cross-reactivity with non-PTH-like factors produced by tumor cells, although the PTH-like factors have sequence homology with parathyroid hormone at the N-terminal rather than the C-terminal end, and the C-terminal assays have also been shown increased circulating concentrations in patients who do not secrete PTH-like factors, such as those with myeloma.[87] It is also possible that in this situation the suppressed parathyroid glands could be releasing biologically inert C-terminal fragments which cross-react with the assays. Lufkin et al[86] found that in their hands the N-terminal assay did not provide enhanced diagnostic usefulness over the C-terminal, mid-region, or intact assays. In fact, the N-terminal assay was not as efficient, either at distinguishing patients with hypercalcemia of malignancy from patients with primary hyperparathyroidism, or at distinguishing patients with primary hyperparathyroidism from normals. Their conclusions were that most assays give similar results, and that the choice of the assay which a clinician uses should be determined by other factors such as cost, relationship with the laboratory, and the efficiency of the laboratory. Recently the commercial PTH assays were assessed by adding synthetic human whole-molecule PTH which was spiked into endogenous survey pools used for the ligand survey of the PTH assay at a concentration of 1000 pg/ml. The assays were compared for assay coefficients of variation and recovery of the hormone by the various assay methods. This survey[94] concluded that there was clearly large heterogeneity in results reported by the available commercial assays. This is hardly surprising, in view of the known fact that PTH circulates as multiple fragments in the circulation.

IRMA

As noted above and in Chapter 4, the new two-site assay or IRMA represents a significant advance and seems likely to replace current radioimmunoassays in the diagnosis of primary hyperparathyroidism in the near future. Comparisons with other radioimmunoassays are limited, but suggest better discrimination between primary hyperparathyroidism and normals and primary hyperparathyroidism and hypercalcemia of malignancy.[92,93]

It is also possible that in time sensitive biological assays for PTH will be developed. The cytochemical assay is clearly very sensitive, but it is difficult to use and is available only in several research laboratories. It utilizes the capacity of PTH to stimulate glucose-6-phosphate dehydrogenase (G6PD) content in renal tubular cells. Another bioassay is the adenylate cyclase assay, using as target cells either renal tubular cells or

osteosarcoma cells. Although these assays have been in the development phase for a number of years, it is unlikely that they will ever be of much use clinically, because the major clinical issue in the differential diagnosis of hypercalcemia is to distinguish the patient with primary hyperparathyroidism from the patient with the hypercalcemia of malignancy. Since many patients with the hypercalcemia of malignancy secrete factors which interact with the PTH receptor, the bioassays will not be able to distinguish PTH from PTH-related molecules which produce similar biological effects and interact with the same receptor.

Treatment of primary hyperparathyroidism

Decision to treat

Once the diagnosis of primary hyperparathyroidism has been made, several important decisions are necessary. The first is whether active treatment is required. This issue has become more important in recent years, since more than half of the patients diagnosed with this common disease are asymptomatic at the time of diagnosis.

Since many patients with primary hyperparathyroidism are asymptomatic, and many of them may never develop features of the disease which will interfere with their lives, not all patients require surgery. Many patients remain asymptomatic for a considerable number of years, some at least for 20 years (although it may be argued that careful history taking will reveal that many patients do, in fact, have symptoms and these often become apparent to the patient when they are relieved; that is, after successful surgery). The problem is that it is not known how to predict at the time of diagnosis which patients will later develop complications from the disease. A large prospective study from the Mayo Clinic of 134 patients indicates that more than 50 per cent of asymptomatic patients will go 5 years without any change in their clinical state.[95] In this study approximately 20 per cent eventually required surgery and 20 per cent were lost to follow-up. It should be noted that some untreated patients feel very uneasy knowing that they have a disease which can be actively treated, and particularly that they have a tumor (albeit benign). The author's own experience indicates that an occasional asymptomatic patient will eventually get into trouble. One case of particular note was a 60-year-old woman who was first recognized as having primary hyperparathyroidism when she presented comatose with a serum calcium of 25 mg/dl (6.25 mmol/l). She had been followed for a number of months by a local physician who did not have the benefit of a serum calcium measurement. While she was followed by him her complaints were of mild aches and pains, and a general feeling of lack of well-being. She presented with an acute parathyroid crisis, could not be resuscitated, and died with progressive renal failure which was found to be related to renal vein thrombosis. Clearly, if she had been diagnosed as having primary hyperparathyroidism in the preceding months, and had been actively treated, then this episode of parathyroid crisis would not have occurred. The problem is that it is difficult to recognize those unusual patients who have such a disastrous clinical course at the time of initial presentation.[14]

As a result of these difficulties in anticipating the natural history of the disease in individual asymptomatic patients at the time of diagnosis, the indications for treatment are arbitrary and should be influenced by factors such as the availability of a competent surgeon who is experienced in this type of surgery, the fitness of the patient to withstand neck exploration, and the ease of follow-up. A history of familial hypercalcemia suggests either multiple endocrine neoplasia or FHH, and surgery should only be embarked on by an expert in the former case, and is best avoided in the latter. In MEN syndromes, four-gland hyperplasia is likely and patients with FHH rarely benefit from surgery. The indications that the author uses for surgery are: (a) symptoms or signs which can be ascribed to hypercalcemia or primary hyperparathyroidism; (b) patients who cannot be followed adequately; (c) corrected serum calcium greater than 11.5 mg/dl (2.9 mmol/l); and (d) patients with parathyroid carcinoma. It is also reasonable to consider surgery strongly in those patients who are at risk for later development of osteoporosis—for example, young or middle-aged Caucasian women with any of the other known risk factors for osteoporosis.

However, the issue of increased risk of bone loss with advancing age and greater propensity for fractures in patients with untreated primary hyperparathyroidism has been questioned recently.[96,97] 174 patients with mild asymptomatic primary hyperparathyroidism were followed without surgical treatment at Henry Ford Hospital. Of the 80 patients who were followed over a number of years, ionized calcium and PTH secretion remained relatively constant and there was no greater decrease in forearm density than could be expected from advancing age. Moreover, there seemed to be no increased risk of vertebral fracture compared with retrospective control groups. Although these results showed that while patients with mild primary hyperparathyroidism may have undergone significant loss of cortical bone by the time of diagnosis, continued bone loss does not appear to occur at an accelerated rate. However, Bilezikian[98] argues that it is possible that the precision of the machines used was not sufficient to permit a significant change in bone mineral density over the duration of this study to be detected. Although the results of Rao and colleagues clearly need to be confirmed by a controlled clinical trial, it is reasonable to conclude that conservative management for mild primary hyperparathyroidism is a reasonable approach.

It should also be realized that parathyroid gland exploration is not a benign procedure. In a retrospective study of the experience from the University hospitals of Cleveland and Case Western Reserve University[99] in the mid-1970s, it was clear that there are risks associated with this surgery. Approximately 9 per cent of patients develop permanent hypoparathyroidism and 8 per cent of patients remain hypercalcemic postoperatively. When this is compared with the long-term results in patients treated conservatively, it is certainly a reasonable course to recommend conservative treatment for mild asymptomatic and uncomplicated primary hyperparathyroidism based on the currently available literature.

Before recommending surgery, it is important to exclude the presence of FHH (detected by urine calcium and careful family history, see Chapter 9), since these patients usually remain hypercalcemic after neck surgery.

Management of the asymptomatic patient

Once a decision has been made not to perform a neck exploration on a patient with asymptomatic primary hyperparathyroidism, a plan of management must be designed to monitor the patient for any change in clinical state. A reasonable approach is a careful evaluation of the patient to exclude, by skeletal radiology, intravenous pyelography, and, if available, photon-beam absorptiometry, the presence of occult bone disease or renal stone disease. Photon-beam absorptiometry of cortical bone using the single-beam approach may be valuable in this situation, since primary hyperparathyroidism does have substantial effects on cortical bone loss. The patient should be re-evaluated every 3–6 months. If the patient develops any evidence of progressive bone loss, new risk factors for osteoporosis, or the presence of renal stones, then surgery is indicated. The patient should be advised to have the serum calcium measured at times of infection or other stress. Asymptomatic patients may progress to a parathyroid crisis when they have become dehydrated. This may be more likely to happen in the elderly patient. Age itself should not be a contraindication for surgery, since elderly patients with this disease do very well when neck exploration is performed by an experienced surgeon.

The author has followed patients with serum calcium values in the range of 12–13 mg/dl who did not wish surgery for a number of years without any change in symptoms or the development of complications. However, it seems empirically likely that those patients with the highest serum calcium concentrations will be more prone to the development of later complications, since many of the features and complications of this disease seem to be directly related to the serum calcium.

Medical therapy

Some patients, particularly elderly ones, are too old for surgery but require medical therapy for control of hypercalcemia. Choice of medical therapy for primary hyperparathyroidism is limited.[100,101] Chronic medical therapy which is available for hypercalcemia due to primary hyperparathyroidism includes oral phosphate, bisphosphonates, and sex steroids. Oral phosphates should only be used in patients with a low serum phosphorus (<3.8 mg/dl) and normal renal function. Oral phosphate therapy in doses up to 2 g/day can be very useful in controlling the serum calcium for years, particularly in patients with mild hypercalcemia. The risk is the development of renal stones, although this risk is probably very small if the patient is carefully followed. The limiting factor with this form of therapy is often troublesome diarrhea. Occasional patients may not have a sustained response to oral phosphate therapy.[102] There is the theoretical risk that the reduction in the plasma calcium leads to increased PTH secretion from the parathyroid adenoma, which may in turn lead to an exacerbation of bone disease, although this complete sequence has never been documented as far as the author is aware. Bisphosphonates have also been used in the treatment of primary hyperparathyroidism, either due to an adenoma or to parathyroid carcinoma. Oral etidronate is not efficacious in this situation.[102] However, clodronate and APD have both been used successfully,[103] although they may not be quite as effective in this situation as they are in malignancy. These drugs may be particularly useful in the control of hypercalcemia in patients with parathyroid carcinoma. Estrogens or progestins are particularly useful in postmenopausal women. This was first suggested more than 10 years ago by Gallagher and Nordin,[104] and was confirmed recently by Marcus et al[105] and by Selby and Peacock.[106] The mechanism of action of estrogens is unknown. It does not appear to be related to changes in PTH secretion. However, the serum calcium falls into the normal range in the majority of patients, and they are asymptomatic. This use of this therapy has been reported only in postmenopausal women. The author had occasion to use it once in an elderly woman with symptomatic hypercalcemia and renal failure, who was not a candidate for surgery. Her serum calcium was maintained in the normal range for 5 years until she died from unrelated causes. Premarin in a dose of 0.625 mg/day for 25 days of each month is a suitable and effective therapy. Other medical therapies which may be considered include corticosteroids, calcitonin, calcitonin and corticosteroids, or plicamycin. Corticosteroids alone are almost always ineffective in primary hyperparathyroidism, and are not worth a trial in the author's view, unless the diagnosis remains in doubt. Calcitonin at best will produce a transient response. The combination of calcitonin and corticosteroids has been effective, particularly in one patient with a parathyroid carcinoma[107] who was maintained normocalcemic with this combination for many months. The author's experience in primary hyperparathyroidism is that calcitonin and corticosteroids in combination produce a partial response, but not a complete one.[108] Plicamycin may be used if no other form of therapy is suitable, and particularly in patients with parathyroid carcinoma.[109] However, its toxicity and the need for repeated injections or infusions limit its usefulness in patients who have benign disease.

Urgent treatment of hypercalcemia

Should hypercalcemia be so extreme that the patient's life is in immediate danger, then the serum calcium concentration must be lowered urgently. There are two possible approaches.

Emergency surgery

Although emergency surgery is still occasionally employed, the risks of anesthesia in a severely hypercalcemic patient, particularly if the surgery is prolonged, are probably high. Since the serum calcium can usually be lowered rapidly by current medical therapy, surgery is usually not indicated until the patient's clinical condition is stabilized and hypercalcemia is under reasonable control.

Emergency medical therapy

This includes rehydration with intravenous isotonic saline, the administration of calcitonin and corticosteroids, plicamycin, phosphate (either oral or intravenous), or intravenous bisphosphonates. Emergency therapy for hypercalcemia is considered in more detail in Chapter 7.

Furosemide may be required in patients who are receiving large amounts of isotonic saline for fluid overload. The author prefers to use vigorous rehydration with intravenous saline together with the combination of calcitonin and glucocorticoids, which are very safe forms of therapy. Plicamycin is usually not safe in this situation, because the patients are severely dehydrated and are likely to have prerenal azotemia. If intravenous bisphosphonates are available, these are a particularly useful form of therapy, although there may be some risk of toxicity in the patient in whom renal function has not been stabilized. Intravenous phosphate may be life-saving, but if the serum calcium is very high this therapy will almost always result in soft-tissue calcium deposition. The author's approach is to use vigorous rehydration with isotonic saline, together with calcitonin and glucocorticoids. If bisphosphonates are available, these are introduced as soon as it is clear that the patient does not have irreversible impairment of renal function.

Surgery

Surgery is the best form of definitive therapy for primary hyperparathyroidism. The indications are listed in Table 8.3. Pre-operative preparation for

Table 8.3   Indications for surgery.

1   Significant clinical features occurring as a consequence of hypercalcemia (for example, renal calculi, symptoms of hypercalcemia, or radiologic bone disease)

2   Corrected serum calcium >11.5 mg/dl (3.9 mmol/l). Serum calcium concentration at which cut-off taken is arbitrary, depends on factors such as availability of experienced and skilled surgeon, psychologic impact on patient, ease with which the patient can be followed

The difficulty in making the decision whether active treatment is necessary is that the natural history of untreated asymptomatic hypercalcemia is unknown

Surgery is contraindicated in patients with FHH

surgery in the patient, in whom a definitive diagnosis has been made, involves excluding FHH by a careful family history and determination of renal calcium excretion, and identifying an adequate and experienced surgeon. The question of whether it is necessary to treat hypercalcemia pre-operatively has never been addressed thoroughly. In the majority of patients it is not necessary, although it is possible that occasional patients with a very high serum calcium would do better if the hypercalcemia were treated medically before they were subjected to anesthesia.

Operative approach

The operative approach at the Massachusetts General Hospital in over 700 cases between 1930 and 1975 has been reviewed recently by Wang.[110,111] The surgical procedure is determined by the pathology. Primary hyperparathyroidism is due to a solitary adenoma in 85 per cent of cases, and hyperplasia in approximately 15 per cent of patients. In rare situations, carcinoma of the parathyroid gland is responsible. Unless a localization technique is used (see page 152), the surgeon does not know before neck exploration which of the parathyroid glands are abnormal, or even whether the patient has one abnormal gland or four. Hyperplasia should be suspected in those patients with a family history or with multiple endocrine neoplasia. Since it is usually not known whether hyperplasia or an adenoma is responsible before neck exploration, the procedure has to allow for all possibilities, including the possibility that an adenoma may be located in an ectopic site, which occurs in about 5–10 per cent of cases. One of the problems is that it may be difficult to distinguish an adenoma from normal parathyroid gland tissue, and sometimes from other normal tissues such as lymph nodes. The operative approach requires an experienced and confident surgeon, together with an expert pathologist to examine the frozen sections. The approach favored by many surgeons is to approach one side of the neck thoroughly, and then both glands on that side of the neck are biopsied. Parathyroid gland tissue can easily be confused with lymphoid tissue or adipose tissue. The identification of glandular tissue should be made by a pathologist on the basis of a frozen section. If a parathyroid adenoma is identified and the other gland on the same side is normal, then the surgeon can exit without going to the other side, since under these circumstances the likelihood is very high that the glands on the other side will be normal. If both glands on one side are normal, then the surgeon should go to the other side of the neck and identify the glands there. However, there are many surgeons who prefer to identify glands on both sides of the neck.

With the advent of new and sensitive assays for circulating PTH, it may be possible to utilize operative measurements of PTH to decide whether further neck exploration is necessary.[112] For example, the new two-site IRMA for PTH can be modified to permit measurement of PTH within a few minutes. Using this approach, a decrease in circulating PTH to the normal range following removal of an apparent adenoma may render continued exploration of the opposite side of the neck unnecessary.

If an adenoma is not found, then the surgeon should examine the retro-esophageal and retro-pharyngeal spaces and the thymic fat pad. If an adenoma is still not found, then the surgeon should perform a thyroid lobectomy on the side of the neck where the parathyroid glands were not detected. The mediastinum is not split at the first operation. If the abnormal gland is not found after this exploration, then the surgeon should exit from the neck. After 2 weeks, an attempt should be made to localize the abnormal gland by techniques such as computed tomography scan, ultrasound, scintigraphic scanning, and, if the appropriate expertise is available, by venous sampling and PTH assay using an N-terminal assay, or by arteriography. Once the adenoma is localized, the neck is re-entered and the adenoma is removed.

Localization procedures should be successful in approximately one-half to two-thirds of patients whose necks have previously been explored. The localization procedures which are available and the indications for their use are discussed in more detail below.

If four-gland hyperplasia is found, then the surgeon should remove three and one-half glands and leave part of the gland which has been least damaged (in other words, the last gland). Others advise leaving approximately 50 mg parathyroid gland tissue.

Hyperparathyroidism is probably more likely to recur in patients with hyperplasia due to MEN Type I than in patients with sporadic four-gland hyperplasia.[113,114] For this reason, the surgeon may wish to perform total parathyroidectomy in patients with MEN Type I followed by an implantation of parathyroid tissue into the forearm musculature where it could be easily removed under local anesthetic at a later date.

The alternative approach is to explore the whole neck and identify all the parathyroid glands. Some would advocate that the procedure should be performed with an ear, nose, and throat surgeon in attendance to identify the recurrent laryngeal nerve. Clips may be left where the parathyroid tissue was found, in case re-exploration be required at a later date.

Most neck explorations (more than 90 per cent) are successful, and patients can return home 4–5 days later.

Postoperative management

Patients recover very quickly from neck exploration. The most common postoperative complication is hypocalcemia. In patients without bone disease before surgery, serum calcium, serum phosphate, urine cAMP, and plasma PTH normalize within minutes to hours of the surgery. Often the patient will have transient hypoparathyroidism for up to 2–3 weeks. This usually reaches a maximum at 4–7 days. The fall in phosphate and cAMP excretion is usually more rapid than the change in serum calcium, and the change in urine cAMP can be observed during the operative period.[115] This may be due to the following.

(a) The 'hungry bone' syndrome. Demineralized bones rapidly absorb calcium, and the patient may be profoundly hypocalcemic for a few days. The duration and severity of hypocalcemia depends on the extent of previous bone disease, and may be prevented by therapy with active metabolites of vitamin D before surgery. The serum phosphorus is usually low.

(b) Postoperative hypocalcemia may persist for up to 6 months. In a few cases this may be due to total-body magnesium depletion and subsequent hypomagnesemia. However, in most patients it is due to damage to the parathyroid glands during surgery, which is sometimes reversible. If function has not returned within 6 months, it can be considered irreversible (see below).

(c) Permanent hypoparathyroidism may occur as a result of parathyroid gland damage during surgery, and most frequently occurs in patients who have four-gland hyperplasia. These patients require lifelong therapy with oral calcium and vitamin D. Since patients with severe bone disease may take 3–6 months to recover from transient hypoparathyroidism, permanent hypoparathyroidism cannot be diagnosed definitively until 6 months after surgery. During this period symptomatic hypocalcemia may require treatment with vitamin D and oral calcium preparations. The bone disease is usually slow to improve, and it may take months for mineral to be restored entirely to the osteopenic skeleton. Cystic areas may remain radiolucent indefinitely. Permanent hypoparathyroidism is a severe side-effect of parathyroid surgery. It occurs most frequently in patients who have undergone re-exploration of the parathyroids, or in patients who have hyperplasia and have had removal of all or part of four glands. Again, the diagnosis should not be made until 6 months after surgery. The therapeutic approach is to treat the hypoparathyroidism with calcium and vitamin D, in the same way that any other patient with hypoparathyroidism would be treated. Some patients do well with autotransplantation of parathyroid tissue removed at the time of surgery into the forearm. If it is anticipated that hypoparathyroidism is possible following difficult surgery, then some parathyroid tissue should be cryo-preserved so that it could be available later for insertion into the forearm to correct this very unfortunate situation.

(d) Other unusual complications occasionally occur in the postoperative period. These include acute pancreatitis (particularly in those cases where the adenoma is large), renal colic, and possibly acute psychosis.[109]

Localization of the parathyroid glands

There are a number of reasons why surgeons would prefer to know where pathologic parathyroid gland tissue is located before surgery in patients with primary hyperparathyroidism. At the time of the first operation the surgeon does not know whether the primary hyperparathyroidism is due to a parathyroid gland adenoma or to hyperplasia. If an adenoma is the cause, then it would be helpful for the surgeon to know whether the adenoma is in an ectopic or atypical location, as occurs in about 5–10 per cent of patients (Figure 8.6). An effective and noninvasive localization procedure, which would give the surgeons this information pre-operatively, would increase the chances of successful surgery. For example, if the surgeon knew the location of an adenoma, then an extensive neck exploration would not be required. Moreover, if the surgeon knew that the patient had primary hyperparathyroidism due to parathyroid gland hyperplasia before the operation, then the surgeon could take steps to insure cryo-preservation of parathyroid gland tissue in case the patient should be rendered hypoparathyroid following removal of most of the functioning tissue. A localization procedure also aids diagnosis in hypercalcemic patients with cancer in whom primary hyperparathyroidism is also suspected. However, the surgeon would be helped most in the patient who requires re-exploration (about 10 per cent of all patients). In these patients the first operation frequently leaves distortions in the anatomy of the thyroid bed, due to scarring, and the diseased parathyroid gland which was not located at the first operation is even more difficult to find.

Figure 8.6

Anatomical location of the parathyroid glands, showing the variability of their location and the difficulties that this provides for the parathyroid surgeon.[110]

Parathyroid IV

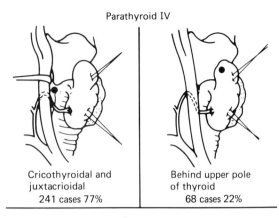

Cricothyroidal and juxtacrioidal
241 cases 77%

Behind upper pole of thyroid
68 cases 22%

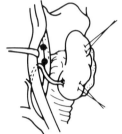

Retropharyngeal and retroesophageal
3 cases 1%

Parathyroid III

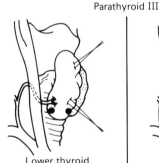

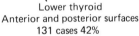

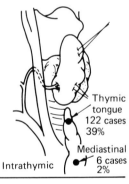

Lower thyroid
Anterior and posterior surfaces
131 cases 42%

Intrathymic

Thymic tongue
122 cases 39%

Mediastinal
6 cases 2%

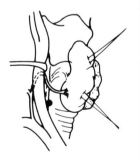

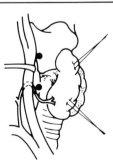

Juxtathyroidal
47 cases 15%

Ectopic
6 cases 2%

The techniques available for localization of the parathyroid glands are selective arteriography, selective venous sampling combined with PTH assay, computed tomography scan, ultrasonography, and technetium scanning. The most widely used techniques until the past few years were selective arteriography and selective venous sampling with PTH measurements.[116–121] These approaches will localize abnormal parathyroid gland tissue in about 75 per cent of patients before the first operation. In necks which have already been explored once, the success of localization of abnormal parathyroid tissue is less, approximately 50–60 per cent. Moreover, these techniques are extremely expensive, and they require an experienced and skilled radiologist, as well as the availability of a reliable PTH radioimmunoassay, preferably one which recognizes the N-terminal portion of the molecule. Selective arteriography is a procedure which carries some risk, and it should not be performed lightly. This technique has resulted not only in infarction of the parathyroids, but also in infarction of the cervical spinal cord and vertebral ischemia due to obstruction of spinal arteries. The blood supply of the parathyroid glands arises from the thyrocervical trunk, which also supplies the spinal cord. Adenomas are usually very vascular, and they show a characteristic blush when contrast is injected.

Recently, other localizing techniques which are noninvasive and more accessible to most clinicians have been used much more widely. These include computer-assisted, thallium-201-technetium-99m scintigraphy, high-resolution computed tomography, and high-resolution real-time ultrasonography. Perhaps the simplest and least-invasive of the newer techniques for identifying abnormal parathyroid gland tissue is high-resolution real-time ultrasonography.[118,122–124]

Ultrasonography

Mallette in particular has been a major proponent of parathyroid ultrasonography as a useful tool in the pre-operative management of primary hyperparathyroidism.[125] This technique is successful in identifying the majority of abnormal glands larger than 1 cm before surgery, although

it is less useful in patients undergoing re-exploration. Real-time high-resolution ultrasonography can locate parathyroid adenomas with a sensitivity of 80–88 per cent and a specificity of 92–96 per cent. In the hands of experienced operators, it should allow visualization of normal parathyroid glands. Since ultrasound can usually differentiate parathyroid adenoma from four-gland hyperplasia, information obtained by preoperative ultrasonography may influence the management of primary hyperparathyroidism. It allows the surgeon to begin dissection on the correct side of the neck, which limits operation time. Moreover, when hyperplasia is diagnosed pre-operatively, a thorough pre-operative evaluation for multiple endocrine neoplasia should be undertaken. In addition, since hyperparathyroidism may recur more frequently in multiple endocrine neoplasia type 1 than in sporadic cases of parathyroid hyperplasia, the surgeon may elect to perform total parathyroidectomy with autologous parathyroid grafting of parathyroid tissue to the forearm musculature. Another advantage of ultrasonography is that it is very effective at identifying intrathyroidal parathyroid glands. Computed tomography with contrast enhancement can identify the majority of abnormal parathyroid glands before the initial operation, and it may be useful in identifying large glands if they are situated in ectopic sites.[126] This approach has improved considerably with third-and fourth-generation scanners, the use of a neck harness, and bolus injection of contrast. It is particularly useful in detecting adenomas at those sites where ultrasonography is ineffective, namely the retro-esophageal, paratracheal, and mediastinal areas. However, it is not effective in detecting adenomas located between the thyroid and the clavicles, and it is useful in less than half of the patients after the first operation. Scintigraphic scanning techniques have recently been improved to localize abnormal parathyroid glands.[127] Scanning the parathyroids using technetium-99m pertechnetate to image the thyroid gland initially, followed by thallium-201 which images both the thyroids and parathyroids, and then using computer-assisted subtraction techniques detects parathyroid adenomas in about two-thirds of cases before initial surgery.[128–130] With this technique, false positives are more frequently encountered, usually due to abnormalities in the thyroid. A recent prospective and blinded comparison of radionuclide, computed tomographic, and sono-graphic localization of parathyroid tumors suggested that scintigraphy and computed tomography are equally sensitive, and very effective when used together.[131]

The major indication for localization of abnormal parathyroid gland tissue until now has been before re-operation following initial failed surgery. The experienced surgeon will locate the abnormal tissue and rapidly cure more than 90 per cent of patients at the time of initial surgery. However, the advent of ultrasonography as a useful localization technique may change this approach. Localization of the diseased gland or glands before initial surgery will reduce operation time and the extent of dissection. In those patients who have had unsuccessful neck explorations, the surgeon benefits enormously by knowing before surgery the location of the abnormal tissue. In these patients a combination of noninvasive methods including computed tomography scan, ultrasonography, and technetium scanning is desirable. A reasonable approach is to use ultrasound first, then scintigraphy, and computed tomography if ultrasound is unrewarding. In the author's view, selective venous sampling and selective arteriography should only be performed in specialized centers by experienced personnel, since they are costly, time-consuming, and, in the case of arteriography, potentially risky.

This topic has been reviewed recently by Winzelberg[132] and in a *Lancet* editorial,[133] and the reader is referred to these reviews for a more extensive bibliography.

Determination of success of surgery during procedure

Serum calcium, phosphate, urine calcium, and urine phosphorus do not change immediately, and their measurement is usually not helpful during the operative period. Urine cAMP changes immediately,[115] but may not be readily measurable and available for the surgeon. If it is possible to measure it, then it can be helpful because it changes within minutes of the removal of a parathyroid gland adenoma. As indicated above, a modified PTH IRMA may be useful in this situation.[112]

Other causes of hypercalcemia

Thyrotoxicosis and hypercalcemia

Hypercalcemia occurs occasionally in patients with thyrotoxicosis from any cause, and thyrotoxicosis should always be considered in the differential diagnosis of hypercalcemia. Usually the hypercalcemia associated with thyrotoxicosis is mild. However, occasionally the hypercalcemia can be more severe and even life-threatening. Although all of the mechanisms involved in the pathophysiology of the hypercalcemia occurring in association with hyperthyroidism have not been clearly delineated, nevertheless there is a lot of information on the effects of biologically active thyroid hormones on calcium and phosphate transport across bone, gut, and the kidney. This topic has been reviewed extensively by Mundy and Raisz,[1] and Auwerx and Bouillon.[2]

Prevalence

The frequency of hypercalcemia in thyrotoxicosis depends on the method of measurement of the serum calcium. When serum ionized calcium concentrations are measured, hypercalcemia is detected more frequently. Some authors have reported frequencies as high as 50 per cent using ionized calcium determinations. Although the author has seen no study to document this, it is suspected that the frequency of hypercalcemia associated with Graves' disease is less frequent than it was 10–15 years ago, just as the frequency of severe Graves' ophthalmopathy seems to have declined in recent years.

When uncorrected serum calcium measurements are made, the frequency with which hypercalcemia is diagnosed is considerably less. Baxter and Bondy[3] reported that nineteen of seventy-seven patients with thyrotoxicosis (a frequency of 22 per cent) had hypercalcemia at some time during the course of their disease. However, they found that two of these nineteen patients had coexistent parathyroid adenomas. In another retrospective study, Gordon et al[4] found that 17 per cent of 121 patients with thyrotoxicosis had an elevation in the serum calcium. Since this was a relatively large number of patients, this study gave these authors the opportunity to correlate their clinical impression of the severity of the thyrotoxicosis with the level of the serum calcium. They were unable to draw correlations between any of the symptoms of thyrotoxicosis that these patients had and the level of the serum calcium. In another study, Farnsworth and Dobyns[5] found that six out of fifty-two patients with hyperthyroidism had clear elevations in serum calcium concentrations. However, it should be remembered that in all of these studies the total serum calcium may give a spuriously low level because the albumin is also increased. Thus, hypercalcemia is almost certain to be more common when 'corrected' serum calcium or ionized serum calcium concentrations are determined.

The great majority of patients with thyrotoxicosis do not has symptomatic hypercalcemia, and usually the serum calcium is <13 mg/dl (3.25 mmol/l). However, occasional patients with severe hypercalcemia have been reported, and there was one reported case of a patient with a total serum calcium of 19.2 mg/dl (4.8 mmol/l) due to thyrotoxicosis. Since severe hypercalcemia is unusual in thyrotoxicosis, when a patient presents with a serum calcium of >13 mg/dl (3.25 mmol/l), other causes of hypercalcemia should be sought carefully.

There have been a number of reports of primary hyperparathyroidism coexisting with thyrotoxicosis. It appears unlikely that there is a real association between these diseases. Rather, it appears that this apparent association is due to the coexistence of two common conditions. The differential diagnosis of hypercalcemia due to primary hyperparathyroidism from hypercalcemia due to thyrotoxicosis is considered below.

decreased mineralization lag time, suggesting an increase in bone turnover. There is no impairment of mineralization, and no evidence of osteomalacia. However, osteopenia is common in patients with hyperthyroidism, particularly those over the age of 50 years. This has frequently been described as 'osteoporosis', although the bone disease is not really that of osteoporosis. Although there have been few studies of photon-beam absorptiometry in thyrotoxicosis, those studies which have been performed suggest that these patients have decreased bone mineral density,[9] and in one study using total-body neutron activation analysis which measured total body calcium, there was a 9 per cent decrease in the total body calcium content in thyrotoxic patients, with an 8 per cent increase after antithyroid therapy had been administered for an average of 9 months.[10]

Pathophysiology

Thyroid hormones cause significant derangements in the transport of calcium across bone, the kidney, and the gut. Thyroid hormones cause a marked increase in bone turnover. In fact, the change in bone turnover is even greater in hyperthyroidism than it is in primary hyperparathyroidism. Changes in bone histology in patients with thyrotoxicosis were noted long before it was recognized that hypercalcemia also occurred in this disease. The changes do resemble in some ways those seen in patients with osteitis fibrosa cystica, the bone disease of primary hyperparathyroidism.[6] However, there are apparently some subtle differences in bone morphology between the two diseases. In one report, Meunier et al[7] suggested that osteoclastic activity was possibly not as great in hyperthyroidism as it was in primary hyperparathyroidism, and Parfitt[8] found that there were differences in the characteristics of the bone. The irregularly woven bone which is characteristic of osteitis fibrosa cystica was not a feature of thyroid bone disease in Parfitt's studies.

The bone disease of thyrotoxicosis is characterized by an increase in osteoid tissue and a

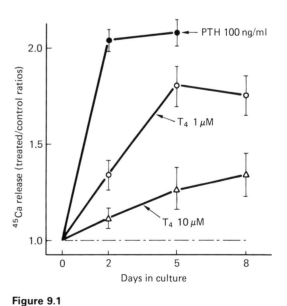

Figure 9.1

Effects of thyroid hormones on bone resorption in organ cultures of fetal rat long bones. (Reproduced with permission, copyright 1976 by the American Society for Clinical Investigation.)[11]

Thus, there is a marked increase in bone turnover in hyperthyroidism. The mechanism is probably related to the direct effects of thyroid hormones on bone. Thyroid hormones have been shown to increase osteoclastic bone resorption directly in vitro, using fetal rat long bones previously labeled with ^{45}Ca (Figure 9.1).[11] The thyroid hormones act fairly slowly. Both thyroxine and tri-iodothyronine exert effects at concentrations equivalent to 10^{-9} M of the free hormone. Resorption of bone was not only slower, but also not as complete as that which is seen when bones are incubated with PTH. Bone resorption stimulated in vitro by thyroxine and tri-iodothyronine was inhibited by cortisol, calcitonin, and phosphate, as well as by propranolol. Other analogs of the thyroid hormones such as reverse T3 and various iodotyrosines were ineffective. However, tri-iodothyroacetic acid, which does have thyroid hormone biological activity in other systems, did exert some bone-resorptive effects. Thus, it is likely that the major mechanism by which osteoclastic bone resorption is stimulated in thyrotoxicosis is by a direct effect of thyroid hormones on the bone-resorption apparatus.

There are other mechanisms by which hyperthyroidism could potentially be associated with an increase in bone resorption. The demonstration that these hormones could inhibit cyclic nucleotide phosphodiesterase in fetal rat calvariae and thereby raise cAMP levels raises the possibility that this could be related to the mechanism of bone resorption.[12] However, structural activity studies using multiple thyroid hormone analogs with varying biological activity showed no relationship between biological activity and cAMP accumulation. Another potential mechanism for increased bone turnover could be via catecholamine mediation. There is abundant evidence that the tissues of patients with hyperthyroidism seem to be more sensitive to the effects of catecholamines than those of euthyroid individuals. In particular, beta-adrenergic receptors seem to be hypersensitive in the presence of thyroid hormone excess.[13] There is no clear relationship between catecholamines and bone resorption, however. On the other hand, the effects of propranolol to inhibit bone resorption stimulated by the thyroid hormones are consistent with this notion.

Another possible mechanism for the increase in bone resorption seen in thyrotoxicosis is production of leukocyte cytokines by immune cells. In Graves' disease it appears likely that immune mechanisms are involved, and these could lead to the production of leukocyte, and particularly lymphocyte, products which stimulate osteoclastic bone resorption. However, leukocyte cultures from patients with Graves' disease have given no indication that this mechanism is responsible, or that there are differences in OAF production between patients with Graves' disease and normals. Moreover, bone resorption and hypercalcemia are also seen in patients with hyperthyroidism which is due to non-autoimmune causes.

The precise mechanism by which bone formation is stimulated remains unknown. When thyroid hormones are added to fetal rat calvarial cultures in vitro and collagen synthesis as a parameter of bone formation is assessed, there is a decrease.[11] An increase would have been expected from the in vivo studies. Similar findings are, of course, found in primary hyperparathyroidism. One explanation for this apparent discrepancy between the effects of thyroid hormones and PTH on bone collagen synthesis in vitro and bone formation in vivo is that local coupling factors that are produced during the resorption process are responsible for the activation and differentiated function of osteoblasts to form new bone, and this overrides the direct effects of the bone-resorbing agent to decrease osteoblast activity.

It is likely that the intestinal absorption of calcium is diminished in patients with hyperthyroidism, from calcium balance studies using measurements of tracer calcium in the blood after an oral load.[14]

There is no gross derangement in vitamin D metabolism in patients with hyperthyroidism. However, there has been one report that 25-hydroxyvitamin D concentrations are decreased in the serum of thyrotoxic patients. Moreover, increased bone turnover could lead to an inhibition of PTH secretion, which could decrease gut absorption of calcium by impairing the formation of 1,25-dihydroxyvitamin D. There is a slight decrease in serum 1,25-dihydroxyvitamin D in patients with hyperthyroidism, possibly related to suppressed PTH and increased serum phosphate.[15]

It has been known for more than 50 years that hypercalciuria is a frequent association of hyperthyroidism.[16] Hypercalciuria is frequent in normocalcemic as well as in hypercalcemic patients. It almost certainly represents the increase in bone resorption which is seen in this condition. It may be due to the compensatory mechanism of the kidneys to impair renal tubular calcium reabsorption, and be due to suppressed secretion of PTH which occurs in patients with increased bone turnover. There is no evidence that renal stone formation is increased in patients with hyperthyroidism. Several explanations are possible. It may be that inhibitors of stone formation are present in the urine of patients with thyrotoxicosis. However, these patients may avoid renal stones because the symptoms of hyperthyroidism bring the patient to medical care and subsequent treatment before renal stones have a chance to occur.

Thus, the overall picture in patients with hypercalcemia due to hyperthyroidism is increased bone turnover with increased osteoclastic bone resorption and new bone formation, decreased renal tubular calcium reabsorption, and probably decreased calcium absorption from the gastrointestinal tract. The most prominent finding is hypercalciuria. This may be the major protective mechanism for preventing more florid and common hypercalcemia.

Clinical manifestations

In the great majority of patients hypercalcemia is asymptomatic. However, the bone disease may be prominent, particularly in older patients. These patients usually present with a radiologic picture of the osteoporosis which occurs in the elderly.[17] If careful radiologic studies are performed on the metacarpal cortex, then thinning is revealed in patients with prolonged hyperthyroidism.[18] Fine striations may be seen in the cortices, which are characteristic of increased states of bone resorption and may even be seen in patients with primary hyperparathyroidism. Subperiosteal resorption of the phalanges is not a characteristic of the bone disease of hyperthyroidism.

It is not clear whether there is a direct correlation between the severity of thyrotoxicosis and the absolute level of the serum calcium. Hypercalcemia has not been reported, to the author's knowledge, in thyroid crisis, although there have not been careful studies of the free thyroid hormone concentrations compared with ionized calcium concentrations in patients with severe thyrotoxicosis.

Serum alkaline phosphatase and urine hydroxyproline are increased in patients with hyperthyroidism.[19] In part, these findings reflect the increase in bone turnover. The increased alkaline phosphatase in the serum may come not only from the skeleton, but also from the gut. Similarly, the source of hydroxyproline found in the urine may be from increased turnover of not only collagen in bone, but also collagen in other sites, such as skin.

Treatment

The major aim of treatment of patients with either hypercalcemia or bone disease in association with hyperthyroidism is to reverse the hyperthyroidism. This is now readily achieved by the use of antithyroid drugs and/or radioactive-iodine therapy. In patients with severe hypercalcemia, it may be necessary to treat the hypercalcemia before it is possible to render the patient euthyroid with antithyroid medications. Under these circumstances several approaches are possible. Propranolol has been shown to reverse thyrotoxic hypercalcemia, although not in every case studied.[20–22] Calcitonin and glucocorticoids in combination could be expected to be very effective in this situation. Although the author is not aware of any reports of the use of bisphosphonates in the treatment of thyrotoxic hypercalcemia, it seems almost certain that they, too, would be effective, since the primary abnormality seems to be an increase in bone resorption. Cortisol, propranolol, calcitonin, and phosphate all block bone resorption stimulated by T_4 in organ culture (Figure 9.2).[11]

Immobilization hypercalcemia

Clinical features

Immobilization leads to profound changes in bone remodeling, which are still mostly unexplained. All patients who are completely immobilized develop hypercalciuria which may last for

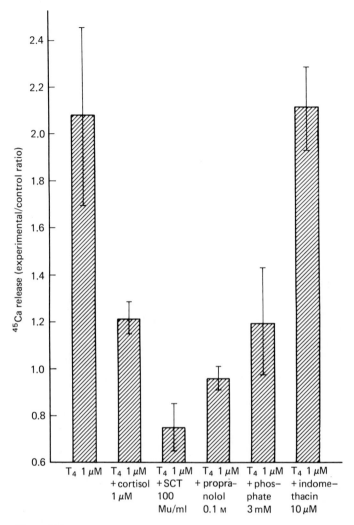

Figure 9.2

Effect of inhibitors on T_4-stimulated bone resorption. The duration of the culture period was 5 days. For each group, one bone was cultured with T_4 and inhibitor and the paired bone was cultured with the appropriate inhibitor alone in control medium. Values are means ± SEM for four pairs of bone cultures. (Reproduced with permission, copyright 1976 by the American Society for Clinical Investigation.)[11]

several months.[23] A small number of these patients, probably less than 10 per cent, develop hypercalcemia which may be quite severe.[24,25] In occasional patients the serum calcium may rise as high as 18 mg/dl (4.5 mmol/l). Hypercalcemia occurs in those immobilized patients who have underlying rates of bone turnover or bone remodeling which are high. Hypercalcemia associated with immobilization occurs most commonly in adolescent boys, particularly those who suffer spinal cord injuries which render them paraplegic or quadriplegic. It has also been described in other circumstances where bone turnover is increased, such as in Paget's disease, and presumably could exacerbate hypercalcemia in conditions such as thyrotoxicosis, primary hyperparathyroidism, or malignancy. However, to the author's knowledge it has not been well-documented in adults with normal underlying rates of bone turnover, and when hypercalcemia occurs in these individuals it is likely that immobilization is an exacerbating factor rather than the underlying cause. Although hypercalcemia occurs most frequently in the completely immobilized adolescent, it can also occur in growing children or adolescents after single-limb fractures.[26]

Hypercalcemia in the immobilized patient may be responsible for characteristic hypercalcemic symptoms such as nausea, vomiting, and confusion, and should be actively treated. Hypercalciuria is also a serious problem, and nephrolithiasis commonly occurs in these patients. Since individuals with spinal cord injuries are also prone to urinary tract infections, these patients frequently develop recurrent renal calculi.

Prolonged immobilization invariably leads to osteopenia, but this is usually asymptomatic. The serum immunoreactive PTH concentration is suppressed.

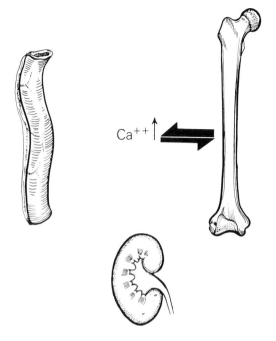

$Ca^{++}\uparrow$

Figure 9.3

Abnormalities in calcium homeostasis in patients with immobilization hypercalcemia, showing increased osteoclastic bone resorption. Little is known of renal tubular calcium reabsorption, but it appears likely that it is not increased, since hypercalciuria is a prominent finding.

Pathophysiology

Hypercalcemia associated with immobilization is due to increased bone resorption relative to bone formation.[27] The mechanism for the increase in osteoclast activity is unknown. It is clear that there is initially an absolute increase in osteoclastic resorption, and probably also bone formation, but then a decline in subsequent new bone formation occurs and the result is osteopenia in all patients, and hypercalcemia in some (Figure 9.3). Thus, immobilization leads to a decrease in the coupling of bone formation to previous bone resorption. The mechanisms by which these disturbances in bone cell function occur are still unknown. Recent work by Lanyon et al[28,29] indicates that mechanical loading influences bone remodeling. These authors suggest that mechanical loading leads to effects on the bone matrix which are transmitted to osteocytes within the matrix. There must be some communication between the osteocytes and cells on or near the bone surface, which leads to changes in osteoclasts and osteoblasts responsible for bone remodeling. Evidence for this concept comes from data obtained by applying graded loads to the transfixed ulnae of turkeys, and studying the effects of these loads on bone remodeling and bone mass.

Immobilization osteopenia is common in all bones relieved of weight-bearing stress. It is seen in individuals placed at complete bed rest, in patients or animals with immobilized limbs associated with nerve section or injury, and in astronauts who are removed from the effects of gravity.

Renal tubular calcium reabsorption is decreased in immobilized patients, presumably reflecting suppressed circulating PTH.[30,31]

Treatment

Vigorous fluid treatment is required to prevent or limit recurrent renal calculi. Hypercalcemia should be treated with agents which inhibit osteoclastic bone resorption. Bisphosphonates, calcitonin and glucocorticoids have all been used successfully. Glucocorticoids are best avoided in osteopenic individuals, particularly those who are prone to infection. Bisphosphonates may be particularly useful for the treatment of the hypercalcemia (see Chapter 7, on the treatment of hypercalcemia).

Sarcoidosis

Sarcoidosis has been known to be associated with hypercalcemia for almost 40 years. The nature of this association was reviewed in a recent editorial by Singer and Adams.[32]

Although hypercalcemia is well known to occur in sarcoidosis, the relative frequency of this association in patients with sarcoidosis has probably been exaggerated. Literature reports of the frequency of hypercalcemia in sarcoidosis have varied from 2 to 60 per cent. The first report of the association described hypercalcemia in eight of eleven patients.[33] Reports in the 1960s suggested that approximately 17 per cent of patients with sarcoid develop hypercalcemia at some stage during the course of the disease,[34,35] although this figure still seems rather high in light of current evidence. When the relationship is considered the other way, that is the frequency of sarcoidosis as a cause of hypercalcemia, then although sarcoid is an important cause to recognize, it is still responsible for less than 5 per cent of all cases of hypercalcemia, probably even in areas of high sarcoid prevalence. Whatever the true frequency, the hypercalcemia is usually of mild degree and is not usually greater than 12 mg/dl.[34] The hypercalcemia may fluctuate, and for this reason may be overlooked in some patients if they are not evaluated on several occasions.[36] However, hypercalciuria occurs at least twice as commonly as hypercalcemia in this disease,[37] suggesting that the abnormality in calcium homeostasis is present much more frequently than is determined by evaluation of the serum calcium alone.

Mechanisms for hypercalcemia

There is strong evidence that calcium absorption from the gastrointestinal tract is increased in patients with sarcoidosis. This has been shown by calcium balance studies[38] and supported by the clinical observations that hypercalcemia can often be relieved by reducing calcium intake in the diet or by the administration of corticosteroids. The increase in gut absorption of calcium was originally ascribed to increased sensitivity to vitamin D. In fact, small doses of vitamin D given to patients with sarcoidosis do result in excessive increases in the serum calcium.[38] Moreover, hypercalcemia in sarcoidosis may be related to exposure to sunlight, and is clearly worse during summer months when skin exposure to ultraviolet light results in the increased synthesis of vitamin D in the skin. However, measurements of vitamin D and 25-hydroxyvitamin D in the serum are normal in patients with sarcoidosis, with or without hypercalcemia.[39] Bell et al[40] were the first to make the observation that there were increased circulating concentrations of 1,25-dihydroxyvitamin D in sarcoidosis, and this is clearly related to the hypercalcemia. This observation has been confirmed by a number of groups.[41–43] Over recent years major interest has focused on the mechanism for this increase in serum 1,25-dihydroxyvitamin D. The best explanation is that the 1,25-dihydroxyvitamin D is synthesized in the sarcoid granuloma tissue. It is clear that 1,25-dihydroxyvitamin D is produced extrarenally, because an anephric individual with sarcoid had high circulating concentrations of 1,25-dihydroxyvitamin D.[44] Recently, it has been found that 1,25-dihydroxyvitamin D can be synthesized both by granuloma tissue and by alveolar macrophages from patients with sarcoidosis in vitro.[45–48]

There are other factors which could be important in the pathophysiology of the hypercalcemia in sarcoid. Osteolytic bone lesions are present in many patients with sarcoidosis. These are usually small punched-out lesions in metacarpal bones and phalanges. Although it is unlikely that the increase in bone resorption associated with these bone lesions is sufficient to cause hypercalcemia, these bone lesions may contribute to the increase in serum calcium. These bone lesions consist mostly of small cysts which are discrete and lattice-like, and which occur predominantly in the phalanges. They are probably related to the sites of sarcoid granulomata, and they may be caused by bone-resorbing cytokines produced by the chronic inflammatory cells associated with the granuloma tissue which can stimulate osteoclast activity such as interleukin-1, tumor necrosis factor, and lymphotoxin (Chapter 3). However, it is likely that there is also a generalized increase in bone resorption in sarcoid, due to the increased circulating 1,25-dihydroxyvitamin D concentrations produced by the granuloma tissue.

There is another mechanism by which hypercalcemia may occur in patients with sarcoidosis. Primary hyperparathyroidism has been described many times in association with this disease.[49,50]

It is likely that the coexistence of these two conditions is coincidental, since primary hyperparathyroidism is a very common disease. Moreover, the serum calcium is likely to be very closely monitored in patients with sarcoidosis, because of the well-known association of this disease with hypercalcemia, and this will lead to early recognition of primary hyperparathyroidism should the two diseases coexist.

The pathophysiology of hypercalcemia in sarcoidosis is shown diagrammatically in Figure 9.4.

Clinical features

Hypercalcemia in sarcoidosis is rarely of severe degree, and usually is not symptomatic. In most patients the serum calcium is less than 12 mg/dl.[34] The hypercalcemia usually occurs in association with increased serum phosphorus concentrations, and together the increase in serum calcium and the increase in serum phosphorus may lead to nephrocalcinosis and impaired renal function.

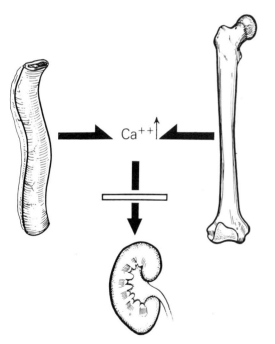

Figure 9.4

Abnormalities in calcium homeostasis in patients with sarcoidosis, showing evidence for increased gut absorption and possibly also increased bone resorption. In many of these patients glomerular filtration is impaired, but this may be secondary to the hypercalcemia and hyperphosphatemia.

Differential diagnosis

In any patient with sarcoidosis and hypercalcemia, it is important to determine whether sarcoid is responsible for the hypercalcemia. Hypercalcemia is rarely the initial presentation in sarcoid. A not uncommon diagnostic dilemma for the physician is to have a patient with known sarcoidosis who develops hypercalcemia. The physician is then faced with considering all of the causes of hypercalcemia which could be operative in this particular patient. If hypercalcemia is not due to sarcoid, then the most likely cause is primary hyperparathyroidism. The best way to diagnose primary hyperparathyroidism is by PTH radioimmunoassay. Patients with primary hyperparathyroidism usually have hypophosphatemia, normal renal function, and a mild hyperchloremic acidosis which, together with the increased plasma immunoreactive PTH concentration, makes this diagnosis easy to make. However, when primary hyperparathyroidism coexists with sarcoidosis, it may occur in a patient with compromised renal function, so parameters such as increased renal phosphate wasting, and hyperchloremic acidosis may not be present. In addition, compromise of renal function may impair clearance of PTH fragments which are recognized by some immunoassays. Since impaired renal function is present in the majority of patients with sarcoidosis and hypercalcemia,[51] the PTH radioimmunoassay will be difficult to interpret, particularly if C-terminal assays are used. These assays measure circulating fragments which are cleared by the kidneys and are likely to accumulate in the presence of renal failure.

The hallmark of hypercalcemia in sarcoidosis is a concurrent increase in the serum calcium and the serum phosphorus, usually in association with nephrocalcinosis and poor renal function in longstanding cases. Serum concentrations of 1,25-dihydroxyvitamin D are also increased. However, serum 1,25-dihydroxyvitamin D may also be increased in patients with primary hyperparathyroidism.

It may be necessary in some patients to perform a steroid suppression test. If hypercalcemia is due to sarcoidosis, 40 mg prednisone given for 10 days will cause a fall in serum calcium. It may take more than 7 days to see the serum calcium decrease to the normal range. However, patients with primary hyperparathyroidism rarely, if ever, respond to corticosteroid therapy with a fall in the serum calcium.

Treatment

Hypercalcemia in some patients with sarcoidosis can be reduced by limiting the calcium intake in the diet to approximately 400–500 mg elemental calcium per day. This may require the use of distilled water. Patients with sarcoidosis should also have limited exposure to sunlight, because of the danger of stimulating vitamin D synthesis following exposure of skin cells to ultraviolet light. If dietary restriction and avoidance of sunlight exposure fail to work, then patients can be treated with corticosteroid therapy.[42,52] The precise mechanism of action of corticosteroid therapy in sarcoidosis is not clear. It is likely that corticosteroids have a direct effect on the

gastrointestinal tract to limit calcium absorption, but in addition they probably interfere with the metabolism of vitamin D in the granuloma tissue. Serum levels of 1,25-dihydroxyvitamin D are reduced in sarcoidosis by glucocorticoid treatment, and they may fall before the decrease in the serum calcium.

Chloroquine has also been used successfully in the treatment of sarcoidosis.[53] Chloroquine reduces serum 1,25-dihydroxyvitamin D levels and urine calcium levels. In a study in two patients reported by O'Leary et al,[54] chloroquine was used in a dose of 500 mg/day (6–6.5 mg/kg body weight per day) for a total cumulative dose of 240 g per patient. One risk associated with this therapy is the potential for retinal damage, but this was avoided by ophthalmologic assessments (fundoscopy and visual field examinations) every 6 months. If necessary, the daily dose can be reduced. There is now widespread experience with this drug in rheumatoid arthritis, so a safe regimen to avoid retinal complications can be chosen.[55,56] Chloroquine can be used in those patients in whom glucocorticoids are unsatisfactory because of the need for longterm therapy or because the dose required is so high that Cushing's syndrome will be produced, with its attendant complications.

Hypercalcemia associated with tuberculosis and other granulomatous diseases

Hypercalcemia occurs occasionally in patients with granulomatous diseases other than sarcoidosis. These include coccidioidomycosis, histoplasmosis, berylliosis, silicone-induced granulomas, and tuberculosis. Since all of these conditions share in common with sarcoidosis the presence of granulomas in affected tissues, it seems likely that the mechanism for hypercalcemia may be similar to that which is likely responsible for hypercalcemia in sarcoidosis, namely increased extrarenal production of 1,25-dihydroxyvitamin D in the granulomatous tissue (see section on sarcoidosis). There have now been several patients with tuberculosis, who have been found to have increased serum 1,25-dihydroxyvitamin D concentrations which co-incided with hypercalcemia.[57–59] In the patient of Felsenfeld et al,[58] hypercalcemia occurred in the presence of active and widely disseminated tuberculosis. This patient was on maintenance dialysis and presumably had no normal renal function. Despite this, the serum 1,25-dihydroxyvitamin D concentration was elevated, and the authors postulated that extrarenal pro-duction of 1,25-dihydroxyvitamin D was respon-sible. The patient described by Gkonos et al[57] was another patient with end-stage renal disease who had high serum 1,25-dihydroxyvitamin D concen-trations.

Thus, this situation does seem similar to the circumstances which are usually present in patients with hypercalcemia due to sarcoidosis, and so the mechanism for the increase in 1,25-dihydroxyvitamin D concentrations in these patients is likely to be the same. In both of these conditions serum 1,25-dihydroxyvitamin D con-centrations may be increased because vitamin D, metabolism, and particularly the conversion of 25-hydroxyvitamin D to 1,25-dihydroxyvitamin D may be less tightly controlled than in normals. In normals the serum 1,25-dihydroxyvitamin D con-centration is closely regulated despite relatively wide variations in serum 25-hydroxyvitamin D concentrations. However, patients with sarcoid-osis show quite marked elevations in serum 1,25-dihydroxyvitamin D in response to increases in the 25-hydroxyvitamin D concentration. This is likely to be the mechanism by which these patients are so very sensitive to changes in vitamin D intake or exposure to sunlight.

Other granulomatous diseases have been as-sociated with hypercalcemia.[60–62] The mechan-ism is probably similar, and as in sarcoidosis and possibly tuberculosis, may be related to 1,25-dihydroxyvitamin D production by granulo-matous tissue (Table 9.1).

Hypercalcemia has occasionally been noted in the acquired immunodeficiency syndrome (AIDS). In two cases reported by Zaloga et al,[63] it was associated with disseminated cytomegalo-virus infection. In these two patients the authors attempted to exclude known causes of hypercal-cemia, such as primary hyperparathyroidism, immobilization, or associated malignancy. Both of the patients were young male adults. Both responded very well to calcitonin therapy,

Table 9.1 Hypercalcemia and granulomatous bone disease.

Diseases associated with bone marrow granulomas	Reports of hypercalcemia
Infections	
Histoplasmosis	+
Blastomycosis	
Coccidioidomycosis	+
Candidiasis	+
Tuberculosis	+
Mononucleosis	
Brucellosis	
Cytomegalovirus	+
Rocky Mountain spotted fever	
Tularemia	
Cryptococcosis	
Saccharomyces	
Malignancy	
Hodgkin's disease	+
Other lymphomas	+
Large-cell lung cancer	+
Mycosis fungoides	
Breast cancer	+
Acute lymphocytic leukemia	+
Drug	
Procainamide	
Ibuprofen	
Phenylbutazone	
Dilantin	
Indomethacin	
Sarcoid	+
Allergic/autoimmune	
Eosinophilic granuloma	+
Farmer's lung	
Felty's syndrome	
SLE	
Primary biliary cirrhosis	
Rheumatoid arthritis	
Silicone	+
Berylliosis	+

although neither survived for more than 2 weeks. It seems that hypercalcemia in AIDS is unusual. When it does occur, it is likely to have a multifac-torial origin. For example, it could be related to chronic granulomatous disease, to adrenal insuf-ficiency or to development of an associated neoplasm. On the other hand, AIDS is becoming such a common disease that it appears likely that its association with hypercalcemia, although unusual, will be seen more commonly.

Vitamin D intoxication

Although vitamin D intoxication has become more frequent over recent years with the advent of the new and more potent vitamin D metabolites for use in modern medical practice,[64] these new compounds have a shorter half-life than their predecessors, and toxicity is of shorter duration once the compound is discontinued. Vitamin D intoxication is seen in patients treated with vitamin D therapy for chronic renal failure and for hypocalcemic conditions such as hypoparathyroidism, rickets, and osteomalacia. It was probably more common in the past, when it was often prescribed for conditions unrelated to calcium or vitamin D metabolism, such as tuberculosis or rheumatoid arthritis. Some patients who take excessive vitamin D are food faddists, but it is sometimes prescribed for disorders such as psoriasis, arthritis and chilblains. In south Texas, many of the patients taking vitamin D have had it prescribed by physicians in Mexico. The recent fad for calcium and vitamin D as potential treatments for osteoporosis and the prevention of age-related bone loss may also lead to some sporadic cases of vitamin D intoxication due to nonprescription vitamin D-containing compounds. Patients vary considerably in their tolerance to large doses of vitamin D, and in how much vitamin D is required to produce intoxication. Most adults probably require more than 50 000 units of vitamin D per day,[65] although vitamin D therapy is notoriously difficult to control. Whether hypercalcemia occurs will depend to some extent on the calcium intake in the diet. When the parent vitamin D compound (cholecalciferol) is used, the toxic effects accumulate and the effects of overdose may take several months to dissipate.

The effects of vitamin D on calcium homeostasis are complex and profound (Chapter 3). Vitamin D and all of its metabolites lead to an increase in serum concentrations of both calcium and phosphate. This is due in part to increased absorption of calcium and phosphate from the gastrointestinal tract, but is also related to increased resorption of bone, particularly when pharmacologic amounts of vitamin D are being administered.[66–68] The relative importance of these effects remains unclear, although most workers ascribe the predominant effect of vitamin D on calcium homeostasis to its effects on the gut. The effects of vitamin D on the gut have been described in detail in a number of recent reviews (Chapter 3). It is now clear that the effects of vitamin D on bone are much more complex than was originally thought. Vitamin D may have effects to stimulate the differentiation of osteoclasts, but it may also have secondary effects to regulate the production of cytokines in the bone micro-environment, which are also involved in the regulation of bone turnover.

Clinical features

The clinical features of vitamin D intoxication have been recognized for more than 50 years, and have been described many times.[64,69–74]

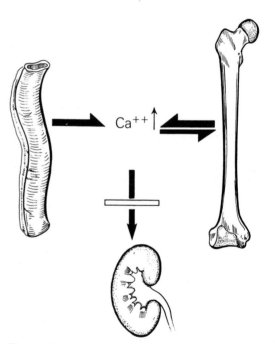

Figure 9.5

Abnormalities in calcium homeostasis seen in patients who are intoxicated with vitamin D or its active metabolites. These patients have increased bone resorption and increased gut absorption of calcium. The pattern of abnormalities in calcium transport are probably identical to those seen in sarcoidosis.

Patients with vitamin D intoxication develop hypercalcemia and hyperphosphatemia, as well as an impairment of renal function. The pathophysiology of hypercalcemia in vitamin D intoxication is shown schematically in Figure 9.5. They frequently have nephrocalcinosis or metastatic calcification in the kidneys as well as in the walls of medium-sized blood vessels, and this accounts for the impairment in glomerular filtration. The renal failure is usually reversible. The soft-tissue calcification in the kidneys may not be visible radiographically unless the patients have been treated with excess vitamin D for prolonged periods, but it still may be responsible for impairment of renal function. These patients frequently develop band keratopathy and conjunctival calcification.

A few years ago it was noticed that patients with chronic renal failure treated with the most active vitamin D metabolites had a deterioration in renal function,[75] raising the question of whether this effect was due directly to the vitamin D metabolite. However, further studies indicated that the impairment in renal function occurred as a consequence of hypercalcemia, rather than the specific vitamin D metabolite which was being administered.[76,77]

If vitamin D intoxication is chronic, then patients may present with painful joints and muscle weakness. The joints frequently include those involved in walking, such as hip, ankle, and knee. The skeletal muscles may be not only weak but also painful, and tendon insertions such as the Achilles and quadriceps insertions may also be tender to touch. The weakness may be sufficiently profound to impair rising from a chair or sitting from a lying position. Radiography may show increased bone formation and osteosclerosis, with periosteal calcification around the long bones. The blood vessels may show evidence of calcification, and there may be ligamentous calcification apparent at sites such as the plantar fascia and iliolumbar ligaments.

There is a well-described animal model which mimics chronic vitamin D intoxication. This is Manchester-wasting disease[78] or enteque seco described in Argentina[79] which occurs in cattle which ingest plants such as *Solanum malacoxylon*.[80] These plants produce a glycoside of 1,25-dihydroxyvitamin D, and chronic ingestion causes hypercalcemia and painful locomotion, characterized by restricted movements due to calcified ligaments and tendons. Patients with vitamin D intoxication and painful calcific tendonitis and periostitis have been reported by Harris[73] and by Davies et al.[69]

Measurements of serum vitamin D metabolites show striking abnormalities. The serum 25-hydroxyvitamin D concentration may be increased by a factor of twenty. There is also a marked, although less dramatic, increase in 1,25-dihydroxyvitamin D concentration, possibly due in part to the marked increase in substrate 25-hydroxyvitamin D.[72] However, it is also possible that there is abnormal regulation of the 1-hydroxylase enzyme system responsible for converting 25-hydroxyvitamin D to 1,25-dihydroxyvitamin D, as there is in sarcoid. In both situations, serum 1,25-dihydroxyvitamin D is readily suppressed by corticosteroid therapy.

Quantitative bone histomorphometry may show striking findings. In osteosclerotic cases there may be an increase in trabecular bone mass, with a marked increase in total osteoid surface, relative osteoid volume, osteoclast numbers and total resorptive surfaces.[69] However, double tetracycline labeling shows an impairment in mineralization, and osteoblasts are spindle-shaped and inconspicuous. These bone abnormalities are difficult to explain. They resemble somewhat the changes in bone morphology seen in vitamin D deficiency. Changes such as this may also occur in patients with aluminum intoxication associated with chronic parenteral nutrition or chronic renal disease.[81,82] The osteoblasts are capable of synthesizing osteoid, but not of causing adequate mineralization. Their spindle-shaped appearance suggests that they are abnormal, but the mechanism for this abnormality remains unclear.

Treatment

Corticosteroid therapy and withdrawal from vitamin D are necessary for the adequate treatment of this condition. Serum calcium can also be lowered rapidly in the short term by administration of calcitonin. Corticosteroids may act by inhibiting calcium absorption from the gut directly,[83] but they may also impair the conversion of vitamin D to its active metabolites.[69] Bone resorption induced by vitamin D metabolites will also be inhibited when corticosteroids are used in pharmacologic doses.[84,85] The duration of hypercalcemia is likely to be much shorter in patients treated with the newer metabolites which have a much shorter half-life than the parent compounds.

Vitamin A intoxication

Vitamin A is a fat-soluble vitamin found in egg yolk, butter fat, and fish oils. Beta-carotene is the major dietary source of vitamin A in plants. Vitamin A is essential for normal bone growth and maturation, but also affects many tissues, influencing cell replication and differentiation. Vitamin A is essential for vision, and its deficiency causes keratomalacia and night-blindness. Deficiency of vitamin A also causes impairment of growth, and excess vitamin A increases bone fragility and may cause hypercalcemia. The

effects of vitamin A and related compounds on bone appear to be unique, and although it is a rare cause of hypercalcemia, it is worth considering in some detail.

Vitamin A is a generic term used for compounds other than carotenoids that exhibit the biological activity of retinol (Figure 9.6). Dietary sources of vitamin A such as β-carotene are converted in the intestinal mucosa to vitamin A, incorporated into chylomicrons, and transported via the lymph to the liver. The liver is the main storage site for vitamin A. Vitamin A is released from the liver with a binding protein called the retinol binding protein, which carries it through the plasma in a large complex with another plasma protein called transthyretin (also called prealbumin).[86] Target tissues take up this complex by recognizing the retinol binding protein. The exact mechanism of action of vitamin A within the cell is still unknown.

The retinoid molecule has many analogs. Analogs of retinoic acid are now widely used in clinical trials for chemoprevention in cancer,

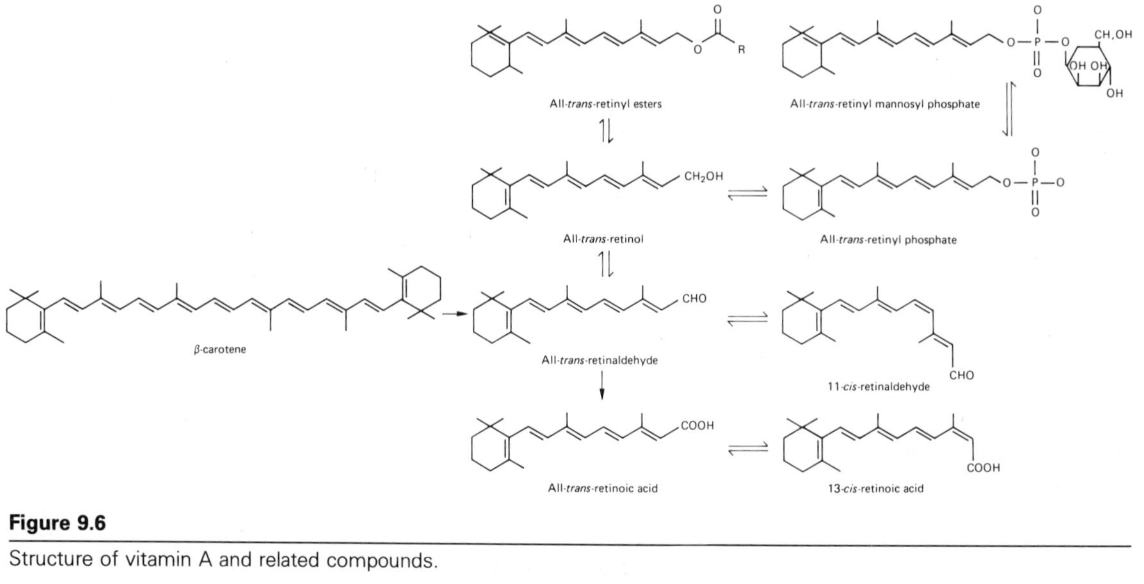

Figure 9.6

Structure of vitamin A and related compounds.

particularly in breast cancer, mycosis fungoides and cancers which are preceded by a premalignant phase (for example, lung and colon). These compounds also have potential efficacy in the treatment of chronic inflammatory diseases such as rheumatoid arthritis, and in dermatologic conditions because of their enhancing effects on cell differentiation and maturation. It is possible that some of their effects are mediated through growth factors such as transforming growth factor β.

One of the active metabolites of vitamin A formed in vivo is retinoic acid. Retinoic acid synthesis occurs in many organs including the liver, although it does not appear to be stored there.[87] Retinoic acid does not seem to have all of the biological effects of vitamin A, since it does not preserve normal vision in vitamin A-deficiency states. Retinoic acid is transported in the circulation bound to albumin. Retinoic acid has a specific cytosolic protein called cellular retinoic acid binding protein.

Recently, the receptors for retinoic acid have been identified and characterized. These receptors belong to a superfamily of receptors which include the receptors of the steroid hormones, thyroid hormones and 1,25-dihydroxyvitamin D.[88] There is considerable homology of all of the proteins in this receptor family. Each of the receptors has parallel domains including steroid–hormone binding domains, DNA binding domains and transcriptional domains. The retinoic acid receptor has now been shown to comprise three isoforms which are 94–97 per cent homologous. These receptors have been referred to as alpha, beta and gamma. The gamma isoform appears to be expressed in the skin and limb buds of developing rodent embryos.

The effects of vitamin A and its analogs on the skeleton have been known for more than 50 years.[89] In the 1920s, vitamin A deficiency was shown to cause decreased endochondral cartilage growth but not to interfere with continued appositional growth. Bone remodeling was decreased. In contrast, vitamin A excess was associated with increased endochondral maturation, increased periosteal osteoclastic resorption and increased susceptibility to fractures. A number of case reports showed that vitamin A intoxication can lead to hypercalcemia.[90]

Vitamin A intoxication is a rare cause of hypercalcemia. Patients who ingest vitamin A excessively are usually food faddists who take megavitamin preparations, or patients who have inadvertently been treated with vitamin A in excessive amounts. Acute vitamin A intoxication was seen frequently in the Arctic explorers who ate the livers of seals and polar bears, which are rich in vitamin A.[91] The subsequent syndrome, which is manifested within a few hours, consists of severe headache, vomiting, extreme lassitude and lethargy, irritability, and peeling of the skin, and is referred to as 'polar bear liver'.

The effects of vitamin A on bone resorption have been known for many years. The Strangeways group at Cambridge University showed that vitamin A stimulates osteoclastic bone resorption in organ culture in association with increased production of cathepsin D.[92–96] This was confirmed by Raisz.[97] Barnicot[98] showed that vitamin A stimulated osteoclastic bone resorption in vivo. Recently, Oreffo et al[99] have shown that vitamin A and its analogues act directly on osteoclasts to stimulate accumulation of tartrate-resistant acid phosphatase and cause changes in microtubule organization within isolated osteoclasts. These effects were seen in cell populations which were essentially homogenous, suggesting that vitamin A is the only known and well-characterized factor which has a direct stimulatory effect on the osteoclast in vitro.

Retinoids also have effects on osteoblastic cells. These factors enhance differentiation of osteosarcoma cells with the osteoblast phenotype, and may increase their responsiveness to other factors such as tumor necrosis factor.[100,101] Similar responses have been seen with isolated rodent calvarial bone cells. Osteoblastic cell differentiation was shown by enhanced expression of alkaline phosphatase.

It has been suggested that vitamin A may have a direct effect on the parathyroid glands, and some of the effects may be related to a secondary hyperparathyroidism. However, where it has been measured, patients with vitamin A intoxication have neither hyperplastic parathyroid glands nor high circulating concentrations of PTH.[90] Moreover, the hypercalcemia of vitamin A intoxication is usually reversed by corticosteroid therapy, even in the presence of continued ingestion of vitamin A. Thus, it appears unlikely that PTH plays a major role in either the bone lesions or the hypercalcemia which occur with vitamin A intoxication. The clinical features of vitamin A excess are readily reversible when vitamin A in the diet is reduced.

The usual consequences of the excessive intake of vitamin A include vague skeletal pains, lethargy, and weight loss with specific skeletal abnormalities.[90] These patients may develop skeletal changes which include increased osteoclastic bone resorption. The radiological changes in the skeleton are also characterized by calcification of ligaments, pronounced osteophyte formation, particularly in the thoracic spine, and osteopenia (Figure 9.7).

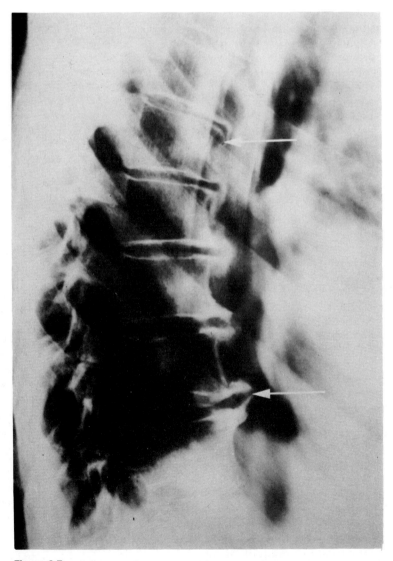

Figure 9.7

Radiologic changes seen in the vertebrae in patients with vitamin A intoxication. Notice the osteophyte formation (arrows).

Idiopathic hypercalcemia of infancy

The idiopathic hypercalcemia of infancy is the term which has been given to hypercalcemia occurring within the first year of life. It clearly represents a grab-bag of conditions including hypercalcemia due to malignancy, primary hyperparathyroidism, familial hypocalciuric hypercalcemia, and the Williams syndrome. Here the hypercalcemia of infancy associated with excess vitamin D supplementation of cow's milk and the William's syndrome are discussed. Idiopathic hypercalcemia of infancy was first reported in the early 1950s in the UK as a minor epidemic which occurred in babies who were fed with cow's milk which had been fortified with vitamin D (reviewed by Nordin[65]). The syndrome was quite a serious one, and it was associated with central nervous system symptoms and signs of hypercalcemia, including lethargy, irritation, hypotonicity, and coma in severe cases. Renal failure was common and the syndrome often had a fatal outcome once this occurred. It was investigated fairly thoroughly, and it became apparent that the clinical features were similar to those which were seen with vitamin D intoxication. In 1957 the amount of vitamin D in fortified baby foods in the UK was reduced, and since then infantile hypercalcemia due to vitamin D intoxication has essentially disappeared. Very few cases were reported from other parts of the world.

The Williams syndrome

The Williams syndrome is an unusual symptom complex which has a number of pathognomonic clinical features, including a characteristic elfin facies, cardiovascular defects, and mild mental retardation (Figure 9.8). A small number of these children develop hypercalcemia during the first year of life. Most, if not all, of the children have been shown to have a defect in vitamin D metabolism which persists.[102] This condition may be confused with other causes of hypercalcemia which occur during the first year of life (see above).

History

Severe infantile hypercalcemia occurring in association with somatic features including growth retardation, mental retardation, and a characteristic facies was first noted in 1951 by Fanconi.[103] In 1961 Williams et al,[104] in New Zealand, reported the presence of supravalvular aortic stenosis associated with a characteristic facial appearance and mental retardation in three children. In a report in the *Lancet*, Black and Bonham Carter[105]

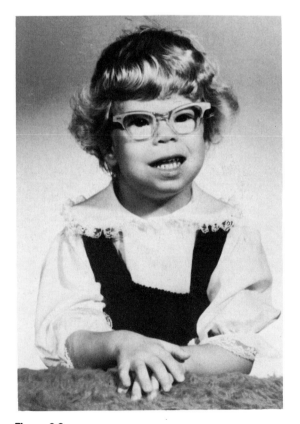

Figure 9.8

Characteristic appearance of a patient with Williams syndrome, showing the elfin facies. The condition is frequently referred to as the 'pixie syndrome'. Hypercalcemia is relatively rare in these patients.

made the connection between the facies seen in this form of supravalvular aortic stenosis and infantile hypercalcemia. The observation made by Black was confirmed the following year in a case report by Garcia et al.[106] For reasons which are not entirely clear, in the USA the syndrome has been known since the early 1960s as the Williams syndrome.

Clinical features

The most striking clinical features are the mental retardation, short stature, and characteristic elfin facies. Children with this syndrome are usually mentally retarded, with an average IQ in the mid-50s, with a range of 40–80. However, they have a characteristic personality. The children are very relaxed and uninhibited. They talk freely to strangers without embarrassment. In fact, they tend to be so outgoing socially that their mild mental retardation may be overlooked and they may be taken for being older than their years. Their personality has been described as a 'cocktail party' personality. In association with mild mental retardation, they often have mild neurologic dysfunction with poor co-ordination.

Ninety per cent of children with this syndrome are small. They are usually around the third percentile and are about 75 per cent of expected height. About two-thirds of the children with this syndrome have mild microcephaly. However, their maturational age equals their chronological age.

Congenital heart disease

About one-third of the children with this syndrome have supravalvular aortic stenosis. However, a great many other cardiovascular abnormalities have been described, some of which are intracardiac and some of which are extracardiac. Some of the children have pulmonary stenosis, some ventricular septal defects, some atrial septal defects, and some valvular aortic stenosis. A small percentage have aortic hypoplasia, and some have peripheral pulmonic stenosis or peripheral arterial stenosis.

Facies

Children with this syndrome have a completely characteristic appearance. Pediatricians and cardiologists who have seen this condition will diagnose a child with Williams syndrome in a crowded room. The facial appearance is often described as the elfin facies, and the syndrome is sometimes called 'the pixie syndrome'. The children are often fair and have a depressed nasal bridge leading to an upturned nose (Figure 9.8). In addition they tend to have a small chin. Other features include medial eyebrow flare, short palpebral fissures, closely set eyes, ocular hypoteleorism, epicanthal folds, periorbital fullness, strabismus (which may need later correction), blue eyes, stellate pattern in the iris, anteverted nares, long philtrum, prominent lips with an open mouth, malar hypoplasia, and sometimes absence of the second lower biscuspid teeth.

Hyperacusis

Hyperacusis is also a common finding in these children. The explanation is not clear. Parents may have noticed that their children are aware of sounds that they themselves cannot hear, and are very aware of ordinary sounds such as a light switch being turned on, an aircraft or helicopter flying overhead, or traffic noises.

There is a miscellaneous group of other somatic abnormalities which may be seen in these children. They may have hallux valgus, hypoplastic nails, and clinodactyly involving the fifth fingers. In addition they are prone to inguinal and umbilical hernias.

The somatic abnormalities which are seen in these children can be confused either with pseudohypoparathyroidism or Noonan's syndrome. Like people with pseudohypoparathyroidism, they may have growth retardation, mental retardation, and occasional abnormalities in calcium metabolism, although of course the abnormalities are opposite to those seen with pseudohypoparathyroidism. Noonan's syndrome is characterized by severe mental retardation, webbed neck, short stature, and severe congenital heart disease. The mental retardation is invariably much more severe than that seen in the Williams syndrome.

Hypercalcemia in Williams syndrome

It is quite clear now that hypercalcemia is a rare event in the child with Williams syndrome. When it does occur, it arises in the first year of life. It is almost certainly due to an increase in the absorption of calcium from the gut, since fecal calcium is decreased and hypercalcemia responds to glucocorticoid therapy. For a number of years it has been suspected that the syndrome is related to an abnormality in vitamin D metabolism. Children with this disorder clearly have increased sensitivity to vitamin D when it is administered orally, and it has been noticed that serum antirachitic activity measured by bioassay is increased. They also have increased circulating concentrations of 25-hydroxyvitamin D after vitamin D therapy. Forbes et al[107] noted that they responded abnormally to intravenous calcium infusion, and Culler et al[108] presented data suggesting these patients have impaired calcitonin secretion. Recently, Taylor et al[102] gave pharmacologic doses of vitamin D_2 for 4 days to children with Williams syndrome, as well as to their siblings and to normal children. They found that the children with Williams syndrome showed an abnormal increase in 25-hydroxyvitamin D in response to vitamin D therapy. The conclusions of this study were that the circulating concentration of 25-hydroxyvitamin D is less tightly regulated in children with Williams syndrome than it is in normal children. Whether this is the full explanation of the abnormality responsible for the development of hypercalcemia in children with Williams syndrome has still to be determined.

The clinical features of this syndrome are summarized in Table 9.2.

Treatment

The hypercalcemia of infancy is usually a self-limiting condition. Some patients subsequently develop renal failure and die. It is therefore important to recognize it early and to treat it appropriately. For any child who is found to have hypercalcemia with the characteristic somatic abnormalities, dietary calcium should be reduced and the child should be treated with corticosteroid therapy. The major differential diagnosis is to exclude other causes of hypercalcemia occurring in this age group. Genetic counseling may be useful. There is no evidence that this condition is inherited, and it has not been reported to occur in families. There is no reason why parents should not have further children if they so desire.

Table 9.2   Clinical features of Williams syndrome.

Mild mental retardation

Elfin (pixie) facies

Short stature

Congenital heart disease (often supravalvular aortic stenosis)

Hyperacusis

Hypercalcemia (uncommon, occurs in first year)

Thiazide diuretics and hypercalcemia

Clinical features

For almost 20 years thiazide diuretics have been known to cause hypercalcemia.[109–111] This relationship between thiazides and the provocation of hypercalcemia has been most obvious in patients who have underlying disorders of calcium homeostasis—in particular, primary hyperparathyroidism.[109–112] One report indicated that twenty of ninety-five patients found to be hypercalcemic were taking thiazides.[112] Of these, fourteen had persistent hypercalcemia after thiazides were discontinued, and all of these were found eventually to have primary hyperparathyroidism confirmed by surgery. Thiazides can also induce hypercalcemia in patients who are otherwise normal,[113,114] although this complication of thiazide therapy in the normal person is fairly unusual, and the increase in total corrected serum calcium is very slight, usually less than 11 mg/dl (3.75 mmol/l). Because thiazides are so likely to produce an increase in the serum calcium in patients with mild primary hyperparathyroidism, the administration of thiazides has been suggested as a provocative test to diagnose this disorder when it is subtle.[115] However, this maneuver has never received wide acceptance, because of its lack of specificity.

Pathophysiology

The mechanism for the increase in serum calcium in patients taking thiazide diuretics remains unclear. In part, it may be due to some degree of contraction of extracellular fluid volume and hemoconcentration, although the pathogenesis of hypercalcemia is clearly more complicated than this. The major mechanism may be due to the effects of thiazides on the renal tubules. Thiazides decrease urine calcium excretion by enhancing renal tubular calcium reabsorption, and this is the rationale for their use in patients with recurrent calcium-containing renal calculi. Both the proximal tubules and the distal tubules have been implicated. In patients taking thiazides chronically, contraction of extracellular fluid volume will lead to enhanced sodium reabsorption in the proximal convoluted tubules, and this will be accompanied by increased proximal tubule calcium reabsorption. This may be part of the reason why serum calcium increases. However, it also appears likely that there is enhanced distal tubular calcium reabsorption. Costanzo and Windhager[116] showed that thiazides, acting from the lumen, enhance distal tubular calcium reabsorption. Parfitt[110] performed a study in seven patients with hypoparathyroidism treated for 1 week with chlorothiazide (200 mg/day) who were also taking vitamin D. From his results he concluded that: (a) the effects of thiazides to decrease renal calcium clearance required the presence of PTH; (b) the effects of PTH and vitamin D metabolites on bone and kidney were enhanced by thiazides; and (c) thiazides have a significant effect (in conjunction with

systemic hormones) on the skeleton. There is some additional fragmentary evidence that thiazides may also have effects on the skeleton. In a study in anephric patients treated with thiazides, Koppel et al[117] noted an increase in the serum calcium. Obviously, in these patients the kidney was playing no role, and altered absorption from the gut was ruled out by studying the patients in a fasting state. Thus, the skeleton was implicated by exclusion.

This effect of diuretics to increase the serum calcium appears to be unique to this class of benzothiadazine (thiazide) diuretics. It has not been noted with other diuretics such as the powerful loop diuretics. Loop diuretics promote calcium excretion, although in the severely dehydrated patient they may further deplete extracellular fluid volume and increase proximal tubule calcium reabsorption indirectly (Chapter 7).

Management

Since many patients who present with hypercalcemia are taking thiazide diuretics, it is frequently a diagnostic dilemma to decide whether the hypercalcemia is due to the thiazides or due to underlying primary hyperparathyroidism or some other disease. The simplest maneuver to resolve this situation is to discontinue thiazides and then to monitor the serum calcium. If the serum calcium is still increased 1 month later, then another cause for hypercalcemia should be sought diligently. As a working guide, the increase in corrected serum calcium in the normal individual due to thiazides should not be greater than 11 mg/dl (3.75 mmol/l). If the serum calcium is greater than 11 mg/dl (3.75 mmol/l) then thiazides should be discontinued and a search for an underlying disease causing hypercalcemia should be undertaken.

The milk-alkali syndrome

Hypercalcemia sometimes occurs in patients taking nonabsorbable antacids such as calcium carbonate in excessive amounts. Historically, this type of hypercalcemia has been referred to as the milk-alkali syndrome. Probably the first description was that of Hardt and Rivers,[118] who found that patients taking a special treatment known as the Sippy regimen for peptic ulcers[119] developed symptoms including altered mental status, irritability, nausea, weakness, and myalgias. These symptoms progressed to prostration, stupor, and eventually coma, but were reversible within 24–48 h of discontinuing the therapy. Cope verified these findings in 1936,[120] and ascribed the clinical features to hypercalcemia.

The milk-alkali syndrome has been thoroughly reviewed recently.[121]

Clinical syndrome

The milk-alkali syndrome is characterized by hypercalcemia, hyperphosphatemia, and alkalosis. Most patients with this syndrome have impaired renal function. The hypercalcemia and impairment of renal function are usually reversed fairly rapidly by discontinuing the antacids. The syndrome has become fairly unusual over the past 20 years, since the fashions in antacid therapy have changed and calcium carbonate is no longer as popular as an antacid. However, over the past few years, with renewed emphasis on oral calcium preparations for the treatment and prevention of age-related bone loss, and the use of large amounts of elemental calcium to avoid peptic ulcer disease and osteoporosis associated with glucocorticoid immunosuppressive therapy following transplantation, the syndrome may have become more prevalent. It can occur acutely when large amounts of antacids and alkalis are ingested, or it can be present on a more chronic basis. The chronic syndrome was well described by Burnett et al,[122] who made a point of the frequency of soft-tissue calcium deposition in the kidneys and nephrocalcinosis. It has been described a number of times since.[123–131]

Pathophysiology

The mechanism for hypercalcemia in the milk-alkali syndrome is complex and still not entirely clear. Ingestion of large amounts of calcium

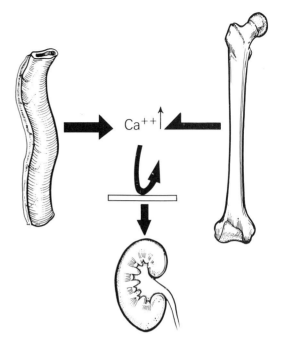

Figure 9.9

Pattern of abnormalities in calcium homeostasis shown in patients with the milk-alkali syndrome. Note that these patients frequently have increased renal tubular calcium reabsorption, probably related to alkalosis, but also mediated in part by the effects of drugs and volume depletion on renal tubular calcium reabsorption. Many of these patients have decreased glomerular filtration as an additional factor to impair renal calcium clearance.

carbonate, certainly greater than 4 g elemental calcium per day, increases the total calcium absorbed from the gastrointestinal tract. However, this alone is unlikely to lead to hypercalcemia unless there is an impairment in renal handling of calcium (Figure 9.9). This impairment in renal handling of calcium could occur in a number of ways. First, glomerular filtration may be impaired in patients with hypercalcemia. Secondly, these patients may become alkalotic, which may in itself impair renal tubular calcium reabsorption.[132–134] Moreover, these patients are frequently treated with drugs which can enhance metabolic alkalosis, such as thiazide diuretics or glucocorticoids, particularly if they are transplant patients. Alkalosis may also be maintained by suppressed circulating PTH concentrations. Patients who are hypercalcemic from causes other than primary hyperparathyroidism can be expected to have decreased circulating PTH concentrations. This will lead to enhanced proximal tubular reabsorption of bicarbonate and loss of sodium chloride. In addition, hypercalcemia itself impairs the renal concentrating ability, which can also lead to volume depletion and increased proximal tubular reabsorption of bicarbonate and alkalosis.

Recently, in a study of almost 300 heart and heart–lung transplant recipients at Stanford University Medical Center, Kapsner et al[135] found that sixty-five patients developed significant hypercalcemia after transplantation. About half of these were alkalotic at the time they developed the hypercalcemia, and approximately one-half had an impairment in renal function. These patients were receiving between 4 and 10 g elemental calcium per day. In most patients the hypercalcemic syndrome was reversed by discontinuing the calcium carbonate therapy, although in several intravenous hydration and forced diuresis were tried because the syndrome was severe.

Treatment of these patients consists of discontinuing the ingestion of calcium and alkali. In most patients the serum calcium rapidly returns to normal and renal function improves. Occasional patients who have had longstanding nephrocalcinosis may have some permanent loss of renal function. In patients with more severe hypercalcemia it may be necessary to lower the serum calcium with more vigorous methods, such as the administration of intravenous saline and the use of calcium-lowering agents such as calcitonin. If hypercalcemia persists, then an underlying hypercalcemic disorder such as primary hyperparathyroidism should be sought.

Familial hypocalciuric hypercalcemia (FHH)

The syndrome of familial hypocalciuric hypercalcemia (also called familial benign hypercalcemia) was first described clearly by Foley et al[136] at the Mayo Clinic in 1972. They noted a benign form of hypercalcemia which occurred in families. The index case was a 5-year-old boy who had ten maternal relatives, including a 76-year-old grandmother, with asymptomatic hypercalcemia. Following these observations, Marx et al[137–140] described the features of this syndrome in considerable detail. They found their patients in relatively large numbers at the NIH, a large referral center where patients were frequently sent following previous failed parathyroid gland surgery. The frequency of familial hypocalciuric hypercalcemia in this particular group (about 10 per cent) is relatively high. They estimate the overall prevalence of FHH to be about the same as that of MEN I.[139–140] The Mayo Clinic experience of this condition was reviewed by Law and Heath.[141]

Clinical features

The clinical features seen in patients with FHH are similar to those which occur in young patients with asymptomatic primary hyperparathyroidism.[137–139,141,142] Familial hypocalciuric hypercalcemia runs a benign course in the great majority of patients. Usually the patients are asymptomatic. However, occasionally patients may present with acute pancreatitis, with pseudogout, or even, surprisingly, with recurrent renal stones.[143,144] However, the prevalence of renal stones may be no greater than that in the general population. There are some differences between the mode of presentation of these patients and patients with other familial forms of hypercalcemia and primary hyperparathyroidism. Patients with FHH are usually young, and the condition is equally prevalent in males and females. Family history reveals an autosomal dominant pattern with complete penetrance. Moreover, in contrast with patients with multiple endocrine neoplasia syndromes, the penetrance is constant with age and these patients have detectable hypercalcemia which is present from early childhood. The important issue in the management of this condition is that it should not be misdiagnosed as primary hyperparathyroidism, because if patients are subjected to neck exploration and parathyroidectomy they rarely show evidence of improvement. A strange and unexplained association has been noted with neonatal primary hyperparathyroidism. It appears that severe neonatal hyperparathyroidism is more common in families with FHH. In some cases this may be related to homozygous expression of FHH in these infants.[145] In other words, it appears that if each two parents have classical FHH, which is autosomal dominant, and they have offspring with homozygous expression of the FHH gene, then these neonates may present with severe neonatal hyperparathyroidism. This was documented in at least one patient, who was the product of a consanguinous mating of two presumed heterozygotes who had mild asymptomatic hypercalcemia. In some cases the severe and dramatic hypercalcemia of the neonatal period may be transient.

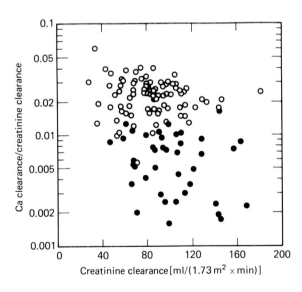

Figure 9.10

Comparison of the renal calcium clearance in patients with primary hyperparathyroidism (O) and patients with FHH (●). Notice that there is almost complete discrimination between the two groups using this parameter. (Reproduced from Aurbach et al, Parathyroid hormone, calcitonin and the calciferols. In: Wilson JD, Foster DW, eds, *Williams' Textbook of Endocrinology* 7th edn (W B Saunders: Philadelphia 1985) 1135–1217.)

Diagnosis

The essential points in the diagnosis are asymptomatic hypercalcemia, low renal excretion of filtered calcium and magnesium, and a history which reveals an autosomal dominant pattern of inheritance with essentially complete penetrance. The hypercalcemia is usually very mild and asymptomatic. Urinary calcium excretion is usually less than 100 mg/24 h (2.5 mmol), in contrast with primary hyperparathyroidism, where it is usually greater than 200 mg/24 h (5.0 mmol). There is, however, slight overlap in urine calcium excretion in the two conditions, and measurements of renal calcium clearance are more discriminating (Figure 9.10). The ratio of calcium clearance to creatinine clearance is usually less than 0.01.[138] A characteristic feature of FHH is an autosomal dominant inheritance pattern with complete penetrance. The condition can be anticipated in half of first-degree relatives.

The condition should be carefully distinguished from other forms of familial hypercalcemia and asymptomatic hypercalcemia occurring in the young (Table 9.3).

Pathophysiology

The pathophysiology of hypercalcemia in FHH is entirely unknown.[146] Plasma PTH, nephrogenous cAMP, and fractional phosphate clearance are not clearly different from those of normals, but in general are lower than those of patients with primary hyperparathyroidism.[138,147] However, these parameters are all abnormal considering the degree of hypercalcemia, which should cause suppression of the normal parathyroid glands. It appears that there must be a defect in divalent-cation handling by the renal tubules, since renal calcium clearance and renal magnesium clearance are low. There is avid reabsorption of calcium and magnesium in the renal tubules in FHH. A similar defect may be present in parathyroid cells, because they apparently do not respond to hypercalcemia by suppressing PTH secretion (Figure 9.11). The parathyroid glands usually, but not always, show normal morphology, and parathyroidectomy rarely lowers the serum calcium to the normal range. Patients who

Table 9.3   Familial causes of primary parathyroid hyperplasia.

Biologic and clinical features	Severe primary hyperparathyroidism in neonates	Hypocalciuric hypocalcemia	Multiple endocrine neoplasia		
			Type 1	Type 2	Type 3
Transmission of syndrome	Probably autosomal recessive	Autosomal dominant	Autosomal dominant	Autosomal dominant	Autosomal dominant
Organs with primary dysfunction	Parathyroid	Parathyroid, kidney	Parathyroid, pancreatic islet, anterior pituitary, adipocyte	Parathyroid, C cell, adrenal medulla	C-cell, adrenal medulla nerve*
Earliest appearance of hypercalcemia	Birth	Birth	15 years	15 years	Rare
Major morbidity	Hypercalcemia, respiratory disease	None	Nephrolithiasis, ulcer	C-cell cancer, pheochromocytoma	C-cell cancer, pheochromocytoma
Fractional renal clearance of calcium	Low	Low	Normal†	Normal†	Normal†
Usual serum calcium level after parathyroidectomy	High	High	Normal	Normal	Not applicable‡

*Primary parathyroid hyperplasia occurs in familial multiple endocrine neoplasia Type 3, but is not manifested clinically.
†Normal fractional clearance with hypercalcemia results in a high filtered load of calcium and hypercalciuria.
‡Not applicable, since parathyroidectomy is unnecessary.

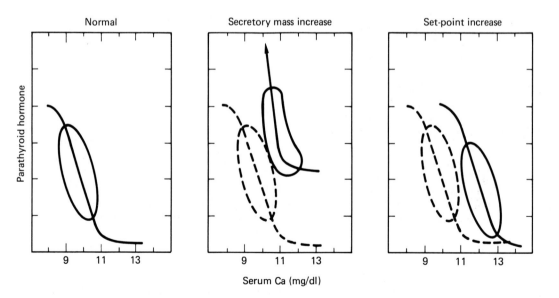

Figure 9.11

Diagram of the theoretical relationship between plasma PTH and plasma calcium. In patients with FHH there is insensitivity to plasma calcium, so there is relatively increased PTH secretion. The set-point for calcium in these patients is therefore increased.[146]

have had total parathyroidectomy with consequent hypoparathyroidism show a persistent
defect in renal calcium clearance despite the
absence of circulating PTH. There are some
similarities between the disturbances in calcium
and magnesium handling in this condition, and
the effects of lithium on calcium homeostasis.
Lithium decreases renal calcium and magnesium
clearance, and can cause increased secretion of
PTH.[148] Lithium may have intracellular effects
which interfere with the capacity of the renal
tubular cell to recognize increased calcium in the
tubular fluid, or the capacity of the parathyroid
cell to recognize hypercalcemia. Lithium behaves
like an agonist for calcium, so that calcium
reabsorption is enhanced in the renal tubules and
PTH secretion is enhanced in the parathyroids.[149]

Figure 9.12 shows a scheme of the possible
pathophysiology of the hypercalcemia in this
condition.

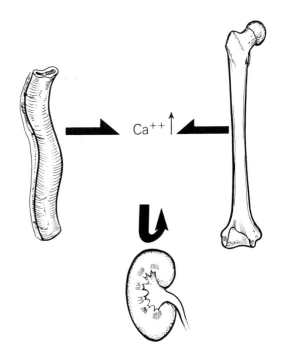

Figure 9.12

Abnormalities in calcium homeostasis seen in patients with
familial hypocalciuric hypercalcemia. This is a very difficult
syndrome to unravel. In addition to increased renal tubular
calcium reabsorption it appears very likely that there are
other changes which have not yet been fully characterized
on gut and bone calcium transport. The complete
pathophysiology of this syndrome has yet to be clarified.

Treatment

The great majority of patients with this condition
require no treatment at all.[150] In fact, many are
better off if hypercalcemia remains unrecognized,
since this discovery may lead to parathyroid
surgery, and this is unlikely to help them. Rarely
are patients who have surgery for this condition
rendered eucalcemic postoperatively. The majority remain hypercalcemic, although total parathyroidectomy will lead to hypocalcemia. Surgery
may be indicated if the patient has pancreatitis,
and in this situation parathyroid tissue should be
cryo-preserved and later autotransplanted.
Surgery may lower the serum calcium sufficiently
to improve symptoms in patients with pancreatitis. One patient that the author saw a number of
years ago was a 35-year-old woman with a 15-
year history of known hypercalcemia and recurrent episodes of pancreatitis, so severe that she
had diabetes mellitus and persistent abdominal
pain. The diagnosis of FHH was made on the basis
of renal calcium handling and family history (her
11-year-old son was also hypercalcemic with low
urine calcium excretion). Her serum calcium was
persistently 14 mg/dl until a subtotal parathyroidectomy was performed by an unsuspect-
ing surgeon who did not recognize the diagnosis.

However, following surgery her serum calcium
was stable at 12 mg/dl, and her episodes of acute-
on-chronic pancreatitis abated.

No medical therapy has yet been shown to
work. Loop diuretics and corticosteroids do not
help. Part of the problem in this condition is that
the primary defect is unknown. The key issue in
the management of these patients is to recognize
that they do not have primary hyperparathyroidism, and thus avoid unnecessary
parathyroidectomy.

Hypercalcemia associated with adrenal insufficiency

Hypercalcemia has occasionally been described in patients with adrenal insufficiency, and particularly in acute adrenal failure.[151-153] However this is an extremely rare association of a very uncommon disease. The mechanism is unknown. Corticosteroids have complicated effects on calcium homeostasis, including causing a decrease in calcium absorption from the gut, which leads to secondary hyperparathyroidism, and complex suppressive effects on the differentiation of both osteoblasts and osteoclasts into fully differentiated cells. They suppress normal bone formation, presumably by impairing the differentiation of osteoblast precursors into mature cells, and similarly inhibit the formation of mature osteoclasts. In pharmacologic doses they are effective inhibitors of osteoclastic bone resorption in vitro and in vivo.[154] It is possible that the major cause of hypercalcemia in patients with acute adrenal failure is increased absorption of calcium from the gut occurring in the absence of glucocorticoids. However, hypercalcemia has been noted on a calcium-free diet, so this explanation is not completely satisfactory. The hypercalcemia in adrenal failure is only a transient phenomenon, and is corrected by glucocorticoid replacement therapy.

Paget's disease

Hypercalcemia is an extremely rare complication of Paget's disease. In fact, it is so rare that the author suspects that in most of the described cases a coexisting condition other than the Paget's disease itself has been the real cause of the hypercalcemia. The most likely cause in most patients who present with hypercalcemia and have underlying Paget's disease is primary hyperparathyroidism, and this should be sought diligently. When hypercalcemia does complicate Paget's disease, it is most likely to occur in a patient with widespread and active disease who is immobilized (for example, following a fracture), which presumably decreases bone formation while bone resorption is increased, possibly as a consequence both of the disease itself and of the immobilized state. In one recent reported case,[155]

it appeared that other causes of hypercalcemia were excluded, and the serum calcium ran in parallel with the state of mobility.

Hypercalcemia in renal failure

Hypercalcemia does not occur frequently as a consequence of renal failure. In fact, the majority of patients with either acute or chronic renal failure are hypocalcemic. Before discussing those circumstances where renal failure leads to hypercalcemia, those situations where hypercalcemia or the disease responsible for hypercalcemia is responsible for renal failure are reviewed.

Renal failure as a consequence of hypercalcemia

Occasionally renal failure can occur as a direct consequence of hypercalcemia.[156] This is most likely to happen in patients who have an increase in both the serum calcium and serum phosphorus, and it is often associated with nephrocalcinosis if the hypercalcemia has been chronic. It is particularly common in patients with vitamin D intoxication, with sarcoidosis or other granulomatous diseases, or the milk-alkali syndrome. Renal failure also occurs frequently in patients with hypercalcemia due to myeloma. In this situation the renal failure is possibly due to 'myeloma kidney' which in turn may be due to Bence Jones nephropathy, uric acid nephropathy, chronic infections, or amyloidosis. Hypercalcemia also occurs occasionally in patients with primary hyperparathyroidism where hypercalcemia has been persistent for long periods, and is complicated by impaired renal function and increases in the serum phosphorus. Renal failure (with nephrocalcinosis) has been a feature of the patients that the author has seen with longstanding hypercalcemia due to parathyroid carcinoma.

Hypercalcemia in acute renal failure

Hypercalcemia occasionally occurs after acute renal failure.[157] When hypercalcemia is present in acute renal failure, it usually occurs during the polyuric phase and probably only in association with acute rhabdomyolysis.[158] About one-third of patients with rhabdomyolysis develop hypercalcemia. In most of these patients there has been a period of hypocalcemia during the phase of acute renal failure and, as renal function begins to recover and polyuria occurs, the serum calcium becomes elevated. Usually the increase in serum calcium is mild and improves spontaneously as the polyuric phase of acute renal failure passes.

However, occasionally hypercalcemia is found without an obvious polyuric phase. The author has noted this several times in patients in whom hypercalcemia was present at the time of admission to hospital. This is likely to occur in a hot climate (such as south Texas during summer) and in a patient who works outdoors and becomes dehydrated following exposure to extreme heat. The hypercalcemia is usually mild and has resolved quickly with improvement of renal function. Other parameters (such as the release of skeletal muscle enzymes or uric acid) will indicate that the patient is undergoing acute rhabdomyolysis. Recovery is usually rapid and complete.

It remains unclear why patients with acute rhabdomyolysis should develop hypercalcemia. Rhabdomyolysis results from skeletal muscle injury associated with release of intracellular muscle cell constituents into the plasma. It is often associated with trauma to muscles, but may be nontraumatic. It was first recorded extensively during World War II, particularly during the Battle of Britain, when many civilians suffered crush injuries when buildings collapsed. It is often associated with alcoholism or occurs in patients who have had a recent alcoholic binge, and may accompany acute alcohol-induced coma. It is also sometimes found in addicts of heroin or other drugs. It is occasionally seen in association with inherited enzyme deficiencies such as carnitine palmityl transferase deficiency. Clinical features consist of nausea and vomiting. It is recognized by an increase in muscle enzymes such as creatinine phosphokinase (CPK) and aldolase, a darkening of the urine which is due to myoglobinuria, with urine which tests positive for blood but not for red cells. There is frequently a marked increase in serum uric acid, due to the release of uric acid from intracellular stores. The serum creatinine increases out of proportion to the blood urea nitrogen. Pigmented granular casts are consistently found in the urinary sediment.

These probably occur due to the interaction of myoglobin with other proteins or cells present in the tubule lumen.

The mechanism responsible for the increase in serum calcium in patients with acute rhabdomyolysis is unknown. A number of suggestions have been made. During the early phase of acute rhabdomyolysis there may be dissolution of calcium deposits which have accumulated in muscle during the hypocalcemic phase. In fact, hypocalcemia may be due to skeletal resistance to PTH (common to all forms of acute renal failure) and is probably related to an increase in the serum phosphorus due to a combination of increased intracellular release of phosphate plus impaired glomerular filtration. The increase in serum phosphorus due to impaired glomerular filtration will lead to a fall in the serum calcium. As serum phosphorus falls during the diuretic phase of acute renal failure, calcium and phosphate salts which have been deposited in soft tissues may be liberated, causing a transient hypercalcemia. However, other suggestions have been made to explain why hypercalcemia occurs in acute rhabdomyolysis. One group has suggested that excessive release of 25-hydroxyvitamin D from injured muscle occurs as a result of acute rhabdomyolysis and leads to increased substrate available for conversion to 1,25-dihydroxyvitamin D as the kidneys recover. Evidence for this mechanism was gathered by Llach et al,[159] who studied six oliguric patients with rhabdomyolysis-induced acute renal failure. All of these patients initially had low levels of serum 1,25-dihydroxyvitamin D associated with hypocalcemia and hyperphosphatemia. During the diuretic phase an increase in serum 1,25-dihydroxyvitamin D was associated with an increase in serum calcium and an increase in PTH. Akmal et al[160] also described this increase in serum 1,25-dihydroxyvitamin D associated with hypercalcemia. They could not identify any known stimulus. In their study neither increased PTH nor low serum phosphate could explain the increase in serum 1,25-dihydroxyvitamin D. The increase in serum calcium and serum 1,25-dihydroxyvitamin D returned to normal later in the diuretic phase. However, Llach et al[159] postulated that the hypercalcemia during the early diuretic phase was due to 1,25-dihydroxyvitamin D, which in turn was due to increased circulating PTH concentrations.

Hypercalcemia associated with chronic renal failure

The majority of patients with chronic renal failure have a decrease in the serum calcium, rather than an increase. Approximately 10–20 per cent develop hypercalcemia. This can occur in a variety of settings. For ease of classification it is simplest to consider two major groups: patients who develop hypercalcemia before renal transplantation, and patients who develop hypercalcemia following renal transplantation.

Hypercalcemia associated with chronic renal failure before renal transplantation

There are a number of mechanisms and different clinical settings which could account for hypercalcemia occurring in patients before renal transplantation. In the past, this type of hypercalcemia has frequently been called 'tertiary hyperparathyroidism'. It has been thought that all of these patients have four-gland parathyroid hyperplasia, but that at least one of the glands becomes autonomous and behaves like an autonomous adenoma.[161] Although this mechanism is a possibility, there has been no clear-cut well-documented case. A more appropriate descriptive term is 'secondary hyperparathyroidism with hypercalcemia'. Hypercalcemia may occur in patients with secondary hyperparathyroidism, because peripheral tissues become more sensitive to PTH. One cause for this may be a decrease in serum phosphate. In some patients hypercalcemia is due to vitamin D intoxication because of the overenthusiastic use of vitamin D or one of its metabolites. In other patients another cause of hypercalcemia may be present. Possibly the most common cause nowadays is aluminum intoxication, but in these patients serum PTH concentrations are not markedly elevated. Some patients with chronic renal failure have a state of low bone turnover (called by some 'aplastic osteomalacia') and increased content of aluminum in the skeleton. These patients are often hypercalcemic. The mechanism for hypercalcemia is unclear, but it may be related to impaired uptake of calcium into the skeleton.

The circumstances under which aluminum causes bone disease and hypercalcemia in chronic renal failure is the following.[162–168] Patients with chronic renal failure who are on dialysis may have excess aluminum intake from the use of aluminum-contaminated dialysis water or the use of aluminum-containing phosphate binders to lower the serum phosphate. However, in some patients no increased intake of aluminum has been identified and the patients may be on a normal diet. These patients frequently have diffuse bone pain and fractures. Bone histology shows osteomalacia which is patchy, with aluminum accumulation along the mineralization front and with little or no uptake of tetracycline and no separation of the tetracycline bands. This type of bone disease is often called aplastic osteomalacia. When the bone aluminum content is measured, it is 20–50 times normal. Features which suggest this condition are the absence of marked increase in the serum PTH, relatively low serum alkaline phosphatase, and resistance to treatment with vitamin D sterols. The frequency is unclear, but it is probably around 20–25 per cent in patients with chronic renal failure. However, this figure is controversial, and some workers in the USA suggest that it is less than 5 per cent. In patients with aluminum bone disease, spontaneous hypercalcemia occurs in approximately 50 per cent. The condition may be associated with encephalopathy, which is probably the explanation for the dementia which occurs in some patients on chronic dialysis.

The evidence that aluminum plays a role in this bone disease is now overwhelming. As indicated above, the aluminum content of the bone is 20–50 times normal. Histochemical studies of aluminum disposition show the presence of aluminum displayed in bone along the mineralization front. In studies in the UK the aluminum content in dialysates correlates with osteomalacia and encephalopathy. Since aluminum is cleared by the kidneys in patients with chronic renal failure or impaired renal function, there will be delayed clearance and the element will accumulate. Moreover, aluminum administration to rats and dogs produces a similar bone disease, and a similar bone disease is seen in patients treated with total parenteral nutrition which is contaminated with aluminum.[81] Some have suggested that this disease represents a disease of progress, namely more adequate treatment of chronic renal failure.

The treatment of aluminum osteodystrophy has

been improved considerably in recent years. The symptoms and bone biopsy appearance often benefit from the use of the chelator desferrioxamine. This chelator removes aluminum by chelating it, so it can be effectively removed by dialysis. Weekly infusions of this agent in patients while they are on dialysis may improve their symptoms as well as the bone appearance. It should be noted that dialysis encephalopathy and vitamin D-resistant osteomalacia in patients with chronic renal failure has been described in non-dialyzed or uremic patients receiving peritoneal dialysis, as well as in dialyzed uremic patients receiving dialysis with aluminum-free dialysate. The source of aluminum in these patients is not known.

Total parenteral nutrition, bone disease, and hypercalcemia

Some patients receiving total parenteral nutrition may develop severe skeletal symptoms, myopathy, and patchy osteomalacia, together with hypercalciuria and hypercalcemia.[81,169–172] Serum 1,25-dihydroxyvitamin D concentrations are low, but 25-hydroxyvitamin D concentrations are normal. Some of these patients develop hypercalcemia. The disorder resolves 1–2 months after discontinuing total parenteral nutrition. The bones contain considerable quantities of aluminum, which presumably enters the body via the protein source casein which is present in the infusate. The bone aluminum content and the histologic features are essentially identical to those seen in uremic patients with low turnover osteomalacia due to aluminum.

Post-transplantation hypercalcemia

A period of hypercalcemia frequently occurs in patients following renal transplantation. This usually does not last for more than 6 months, and it is usually of mild degree with a serum calcium not increasing above 12.5 mg/dl. This period of hypercalcemia is caused by hyperparathyroidism occurring as a consequence of slow involution of the hyperplastic parathyroid glands. The parathyroid glands become markedly hyperplastic in patients with chronic renal failure, but the skeleton is relatively resistant to the effects of increased circulating PTH. The patients are hypocalcemic in spite of increased PTH secretion, probably as a consequence of impaired gut absorption of calcium, decreased production of 1,25-dihydroxyvitamin D, and hyperphosphatemia. The mechanism for skeletal resistance to PTH is not known, but it may be related to high phosphate concentrations occurring as a consequence of impaired glomerular filtration, or lack of 1,25-dihydroxyvitamin D which may be required for PTH to produce significant effects on bone and kidney (Chapter 2). Whatever the reason, when renal failure is corrected the skeleton becomes sensitive again to the effects of high PTH levels. This period following transplantation can be a dangerous period because hypercalcemia may be severely toxic to the transplanted kidney.

Hypocalcemia—causes, clinical features, differential diagnosis

Causes and clinical features

Hypocalcemia is a less common clinical problem than hypercalcemia, and the causes are fewer. However, it can be extremely difficult to manage properly, particularly when the cause is hypoparathyroidism, and it may be responsible for a number of disabling symptoms and signs. The study of some patients with unusual forms of hypocalcemia has led to special insights into the molecular mechanisms of action of calciotropic hormones such as PTH and 1,25-dihydroxyvitamin D.

The total serum calcium concentration rarely drops below 5 mg/dl in humans, despite the complete absence of the parathyroid glands and very low concentrations of 1,25-dihydroxyvitamin D. Parathyroid hormone is probably the major regulator of extracellular fluid calcium homeostasis (Chapter 1). However, since serum calcium rarely falls below 5.0 mg/dl in a completely hypoparathyroid patient, it is therefore clear that there must be other mechanisms for maintaining the serum calcium independently of this hormone. As indicated in Chapter 1, it appears likely that these other mechanisms include the exchange of calcium between the bone fluid, which is in intimate contact with the mineral surface of bone, and the extracellular fluid. This is fortunate, because although a serum calcium of 5 mg/dl (1.25 mmol/l) is still compatible with life, further decreases impair neuromuscular function and critical cell functions so that continued viability is severely threatened.

In this chapter, the major causes (Table 10.1) and clinical features of hypocalcemia are reviewed, followed by a more detailed review of hypocalcemia associated with hypoparathyroidism, pseudohypoparathyroidism, and other causes.

Table 10.1 Causes of hypocalcemia.

Hypoparathyroidism
   Idiopathic
      congenital absence of parathyroid glands
      branchial dysembryogenesis
      associated with multiple endocrine deficiencies
      isolated late onset
   Acquired (post-surgical, infiltrative diseases)
   Pseudohypoparathyroidism (Table 10.3)

Vitamin D deficiency

Malignant disease

Malignant disease

Hypomagnesemia

Toxic-shock syndrome

Neonatal hypocalcemia

Pancreatitis

Renal failure

Table 10.2   Clinical features of hypocalcemia.

1 Neurologic
   (a) Peripheral—irritability (tetany)
   (b) Central
       irritability, seizures
       intracranial calcification
       papilledema
       mental changes

2 Cataracts

3 Abnormal dentition

4 Cardiovascular
   (a) Abnormal EKG (with prolonged QT interval)
   (b) Susceptibility to congestive heart failure and digitalis
       resistance

Clinical manifestations of hypocalcemia (Table 10.2)

These have been described in detail in a number of extensive reviews.[1–3] The reader is referred to these articles for more detailed information and references.

Neuromuscular manifestations

Tetany

Tetany is the term given to spontaneous muscle contractions which occur in patients who are hypocalcemic. The spontaneous muscle contractions are due to increased neural excitability, which occurs as a consequence of the low concentration of extracellular fluid calcium reducing the threshold for excitation for the neural tissue. The result is repetitive responses to a single stimulus, followed by continuous activity. The clinical manifestations of tetany occur as a result of spontaneous discharges of both sensory and motor fibers in peripheral nerves. An attack of tetany is characteristic. It is usually preceded by a sensory syndrome causing perioral paresthesias. These proceed to carpopedal spasm, which usually involves the hands rather than the feet. A characteristic deformity occurs, consisting of abduction of the thumb, extension of the interphalangeal joints, compression of the fingers, and hollowing of the palm. This deformity is frequently called obstetrician's hand (*main d'accoucheur*). Tetany is rarely dangerous, but it is very frightening to the sufferer. In milder forms a tetanic attack may occur only when it is provoked. The physician can provoke tetanic signs by eliciting Chvostek's sign (a sharp tap over the facial nerve in front of the ear just below the zygomatic process or between the zygomatic arch and the corner of the mouth). The result is twitching of the facial and upper lip muscles. This sign may be seen in 10 per cent of normals. Another test is Trousseau's sign, which can be provoked by increasing the blood pressure for 3 min. A sphygmomanometer cuff is applied to the upper arm at 10 mmHg above the systolic blood pressure for 3 min, and the consequence is a typical episode of carpopedal spasm. Trousseau's sign, unlike Chvostek's sign, should not occur in normal individuals.

There are some variations in expression of the tetanic attack. Some individuals complain of muscle cramps or mild paresthesias rather than overt episodes of carpal spasm. Laryngeal stridor is occasionally seen due to spasm of the muscles of the larynx. Tetany is more likely to occur in patients with acute hypocalcemia, rather than those who have a stable serum calcium which is low for a longer period and of slower onset. Sometimes tetany is provoked by emotion, menstruation, pregnancy, vomiting, or exercise. Tetany is not specific for hypocalcemia, but it can also be seen in metabolic alkalosis, hyperkalemia, hypokalemia, hypomagnesemia, and hyperventilation.

Seizure disorders

Hypocalcemia occasionally precipitates seizures in an epileptic and may even cause a type of seizure disorder in normals, which resembles a grand mal seizure in some ways. This is more likely to occur in children. The disorder varies a little from a grand mal seizure, because it consists essentially of a generalized tetanic response followed by prolonged clonic spasms. Loss of consciousness, aura, incontinence, tongue-biting, and the postepileptic confusional state do not occur. This type of disorder produces a characteristic tracing on the electroencephalogram (EEG), including an irregularity and fragmentation of the postcentral background activity, the presence of theta- rather than delta-waves, and increased low-voltage fast activity. There are bursts of high-voltage slow waves which are paroxysmal and are more prominent when the serum calcium is low. This type of seizure disorder may be associated with singular episodes. There are frequently other central nervous system manifestations of hypocalcemia, including papilledema and calcification of the basal ganglia. Papilledema is most frequent in longstanding untreated disease. There are no symptoms and the cause is unknown. It is not specific for idiopathic hypoparathyroidism, and it may occur in pseudohypoparathyroidism or post-surgical hypoparathyroidism. It resolves completely with treatment of the hypocalcemia. Basal ganglia and extrapyramidal calcification occur in and around the walls of small blood vessels. This is a manifestation of chronic hypocalcemia, and it occurs in all varieties of hypoparathyroidism and pseudohypoparathyroidism of sufficiently long duration. These patients may also have a wide range of mental changes and psychiatric disturbances. These can include psychoneurosis, psychosis, and organic brain syndromes.

Cardiac manifestations

Patients with hypocalcemia have characteristic electrocardiographic changes. There is prolongation of the QTc and ST intervals. With more severe decreases in the serum calcium patients may develop arrhythmias, heart block, and congestive failure. In patients with hypoparathyroidism correction of hypocalcemia will improve both the cardiac output and the blood pressure.

Skeletal manifestations

Changes in bone are not well described in patients with hypoparathyroidism. Although there have been reports of increased cortical thickness and trabecular bone density, these are not well documented. In some patients with pseudohypoparathyroidism the skeletal features of secondary hyperparathyroidism may occur, and some patients may develop features of osteomalacia associated with vitamin D deficiency. However, these cases are unusual.

Ocular findings

Cataracts are frequent in patients with chronic hypocalcemia. These occur in 28 per cent of patients who have been hypocalcemic for more than 4 years. They are probably due to interference with the normal hydration of the lens. In severe cases the cataracts may become confluent and cause blindness. They are not specific for hypoparathyroidism, and may occur in a number of other metabolic disorders. However, they are distinct from senile cataracts, and are not necessarily confined, unlike senile cataracts, to one segment of the lens.

Ectodermal abnormalities

The ectodermal abnormalities which occur may be related more to the type of hypoparathyroidism rather than to hypocalcemia. They are most likely in patients with idiopathic hypoparathyroidism. In patients with these abnormalities the skin is dry and scaly, the nails are brittle, the hair is coarse, and the eyebrows are sparse. Some patients also have moniliasis, which involves the gastrointestinal tract. These changes must be distinguished from those associated with mucocutaneous candidiasis, which occurs in some patients with idiopathic hypoparathyroidism. If hypocalcemia occurs in infancy, then abnormalities in tooth development occur. These include failure of eruption, enamel hypoplasia, irregularity of the dentin, and liability to caries. The etiology of the ectodermal abnormalities is not known.

In patients with longstanding hypoparathyroidism there may be intestinal malabsorption and steatorrhea. Malabsorption is corrected when hypocalcemia is treated. Gingival biopsy shows partial villous atrophy, which is not responsive to a gluten-free diet, but is reversible when hypocalcemia is corrected with vitamin D. Presumably, normal function of intestinal villi is dependent on serum calcium and/or vitamin D.

Hypocalcemia associated with vitamin D deficiency

Hypocalcemia may occur as a consequence of vitamin D deficiency. Vitamin D deficiency may be due to absolute deficiency or due to an abnormality in metabolism of vitamin D to the biologically more active polar metabolites. The biological effects and metabolism of vitamin D are considered in Chapter 3. Osteomalacia is reviewed in more detail in Frame and Parfitt.[4]

Causes of vitamin D deficiency (Table 10.3)

Dietary lack

Dietary lack of vitamin D is rare in the USA because of supplementation of food and milk with vitamin D. However, it still occurs occasionally, particularly in the elderly and alcoholics who eat poorly, and who may have other disturbances in vitamin D metabolism and poor exposure to sunlight. The consequences are most profound during growth, since it is at this age that there are increased requirements for calcium and the delay in bone mineralization causes the most profound

Table 10.3   Causes of vitamin D deficiency.

1 Dietary lack

2 Poor sunlight exposure

3 Gut malabsorption

4 Anticonvulsant therapy

5 Chronic liver disease

6 Chronic renal disease

7 Vitamin D-dependent osteomalacia
   Type I inherited deficiency of the 1-hydroxylase enzyme
   Type II inherited inability of target organs to respond to vitamin D metabolites

clinical features. Vitamin D deficiency is more common in the UK, particularly among Asian immigrants who eat poorly and ingest diets with a high content of phytates which may bind calcium in the gut and hinder its absorption.

Malabsorption

Vitamin D is a fat-soluble vitamin, and disorders which lead to steatorrhea are therefore associated with vitamin D deficiency. This is seen frequently in patients with hepatobiliary and gastrointestinal disorders such as celiac disease, cystic fibrosis, and chronic pancreatitis. Of course, in patients with gastrointestinal disorders which lead to malabsorption, there may also be abnormalities in the intestinal epithelium which reduce the efficacy of vitamin D metabolites in stimulating the absorption of calcium across the epithelium. Vitamin D deficiency and hypocalcemia also occur in patients with primary biliary cirrhosis, short bowel syndromes, and in those who have had prior partial or total gastrectomies. However, these disorders are often associated with osteoporosis, and the mechanisms responsible for bone disease and hypocalcemia may be more than simple vitamin D deficiency. In this group of disorders, the enterohepatic circulation of 25-hydroxyvitamin D may also be disordered, leading to vitamin D deficiency. Evidence suggests that there are different absorptive mechanisms in the gut for vitamin D and its more polar metabolites, and so patients with malabsorption may be more responsive to 25-hydroxyvitamin D and 1,25-dihydroxyvitamin D than they are to the parent compound.

Anticonvulsant therapy

Vitamin D deficiency and hypocalcemia may occur in patients taking anticonvulsant therapy for long periods. Anticonvulsants such as phenytoin, phenobarbital, and glutethimide have been associated with rickets/osteomalacia. However, many of these patients are children who have seizure disorders associated with mental retardation. As a consequence, long-term institutionalization is common, and these patients may eat poorly and have poor exposure to sunlight. Thus, multiple mechanisms may be responsible for vitamin D deficiency. There is clear evidence that serum 25-hydroxyvitamin D concentrations are decreased in patients taking anticonvulsants.[5] The mechanism for decreased serum 25-hydroxyvitamin D concentrations is uncertain. One possibility is that anticonvulsant drugs induce hepatic microsomal enzymes which metabolize vitamin D to biologically inactive degradation products.[6] Alternatively, there may be decreased synthesis of 25-hydroxyvitamin D. Anticonvulsants may also have other effects independent of vitamin D metabolism. These agents may alter membrane stability, interfere with bone resorption, and influence the enterohepatic circulation of various drugs.[7]

Chronic renal disease

The proximal tubule of the kidney is the major site for synthesis of the most biologically active vitamin D sterol, 1,25-dihydroxyvitamin D. In patients with chronic renal disease, 1,25-dihydroxyvitamin D concentrations are markedly reduced (although still detectable). This is probably due not only to decreased functioning renal tubular cells capable of producing 1,25-dihydroxyvitamin D, but also because of the increased circulating phosphate concentrations which suppress the renal 1-hydroxylase and which occur as a consequence of impaired glomerular function. Patients with chronic renal failure also develop hypocalcemia because the uremic state directly impairs calcium absorption from the gut.

Vitamin D-dependent osteomalacia

There is a group of syndromes associated with either abnormalities in conversion of 25-hydroxyvitamin D to 1,25-dihydroxyvitamin D, or insensitivity of target organs to active vitamin D metabolites. In Type I vitamin D-dependent osteomalacia, there is deficiency of the 1-hydroxylase enzyme necessary for conversion of 25-hydroxyvitamin D to 1,25-dihydroxyvitamin D.[8] This is an autosomal recessive disorder which is characterized by rickets with low or undetectable serum 1,25-dihydroxyvitamin D concentrations. These patients show no increase in serum 1,25-dihydroxyvitamin D when treated with 25-hydroxyvitamin D, but respond very well to treatment with 1,25-dihydroxyvitamin D. The pre-

cise nature of the molecular defect responsible for this disorder has not been clarified. In Type II vitamin D dependent osteomalacia, circulating concentrations of 1,25-dihydroxyvitamin D are increased in the presence of rickets, and the patients have target organ resistance to the sterol.[9,10] A number of kindreds with this condition have been described, and it is clear that there are different subgroups of this disorder, probably reflecting different molecular mechanisms of target organ resistance. Some of these patients have ectodermal abnormalities including total alopecia, loss of the eyebrows, and dental anomalies including hypoplasia with cone-shaped teeth due to enamel deformities.

Abnormalities in calcium homeostasis

In patients with vitamin D deficiency but without chronic renal disease, the characteristic abnormalities are hypocalcemia associated with severe hypophosphatemia and increased serum alkaline phosphatase. Hypocalcemia occurs as a consequence of impaired absorption of calcium from the gut and decreased bone resorption. Hypophosphatemia is due to impaired gut absorption of phosphate, but also occurs because of secondary hyperparathyroidism, which is a result of hypocalcemia, and leads to renal phosphate wasting. Urine calcium excretion is reduced, first because the filtered load is decreased, and secondly because of the effects of PTH on the distal tubule of the kidney to enhance renal tubular calcium reabsorption. Serum 25-hydroxyvitamin D is markedly reduced.

In patients with hypocalcemia occurring as a consequence of chronic renal disease, the serum phosphate is usually increased rather than decreased. This occurs because impairment of glomerular filtration leads to decreased filtered phosphate.

Clinical features of vitamin D deficiency

Vitamin D deficiency causes rickets in children and osteomalacia in adults. This disease is characterized by bone pain, susceptibility to fracture, and subsequent deformity. The bone is more fragile than normal bone and subject to deformity under stress. Children are more likely to suffer from attacks of tetany than are adults. Vitamin D deficiency is also accompanied by muscle weakness and tenderness, particularly involving the proximal muscles of the lower extremities associated with hypotonia. The bone disease is characterized by impaired mineralization of newly formed bone. The result is excess unmineralized bone matrix (osteoid tissue). Before puberty, the disease is characterized by failure of calcification of the cartilage at the growth plate, namely, decreased endochondral ossification. This condition is known as rickets.

Osteomalacia can be diagnosed histologically by a bone biopsy using undecalcified sections and double labeling by tetracycline with quantitative histomorphometry. There are a number of striking histologic parameters which are found in this disorder, including an increase in osteoid volume, an increase in bone surfaces covered with osteoid which may be relatively resistant to osteoclastic resorption, an increase in osteoid seam width and a variable decrease in bone surfaces covered by osteoblasts. If tetracycline has been taken, this will fluoresce in biopsy sections at the site of mineralization (the mineralization front) and when two courses are taken separated by about 12 days, the distance between labels in undecalcified sections allows calculations of the mineral appositional rate. This is markedly decreased in patients with osteomalacia. Sometimes the single bands may be diffuse and indistinct in osteomalacia.

Diagnosis

Vitamin D deficiency with normal renal function is characterized by hypocalcemia and severe hypophosphatemia. Serum alkaline phosphatase is usually increased. In most cases, serum 25-hydroxyvitamin D will be severely decreased (usually less than 5 ng/ml). X-rays may show the characteristic features of osteomalacia including pseudofractures and osteopenia. Pseudofractures are lines of rarefaction which occur at sites where arteries cross bone, such as below the glenoid forsa, the scapula, the pubic rami and the ribs. In children, rickets is characterized by irregular expansion of the growth plate which is readily detectable radiologically. The growth plate is widened, cupped and has an irregular appearance at the end of the metaphysis. The widening of the growth plate can be felt by palpation at the wrist and around the rib cage—the rickety rosary.

Treatment

Vitamin D deficiency can be corrected by oral calcium and a form of vitamin D. The form of vitamin D should depend on the nature of the defect in vitamin D metabolism. Where there is a defect in metabolism, 1,25-dihydroxyvitamin D or an active metabolite is preferred. The available vitamin D preparations are shown in Table 11.2

Patients may develop rickets or osteomalacia for causes other than vitamin D deficiency. Phosphate deficiency or aluminum intoxication both lead to osteomalacia. Phosphate deficiency occurs most commonly as in inherited condition which is usually sex-linked and associated with renal phosphate wasting. It can also occur in patients with certain benign, slowly growing mesenchymal tumors, and is again due to renal phosphate wasting. Etidronate therapy for over 3 months in doses greater than 10 mg/kg body weight per day causes defective mineralization (Chapter 7). Aluminum intoxication occurs usually in patients with chronic renal failure or patients taking chronic parenteral nutrition and is associated with an aplastic form of osteomalacia, characterized by low bone turnover and a decreased bone–osteoblast interface. Patients receiving chronic parenteral nutrition who become osteomalacic may have been given preparations in which aluminum is contaminating the casein hydrolysate (Chapter 9).

Hypoparathyroidism

Hypoparathyroidism may be due to diseases in the parathyroid gland causing impaired synthesis and secretion of PTH, or to target organ resistance to the effects of PTH, the disease called pseudohypoparathyroidism. Both of these conditions are heterogeneous and consist of distinct entities. They are discussed separately below.

Idiopathic hypoparathyroidism

Idiopathic hypoparathyroidism is the term given, possibly inappropriately, to diseases of the parathyroid gland associated with insufficient PTH secretion which are not secondary or acquired. The term encompasses a number of separate but indistinct clinical entities.

Congenital absence of the parathyroid glands

In individuals without parathyroid glands hypocalcemia is manifested at birth. The entity is usually sporadic, although occasional familial cases have been reported. This variety is very rare.

Branchial dysembryogenesis

In this group of conditions hypocalcemia is also usually evident during the neonatal period. Hypoparathyroidism is due to an abnormality in embryologic development of the branchial clefts. There are frequently other branchial cleft abnormalities, most commonly a disorder of the third and fourth branchial clefts which causes the DiGeorge syndrome associated with absence of the thymus. Sometimes there are defects in the first and fifth branchial clefts, leading to characteristic facial abnormalities and defects in the aortic arch or in the heart. The characteristic facial defects include a small chin (micrognathia), slanted eyes, low-set and notched ears, and hyperteleorism. The cardiac abnormalities include tetralogy of Fallot, as well as a wide variety of other congenital aortic arch anomalies. The absence of the thymus leads to a profound defect in cell-mediated immunity. Sometimes the defects in the heart and thymus are not so severe, and these defects are occasionally compatible with survival into adulthood.

Idiopathic hypoparathyroidism associated with multiple endocrine deficiencies

This is probably the most common variety of idiopathic hypoparathyroidism, or at least the variety which has received the most attention.

The syndrome is called Type I polyglandular autoimmune syndrome.[11] The condition is usually sporadic, although it is occasionally familial. There is no strict pattern of inheritance, although most patients with the familial variety seem to have an autosomal recessive pattern. More than 100 total cases have been reported in the medical literature. Patients with this condition in its complete form (Figure 10.1) suffer from hypoparathyroidism, chronic mucocutaneous candidiasis, primary adrenal insufficiency, primary hypogonadism, diabetes mellitus, and pernicious anemia. The condition usually begins with

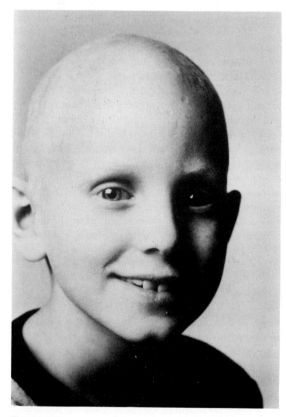

Figure 10.1

Characteristic facies in a patient with Type I polyglandular autoimmune syndrome. These patients are also liable to hypoparathyroidism, hypoadrenalism, and other endocrine deficiencies. They frequently have chronic mucocutaneous candidiasis and occasionally alopecia areata.

candidiasis which has its onset by the age of 4 years. Idiopathic hypoparathyroidism presents a few years later, usually before the age of 10 years, and then Addison's disease within the next few years, usually during adolescence. Occasionally these patients develop alopecia areata and vitiligo, associated with antibodies to melanocytes. The pathophysiology of the chronic cutaneous candidiasis syndrome is unknown. It is presumably related to an impairment in cell-mediated immunity. There is no direct association between the candidiasis and the serum calcium. It presents in childhood and affects the skin, nails, oropharynx, and vagina. Systemic candidiasis does not occur unless the patients have other predisposing factors such as diabetes mellitus.

These patients frequently have autoantibodies to the parathyroid gland cells. However, the significance of these autoantibodies is unknown. They could be primarily responsible for the disease, or they could merely represent epiphenomena.

Isolated late-onset hypoparathyroidism

This is not a distinct clinical entity, but rather represents those cases of idiopathic hypoparathyroidism described above of milder nature which did not present during childhood. Patients who present with idiopathic hypoparathyroidism in adult life occasionally have parathyroid gland autoantibodies. These cases are usually sporadic (but sometimes inherited), and they usually do not have the other endocrine abnormalities associated with Type I polyglandular autoimmune syndrome. These patients may present with symptoms at any time during adult life up to the age of 80 years.

Diagnosis of hypoparathyroidism

(a) The diagnosis of hypoparathyroidism is suggested by measurements of the serum calcium and phosphorus, which show hypocalcemia in association with hyperphosphatemia. Patients presenting with hypocalcemia, hyperphosphatemia, and normal renal

function are most likely to have either idiopathic or acquired hypoparathyroidism, or pseudohypoparathyroidism.

(b) Measurement of serum immunoreactive PTH should distinguish between hypoparathyroidism and pseudohypoparathyroidism. The serum PTH in patients with hypoparathyroidism should be undetectable, although in many assays a measurement in the low normal range may be obtained since few PTH assays reliably distinguish between low and normal concentrations. In pseudohypoparathyroidism, serum PTH should be increased because the parathyroid glands will be hyperplastic as they respond normally to the hypocalcemia caused by peripheral tissue resistance to PTH.

(c) The Ellsworth–Howard test and its variants. The distinction between hypoparathyroidism and pseudohypoparathyroidism can be made by an injection or infusion of PTH followed by measurement of renal phosphate excretion, renal calcium excretion, or, better still, urinary cAMP. In a patient with normal kidneys, PTH should cause a brisk increase in urine phosphate excretion, an increase in the serum calcium, and a striking increase in urinary cAMP. In patients with pseudohypoparathyroidism, these responses will be impaired or absent (see below). It should be noted that in patients with hypoparathyroidism, the absolute urine calcium and urine phosphorus may be in the normal range because of changes in the filtered load, but the fractional excretion will be altered. This is particularly applicable to urine calcium excretion. Because the filtered load is less, the absence of the effect of PTH on the distal tubules of the kidney to cause renal tubular calcium reabsorption may be masked unless fractional calcium excretion is calculated.

The Ellsworth–Howard test (or variations on it measuring cAMP) is performed less often nowadays. Parathyroid extract, which was formerly used in this test, has been removed from the market and replaced by human synthetic PTH 1–34. However, improvements in PTH assays may make this test obsolete.

Differential diagnosis of hypoparathyroidism

The major differential when patients present with hypocalcemia and hyperphosphatemia is to distinguish hypoparathyroidism from pseudohypoparathyroidism. However, within the category of hypoparathyroidism are not only the idiopathic varieties described above, but also hypoparathyroidism due to acquired causes. This includes surgical hypoparathyroidism associated with removal of the parathyroid glands inadvertently during thyroid or parathyroid surgery, and other infiltrative and destructive conditions of the parathyroid glands, including hemochromatosis, previous radioactive-iodine therapy for thyroid disease, neoplastic infiltration, which occurs rarely but has been described, and infiltrations with chronic inflammatory diseases such as tuberculosis, sarcoidosis, and amyloidosis. Acquired but reversible hypoparathyroidism may also be associated with hypomagnesemia (described below) and is occasionally seen following parathyroid gland surgery, and in infants born to mothers who have primary hyperparathyroidism (see below).

Post-surgical hypoparathyroidism

Clinical circumstances

Hypoparathyroidism occurring as a complication of surgery to the head and neck is now less frequent since this type of surgery has become less popular. The frequency of postoperative hypoparathyroidism depends not only on the skill and experience of the surgeon, but also the reason for operation. It is most likely to occur in three sets of circumstances. The most common of these is following thyroid surgery for Graves' disease or thyroid cancer. Surgery for Graves' disease has become much less frequent since radioactive iodine has become more established as a safe and effective form of treatment for Graves' hyperthyroidism. At the same time, since surgeons have less experience in performing subtotal thyroidectomy, the relative frequency of complications such as postoperative hypothyroidism or recurrent laryngeal nerve palsy has become more common. The usual operative

approach is to remove three-quarters to seven-eighths of the thyroid, but avoidance of the common complications requires considerable skill, particularly in situations where the thyroid gland is very vascular and enlarged, and where the parathyroid glands are in ectopic or atypical locations. The frequency of hypoparathyroidism is much higher when the surgery is for cancer, particularly where total thyroidectomy is indicated. Total thyroidectomy is more likely to be required in patients with medullary carcinoma. Total thyroidectomy is now used infrequently for papillary or follicular cancer, and anaplastic carcinoma is usually inoperable. When total thyroidectomy is performed, the frequency of postoperative hypoparathyroidism is greatly enhanced, and of course may be anticipated when it is necessary to remove all of the neoplastic tissue *en bloc*. The second type of surgery frequently associated with hypoparathyroidism is head and neck cancer surgery, in which radical removal of tumor tissue *en bloc* requires removal of the thyroid and adjacent structures. Post-surgical hypoparathyroidism is also common following surgery for parathyroid gland hyperplasia. This can be anticipated in three situations: in patients with familial multiple endocrine neoplasia syndromes; in patients with familial hypocalciuric hypercalcemia; and in patients with secondary hyperparathyroidism associated with renal failure. In patients with parathyroid gland hyperplasia it is very difficult to remove the precise amount of parathyroid gland tissue to leave the patient normocalcemic. The chances of postoperative hypoparathyroidism are high, and this complication should be anticipated. Postoperative hypocalcemia is particularly likely to occur in patients who have renal failure, who are magnesium deficient, or who have extensive bone lesions due to osteitis fibrosa before surgery. These patients develop the 'hungry bone' syndrome.

Postoperative course

Patients whose parathyroid glands are inadvertently removed during surgery are clearly hypocalcemic within 24 h. This hypocalcemia is still present after 1 week. In the usual clinical situation the clinician is not certain in the first week whether hypoparathyroidism is permanent or will eventually resolve. In some patients who are initially hypocalcemic after neck surgery, the hypocalcemia may be only transient. In these there is presumably ischemic vascular damage to the parathyroid glands suffered at the time of surgery. This type of hypocalcemia often improves spontaneously within 2 weeks, although it may take up to 6 months. Some of these patients may recover normal serum calcium levels, but are liable to intermittent hypocalcemia during periods of stress. These patients presumably suffer from impaired parathyroid gland reserve. Prolonged hypocalcemia is particularly likely to occur in patients with bone disease due to osteitis fibrosa cystica before surgery. The mechanism responsible for this type of hypocalcemia is not clear, but it may be due to enhanced uptake of calcium into the skeleton. This syndrome is known as the 'hungry bone' syndrome, and may be improved by the pre- and postoperative administration of active metabolites of vitamin D in patients who appear predisposed.[12]

In general, postoperative hypoparathyroidism is easier to manage than idiopathic hypoparathyroidism. The hypocalcemia remains fairly constant. Moreover, the patients are otherwise well and do not have other complications such as the endocrine gland abnormalities seen in many patients with idiopathic hypoparathyroidism, including adrenal failure, gonadal failure, or diabetes. Of course, the clinical situation will be more difficult to manage if the patients also have bilateral recurrent laryngeal nerve palsy.

Management

Management of patients with post-surgical hypoparathyroidism follows the same principles as that of patients with idiopathic hypoparathyroidism, although in the former situation the occurrence of postoperative hypoparathyroidism can often be anticipated before surgery. Consequently, healthy parathyroid gland tissue removed at surgery should be saved for cryopreservation and later autotransplanted into the musculature of the forearm. If hypoparathyroidism appears certain to occur after surgery, then the surgeon may implant healthy parathyroid gland tissue into the forearm musculature at the time of surgery. Autotransplantation of parathyroid gland tissue is discussed in more detail in Chapter 11.

Pseudohypoparathyroidism

Pseudohypoparathyroidism is a heterogeneous syndrome which appears to be caused by a combination of separate defects in the target-cell response to PTH. The condition is characterized by hypocalcemia and tissue resistance to the effects of increased circulating concentrations of PTH. Many patients also have a number of somatic abnormalities, which are referred to collectively as the phenotype of Albright's hereditary osteodystrophy. Other hormone resistance syndromes are also frequently present. This interesting condition has been thoroughly reviewed by Drezner and Neelon,[13] with an extensive bibliography of this condition.

Classification (Table 10.4)

Type I

These are patients with pseudohypoparathyroidism who have a decreased urine cAMP response to PTH. There are several subsets of this category.

Ia. These patients have decreased urine cAMP responses to PTH, decreased G_s units in red blood cells (G_s is the membrane protein complex which

Table 10.4 Classification of types of pseudohypoparathyroidism.

Type I	Decreased urine cAMP response to PTH Type Ia—decreased G_s units Type Ib—normal G_s units
Type II	Normal urine cAMP response to PTH, decreased phosphaturic response to PTH
Others	Pseudohypoparathyroidism with osteitis fibrosa Pseudopseudohypoparathyroidism

forms the functional link between the receptor and the catalytic unit of the adenylate cyclase enzyme), the somatic features of the pseudohypoparathyroidism phenotype, and multiple hormone resistance syndromes.

Ib. In these patients there is also decreased urine cAMP responsiveness to PTH. However, they have normal levels of the G_s units in their red blood cells. These patients may sometimes have the phenotype of Albright's hereditary osteodystrophy, although most appear normal.

Type II

A few patients have been described who have a normal cAMP response to PTH and normal G_s units in their red cells.[13] They may or may not have the phenotype of Albright's hereditary osteodystrophy. These patients, despite their normal cAMP response to PTH, have a subnormal phosphaturic response to PTH unless the serum calcium is corrected. It is therefore postulated that in these patients there is a defect between the generation of cAMP in the renal tubular cell and the effects of this increase in cAMP on phosphate reabsorption.

Pseudopseudohypoparathyroidism

These patients have normocalcemia and a normal cAMP response to PTH, but the physical appearance and somatic abnormalities which are characteristic of Albright's hereditary osteodystrophy.[14] Some have decreased G_s units, and they frequently have relatives with pseudohypoparathyroidism. These patients are difficult to diagnose. To make the diagnosis in these patients, they should have a first-degree relative with pseudohypoparathyroidism and obvious phenotypic features of the Albright's hereditary osteodystrophy.

Clinical manifestations of pseudohypoparathyroidism

Pseudohypoparathyroidism is characterized by two major features: hypocalcemia and the phenotype of Albright's hereditary osteodystrophy.

Hypocalcemia

The clinical manifestations of the hypocalcemia are the same as those due to hypocalcemia from other causes. They are characterized by neurologic features, cataracts, abnormal dentition, and cardiovascular abnormalities. They are described in detail earlier in this chapter.

Pseudohypoparathyroidism associated with metabolic bone disease

Occasional patients with pseudohypoparathyroidism also have osteitis fibrosa cystica, and occasionally osteomalacia. This has been recognized for more than 25 years. In these patients, measurement of serum 1,25-dihydroxyvitamin D is decreased. Therapy with 1,25-dihydroxyvitamin D reverses the bone changes and also improves the calcemic response to parathyroid extract, although the renal non-responsiveness to PTH (which can be assessed by changes in urine cAMP and phosphate excretion) persists. The osteitis fibrosa cystica is clearly due to the excess action of PTH on bone, so in these patients it is apparent that PTH action on bones is not substantially impaired. Some have considered that this syndrome may represent a variant of pseudohypoparathyroidism where the bone cell response to PTH is normal but the kidney is unresponsive. Bone lesions can be improved by restoring the serum calcium to the normal range with vitamin D therapy to suppress PTH levels in the circulation.

Phenotype of Albright's hereditary osteodystrophy

In 1942 Albright et al[15] first described the syndrome of pseudohypoparathyroidism as an example of a hormone resistance syndrome, in this case resistance of peripheral tissues to PTH. These patients also had a rather characteristic physical appearance (Figure 10.2). He later described several patients who had the phenotype of pseudohypoparathyroidism, but no apparent abnormalities in calcium homeostasis. He called this syndrome pseudo-pseudohypoparathyroidism. The somatic abnormalities are characterized by a round facies, short stature, thick neck, obesity, barrel chest, short phalanges,

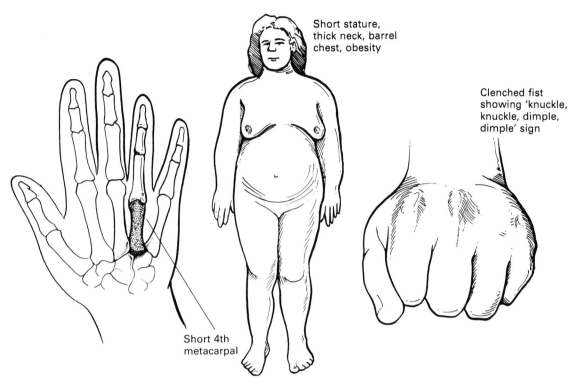

Short stature, thick neck, barrel chest, obesity

Clenched fist showing 'knuckle, knuckle, dimple, dimple' sign

Short 4th metacarpal

Figure 10.2

Characteristic somatic abnormalities of a patient with Albright's hereditary osteodystrophy.

short metacarpals and metatarsals, frequent mental deficiency, and olfactory impairment. These somatic features are variable. Many patients may have a phenotype which resembles that of Turner's syndrome. The shortening of the third and fourth metacarpals results in dimples over the fourth and fifth knuckles when the hand is clenched (Figure 10.3). The mental retardation is apparently not due to the effects of prolonged hypocalcemia on the central nervous system during early life, or to hypothyroidism.

Mode of inheritance

The mode of inheritance is still not clear. The condition does occur twice as frequently in females as it does in males. Early studies suggested that it was inherited as an X-linked trait because there was apparent female-to-male preponderance of affected individuals and lack of clear-cut evidence of male-to-male transmission. However, more recent reports suggest that an autosomal mode of inheritance may be more usual, since

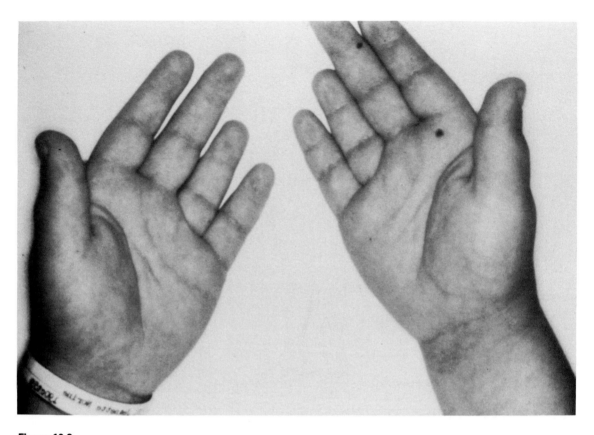

Figure 10.3

Hands of a patient with Albright's hereditary osteodystrophy, showing short fourth and fifth metacarpals.

father-to-son transmission has been noted.[16] Recently, Levine et al[17] performed pedigree analysis using measurement of the G_s units in patients with the most common form of pseudohypoparathyroidism (Type Ia) and from family members with Albright's hereditary osteodystrophy alone. They found that patients with pseudopseudohypoparathyroidism had reductions in red cell membrane G_s comparable with those in the patients with Type Ia. The pattern of inheritance was consistent either with autosomal dominant or with X-linked patterns of transmission.

Pathophysiology

Albright suggested that pseudohypoparathyroidism was a hormone resistance syndrome, similar to the Se(a)bright bantam where the male bantam fails to develop feathering like the normal male,

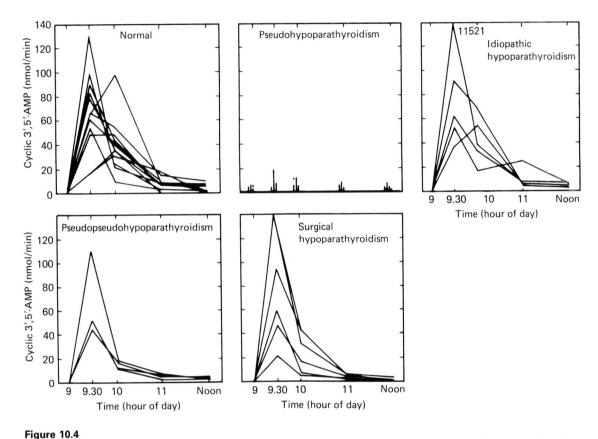

Figure 10.4

Effects of PTH on urinary cAMP excretion in patients with hypoparathyroidism and pseudohypoparathyroidism compared with normals. Note that the patients with pseudohypoparathyroidism show no response to PTH, whereas normals and patients with hypoparathyroidism show a brisk response. Similar, but less dramatic, responses are seen when fractional excretion of calcium (decreased) and fractional excretion of phosphate (increased) are measured. Reprinted with permission.[19]

but rather more like the female. In fact, more recently George et al[18] have shown that the male Sebright bantam's feathering is not due to resistance to androgens, but rather is due to an aromatase in the skin which converts circulating androgens to estrogens. However, Albright's concept that pseudohypoparathyroidism was a hormone resistance syndrome has proven to be correct. The early observations were based on failure of these patients to respond to infusions of parathyroid extract by increasing fractional excretion of phosphate or decreasing fractional excretion of calcium. Chase et al[19] showed that these patients had impaired responsiveness to PTH in generating increased urinary cAMP excretion (Figure 10.4). Although it is clear that the syndrome is due to resistance to the peripheral effects of PTH, the molecular basis for this underlying defect remains unclear. There are a number of possibilities.

Abnormal G$_s$ protein

This was first suggested by Drezner and colleagues,[20,21] based on their studies on a patient with pseudohypoparathyroidism in whom they were able to perform in vitro studies on renal cortical plasma membranes. They examined the responsiveness of this renal tissue to PTH as measured by adenylate cyclase activity, and found that PTH responsiveness was low compared with normal controls when guanine triphosphate (GTP) was excluded from the incubation medium, but was normal when GTP was added to the incubation medium. They postulated, based on this single study, that an abnormality in the G$_s$ to which GTP binds was responsible for the decrease in PTH responsiveness. This suggestion was confirmed independently by Bourne and Spiegel, who used a complementation assay in which they could measure levels of functional G$_s$ units. They examined red cell membranes from patients with pseudohypoparathyroidism, and found that most patients (but not all) had a deficiency in G$_s$ protein

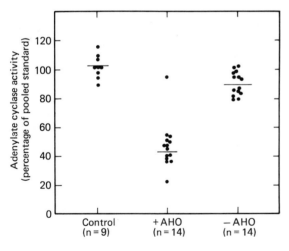

Figure 10.5

Measurement of G$_s$ unit activity in red cells in patients with the phenotype of Albright's hereditary osteodystrophy (AHO) or without it. (Reproduced with permission from Spiegel et al, Deficiency of hormone receptor-adenylate cyclase coupling protein: basis for hormone resistance in pseudohypoparathyroidism, *Am J Physiol* (1982), **243**: E37–42.)

in the red cells (Figure 10.5). Since that time, deficiencies in the G$_s$ protein have been found in other cells of patients with this disease, including platelets, fibroblasts, transformed lymphocytes, and kidney tissue.[22–27] The deficiency in G$_s$ is universal in patients with Type Ia pseudohypoparathyroidism, and is found in many patients with pseudopseudohypoparathyroidism. However, patients with Type Ib and Type II pseudohypoparathyroidism have normal G$_s$ units. Clearly, then, a deficiency in G$_s$ subunits cannot explain all of the features of the disease. Deficiency in the G$_s$ units correlates well with some features of the disease, in particular with mental deficiency and impaired olfaction.[28] It also helps to explain why some patients have other hormone resistance syndromes.

The specific defect in Type Ia pseudohypoparathyroidism is in the alpha subunit of the guanine nucleotide-binding protein (G$_s\alpha$). The deficiency in G$_s\alpha$ is present in target tissues for PTH such as bone and kidney, but is also present in non-target tissues such as fibroblasts and red blood cells. In most patients, messenger RNA for the α subunit is decreased but Southern analysis shows no detectable abnormality in the gene which encodes the α subunit.[29,30] There is approximately 15 per cent reduction in mRNA content for G$_s\alpha$. Restriction endonuclease analysis of genomic DNA from patients with Type Ia pseudohypoparathyroidism has not provided evidence of abnormalities in the G$_s\alpha$ gene such as deletions or rearrangements which could lead to altered gene expression.[29,30]

Abnormal PTH

Some reports have suggested that there is a discrepancy between immunoreactive PTH concentrations and bioactive PTH in the serum of patients with pseudohypoparathyroidism.[31,32] These findings suggest that an abnormality in PTH structure could account for some of the apparent hormone resistance that is present in these patients. These workers have postulated that the plasma of some patients with pseudohypoparathyroidism contains an inhibitor of PTH activity which is released by the parathyroid glands. When PTH was added to the plasma of a patient, they found difficulty in recovering bioactive PTH compared with the addition of PTH to normal plasma. They suggest that the abnormal PTH acts as an inhibitor of PTH effects.

Goltzman et al[33] have recently reviewed the case for release of altered forms of PTH in patients with pseudohypoparathyroidism. Using a sensitive cytochemical assay for PTH, they detected discrepancies between biologically active PTH and immunoreactive PTH in patients with pseudohypoparathyroidism. These altered forms of PTH may act as endogenous inhibitors of PTH action. These workers postulate that an alteration in PTH synthesis and processing within the parathyroid gland cell may occur as a consequence of an abnormality in $G_s\alpha$ within the parathyroid cell. Using a model of parathyroid hormone resistance in rats rendered vitamin D deficient (and as a consequence, hypocalcemic, with undetectable 1,25-dihydroxyvitamin D as well as increased circulating PTH), they found that these rats exhibited a decrease in $G_s\alpha$ activity in target tissues for PTH. This raises the possibility that increased circulating PTH may contribute to the reduction in $G_s\alpha$ in target tissues. As a consequence, this may first further reduce biological responses to PTH in target organs, and second, lead to altered forms of PTH being released from the parathyroid gland.

Defective cytosolic response to increased cAMP

The few patients (about twenty to date) who have been described with the rare Type II syndrome have a normal cAMP response to PTH but fail to show the metabolic effects of PTH such as a change in renal phosphate excretion. In these patients there is presumably a defect between the generation of cAMP within the cell and the mediation of its metabolic effects. It is possible that a similar defect is present in other patients with pseudohypoparathyroidism, but is not present as the sole defect. The defect in the phosphaturic response to PTH in these patients is corrected when the serum calcium is normalized.

Associated hormone resistance syndromes

Many patients with pseudohypoparathyroidism have associated hormone resistance syndromes.

The most common is probably hypothyroidism due to thyroid stimulating hormone (TSH) unresponsiveness. Even those patients who have normal thyroid function tests may have a supernormal TSH response to a thyrotropin releasing hormone (TRH) infusion. Many patients have hypogonadism, due to resistance to follicle stimulating hormone (FSH) and luteinizing hormone (LH). Most of the female patients seen with this condition have an increased frequency of menstrual irregularity, infertility, and poor sexual development. No hypogonadal features have been noted in men, although this has not been studied in detail. These patients may have decreased prolactin responses to TRH infusions and decreased hepatic glucose output in response to glucagon infusions. It has also been suggested that some patients may develop nephrogenic diabetes insipidus because of impaired responsiveness to antidiuretic hormone.

Natural history of the disease

This disease runs a variable course for reasons which are not clear. Some patients (approximately 5 per cent) with pseudohypoparathyroidism may have periods of normocalcemia. The patients with the most severe hypocalcemia tend to remain hypocalcemic, but those with mild hypocalcemia may show fluctuations in the serum calcium. These fluctuations in serum calcium are not necessarily correlated with PTH concentrations.[34–36] This is particularly likely to occur early in the course of the disease. The reasons are not known.

Treatment of pseudohypoparathyroidism

The treatment of pseudohypoparathyroidism is the same as that for any other form of hypoparathyroidism. Vitamin D and calcium supplements are required.

Hypocalcemia associated with malignant disease

Hypocalcemia with a low ionized or corrected serum calcium occurs in malignant disease as a consequence of accelerated bone formation, presumably caused by tumor cells producing a growth factor which stimulates osteoblasts to lay down new bone. As a result there is markedly enhanced entry of calcium into the skeleton, and the patients may develop profound hypocalcemia. This is seen most frequently in solid tumors such as prostatic cancer and breast cancer, but has even been reported in acute leukemia.[37,38] The reported frequency in prostate cancer is 31–45 per cent, and 13 per cent in lung or breast cancer. These patients frequently have evidence of osteoblastic metastases or osteosclerosis on skeletal X-ray photographs and marked elevations in the serum alkaline phosphatase, and bone biopsies when taken show evidence of increased osteoblast numbers, new bone formation occurring on bone surfaces without evidence of previous resorption and hypocalcemia and hypophosphatemia. In a recent case report,[38] a patient with acute leukemia had marked hypocalcemia and hypophosphatemia, and the hypocalcemia resolved when the leukemia was brought under control with cytotoxic drug therapy. Occasionally, new bone formation and hypocalcemia occurs in myeloma (Chapter 6). The cause here is also unknown. Therapy of the hypocalcemia is symptomatic with either oral or intravenous calcium and vitamin D. These patients may have appropriately elevated serum PTH, which excludes the diagnosis of hypoparathyroidism, and hypophosphatemia without evidence of renal phosphate wasting. No detectable abnormalities in vitamin D metabolism have been found.

Another potential cause of hypocalcemia associated with malignancy is due to tumor lysis, which may occur as a consequence of successful therapy of neoplastic disease.[39] This is most frequently seen in the hematologic malignancies, and it has been particularly noted in Burkitt's lymphoma. When tumor cells are lysed, they release intracellular ions such as phosphate and potassium. The hyperphosphatemia can lead to hypocalcemia, because a sudden increase in the serum phosphate will lead to a rapid fall in serum calcium, just as occurs when phosphate is infused intravenously (Chapter 7). In this syndrome hypocalcemia occurs in association with hyperphosphatemia, unlike the condition described above. These patients frequently develop renal failure associated with hyperuricemia.

Patients with neoplastic disease can also develop oncogenic osteomalacia due to renal phosphate wasting. These patients have normal serum calcium measurements and therefore should not be confused with patients who have osteomalacia on the basis of vitamin D deficiency.

Hypocalcemia and magnesium deficiency

The association of hypocalcemia and magnesium deficiency was first described by Heaton and Fourman in 1965.[40] It is a complicated association, since both electrolyte disturbances may occur together, and the symptoms of each are similar. It is extremely important to recognize magnesium deficiency in a patient with hypocalcemia, since the symptoms may not be reversed and the hypocalcemia not corrected adequately until magnesium repletion is complete.

Clinical features

When hypocalcemia occurs in patients with magnesium deficiency, the decrease in serum calcium is usually profound. It is often a complex situation associated with other deficiencies. Frequently the patient is an alcoholic with malabsorption, and may have coexistent osteomalacia and vitamin D deficiency. Symptoms due to magnesium deficiency include anorexia, nausea, tremor, mood swings, tetany, and seizures. Identical symptoms may be found in patients with hypocalcemia. However, these symptoms and signs may occur in patients with magnesium deficiency who are not hypocalcemic, and thus magnesium deficiency alone can be responsible. The important point is that these symptoms will not be reversed until the magnesium deficit has been corrected.

Hypokalemia is also a frequent association of magnesium deficiency, and it may be responsible for similar neuromuscular and cardiac effects.

Patients with magnesium deficiency are particularly liable to digoxin toxicity and sudden death from cardiac arrhythmias. These arrhythmias may be enhanced by associated hypokalemia or hypocalcemia. Arrhythmias which occur frequently are atrial premature contractions, sinus or nodal tachycardia, prolongation of the PR and QT intervals, and flattening of the T-waves. Ventricular premature contractions, ventricular tachycardia, and ventricular fibrillation may also occur.

The relationship between magnesium deficiency and digoxin toxicity is not entirely clear. Magnesium deficiency may increase the uptake of digoxin by myocardial cells, and the Na–K-ATPase pump may be impaired by both magnesium deficiency and cardiac glycosides which would cause reduction in intracellular potassium content.

Causes of magnesium deficiency

There are numerous cases of magnesium deficiency. These are listed in Table 10.5. In general, magnesium deficiency may occur because of internal shifts, decreased absorption of magnesium from the gut, or renal losses due to intrinsic tubular disorders or the effects of drugs.

Table 10.5   Causes of magnesium deficiency.

1 Renal loss
   (a) Intrinsic tubular disorders; for example, Bartter's syndrome, primary renal Mg washing, renal tubular acidosis, diuretic phase of acute renal failure
   (b) Drugs; for example, loop diuretics, aminoglycosides, alcohol
   (c) Mineralocorticoid excess

2 Decreased gut absorption
   (a) Malabsorption
   (b) Decreased intake

3 Internal shifts
   Refeeding after protein-calorie malnutrition, intravenous glucose, hungry bone syndrome

The most common cause, particularly when associated with hypocalcemia, is alcoholism. In this situation multiple factors may be involved. These include poor dietary intake, malabsorption and diarrhea, renal losses due to starvation ketosis, and the effects of acute ethanol ingestion. Alcohol may act on the renal tubules to cause renal tubular magnesium wasting.

Diagnosis

In any patient with hypocalcemia, the serum magnesium should be measured. If it is less than 0.5 mmol/l or 1.2 mg/dl, then symptoms may at least in part be due to magnesium deficiency. Patients with severe life threatening magnesium deficiency have a serum magnesium less than 0.25 mmol/l. A low 24-h urine magnesium is a reliable measure of total body magnesium deficiency in a patient in whom renal mechanisms are not responsible for the loss.

Mechanisms

The cause of hypocalcemia in patients with magnesium deficiency is not entirely clear. Two mechanisms have been suggested. Magnesium deficiency decreases PTH secretion from the parathyroid glands both in vitro and in vivo. Particularly in chronic situations, PTH secretion is diminished when magnesium is deficient and increased when magnesium is supplemented. However, magnesium deficiency may also interfere with the effects of PTH on peripheral tissues, and in particular bone and the kidneys. There are many studies in rats, dogs, and humans showing impaired PTH responsiveness in bones and kidneys in states of magnesium depletion. This has also been demonstrated both in vitro and in vivo.

It is frequently very difficult in the clinical situation to decide which of these mechanisms is likely to be the most important. The clinical circumstances surrounding magnesium deficiency are usually complex and heterogeneous. It is likely that in many patients both mechanisms (decreased secretion of PTH and impaired effectiveness of PTH on peripheral tissues) are involved.

Treatment

Where oral replacement with magnesium is appropriate, there are several different preparations from which to choose. These include magnesium glycolate, magnesium oxide, and milk of magnesia (magnesium hydroxide). The problem with all of these preparations is that they may induce diarrhea. Possibly magnesium oxide is the most suitable preparation. Doses of 250–500 mg magnesium oxide four times per day are usually well tolerated.

When severe deficits are present and urgent therapy is required, parenteral magnesium sulfate may be administered in a 50 per cent solution in 2 ml ampules (1 ml contains 4 mmol) at a maximum rate of 20 ml each 24 h. Magnesium sulfate is administered intravenously in dextrose and water. Whenever renal failure is present, the dose must be reduced. Serum concentrations of magnesium should be kept between 1 and 1.25 mmol/l during replacement. If hypokalemia and hypocalcemia are also present, then the magnesium deficiency should be corrected first and then adequate potassium and calcium replacement achieved.

The toxic-shock syndrome

The toxic-shock syndrome was described about 10 years ago in young women who use tampons during menstruation, and it is related to a toxin produced by certain strains of staphylococci. These patients develop shock characterized by hypotension and circulatory failure, high fevers, liver and renal dysfunction, and erythroderma. These patients frequently develop hypocalcemia which may last up to several days.[41] The mechanism responsible for the hypocalcemia is not clear. In some it may be related to acute renal failure and markedly decreased serum albumin. Some patients have elevations in serum calcitonin. Another possibility is that the toxin produced by the bacteria induces the production of mediators which lead to hypocalcemia. The author has noticed this with infusions of large amounts of tumor necrosis factor, which is thought to be a mediator of endotoxic shock, and interleukin-1. Patients with bacteremic shock are frequently hypocalcemic.[42] In the toxic-shock syndrome, hypocalcemia should be treated symptomatically.

Neonatal hypocalcemia

Neonatal hypocalcemia has been noted in the first week of postnatal life, and it occurs most often in premature children. It also occurs frequently in children born to diabetic mothers and who suffer from the respiratory distress syndrome. The mechanisms are unclear, but may be related to relatively low serum PTH and high serum calcitonin in the infant's blood in the first 2 or 3 days of life in response to active placental transfer of calcium from the mother to the fetus. These infants are also often hypomagnesemic, which is another factor which may impair the infant's capacity to secrete PTH and enhance hypocalcemia. Magnesium depletion may also explain why infants born to mothers with primary hyperparathyroidism may have severe neonatal hypocalcemia, usually occurring after the first week of life and probably related to transient suppression of PTH secretion in the infant.

Hypocalcemia associated with pancreatitis

Hypocalcemia is occasionally associated with pancreatitis. The mechanisms may be multifactorial and related to the association of acute pancreatitis with acute renal failure, or increased circulating concentrations of calcitonin which lower the serum calcium, or possibly the release of pancreatic enzymes which cause calcium deposition at sites of fat necrosis to form insoluble calcium soaps. However, the amount of calcium deposited in areas of fat necrosis is unlikely to be sufficient to account for the hypocalcemia. Glucagon release has also been implicated as a potential mechanism. Whenever patients with acute pancreatitis are normocalcemic or hypercalcemic, then primary hyperparathyroidism should be considered, since occasionally primary hyperparathyroidism can present as acute pancreatitis (Chapter 8).

Other causes of hypocalcemia

Vitamin D deficiency and chronic renal disease are important causes of hypocalcemia, which will be the subjects of further volumes in this series, and have not therefore been considered in detail here.

Treatment of hypocalcemia

Although the principles of managment of hypocalcemia, and in particular hypoparathyroidism, have not changed in the past 40 years, the clinician is now faced with a variety of therapeutic alternatives from which to choose. The goals of treatment are: (a) to increase the serum calcium into the range in which symptoms and complications directly related to hypocalcemia do not occur; (b) to prevent skeletal deformity due to secondary hyperparathyroidism, in those situations where chronic hypocalcemia is present; (c) to avoid hypercalciuria and hypercalcemia and their consequences; and (b) in children, to promote normal growth and development. It is extremely difficult to achieve all of these goals in an individual patient. Moreover, treating the patient who has longstanding hypocalcemia, particularly if it is due to hypoparathyroidism, can be a frustrating and difficult experience which requires the co-operation of the patient and tolerance by the physician. Of course, the major problem is that the primary hormonal deficiency is PTH lack, but there is no suitable PTH preparation available for replacement therapy. The mainstay of therapy is still oral calcium supplementation together with a form of vitamin D. However, there are now numerous forms of both vitamin D metabolites and oral calcium preparations to choose from, and this choice will be influenced by factors such as cost, previous experience, and preference of both the clinician and the patient. This topic has been reviewed by Avioli and others.[1–3]

Indications for treatment

Most patients with hypocalcemia present with a corrected serum calcium between 5 and 9 mg/dl (1.25 and 2.25 mmol/l). Rarely will the serum calcium be less than 5 mg/dl. However, it may fluctuate between 5 and 9 mg/dl (1.25 and 2.25 mmol/l), depending on the diet and the compliance of the patient with the therapeutic regimen. Treatment is particularly difficult when hypocalcemia is due to hypoparathyroidism. The reason is that since PTH itself is not available as a form of therapy, the serum calcium must be increased by the use of oral calcium together with a form of vitamin D. Since vitamin D therapy will increase the serum calcium by increasing gut absorption, but will not have the pronounced effect of PTH on the renal tubules to promote renal tubular calcium reabsorption, the consequence of an increase in the serum calcium by administration of vitamin D or an active metabolite in the absence of PTH is hypercalciuria. For this reason, when vitamin D is used to increase the serum calcium of patients with PTH deficiency, the aim is to bring the serum calcium back towards the normal range, but not to increase the urine calcium to a level where it is likely to lead to renal complications, renal stones, or nephrocalcinosis.

The indications for treatment of hypocalcemia are symptoms and signs which can be ascribed to hypocalcemia such as tetany, paraesthesiae, seizures, and evidence of neuromuscular irritability which can be induced by maneuvers which provoke Chvostek's sign or Trousseau's sign (Chapter 10). In addition, the author believes it is worthwhile actively treating patients who have a corrected serum calcium of <7 mg/dl, even if they have no symptoms or signs which can be ascribed to hypocalcemia.

Urgent treatment of hypocalcemia

Hypocalcemia may be associated with severe symptoms and be acutely life-threatening when severe. Conversely, it may be chronic and mild, and be associated with more minor but nevertheless irritating symptoms. The management of these situations differ. Life-threatening hypocalcemia is most likely to occur immediately after a surgical operation in which the parathyroid glands are ablated and hypoparathyroidism occurs abruptly. It may also occur in the patient with a tumor associated with widespread osteoblastic metastases. It is likely that the symptoms will be severe in these cases because the hypocalcemia develops rapidly. In these circumstances the serum calcium must be corrected rapidly. Calcium chloride (a 10 per cent solution containing 9 mg/ml elemental calcium) should be infused intravenously (10–30 ml over 15 min). While calcium chloride is being administered, the patient should have continuous cardiac monitoring by EKG, particularly if they are already digitalized. Calcium gluconate can also be administered by infusion, in 5 per cent glucose in water over 4 h to provide 10–15 mg calcium per kilogram body weight. The major side-effects are the precipitation of digitalis toxicity, hence the reason for cardiac monitoring, and irritation of veins if the solution is more concentrated than that recommended above. In patients who can swallow, oral calcium is begun immediately, in doses of 200 mg elemental calcium every 2 h. For patients who cannot take oral medications, continuous intravenous calcium can be administered by adding 10 ml 10 per cent solution to 500 ml isotonic saline given every 6 h.

Chronic treatment of hypocalcemia

The principles of the chronic treatment of hypocalcemia are to give sufficient oral calcium and vitamin D that the absorption of calcium from the gut is increased sufficiently to increase the serum calcium towards the normal range and to render the patient asymptomatic. Most patients will be asymptomatic if the corrected serum calcium is >8 mg/dl (2 mmol/l). Many different preparations of oral calcium are available. Some of these are listed in Table 11.1. Perhaps the most popular

Table 11.1 Oral calcium preparations for the treatment of chronic hypocalcemia.

Salt	Tablet size (mg)	Elemental Ca (%)	*Required dose/day (tablets)
Calcium gluconate	500	9	22
Calcium lactate	325	13	24
Calcium carbonate	1250	40	2
Calcium citrate	1000	24	4

*For 1 g elemental calcium.

preparation currently available is calcium carbonate, which is also used by many physicians for its potential beneficial effects on bone mass. However, calcium gluconate or calcium citrate are equally suitable. The major side-effects of these oral calcium preparations are bad taste, gastrointestinal upsets, and failure of absorption. The potential for calcium carbonate to cause the milk-alkali syndrome is reviewed in Chapter 9.

Vitamin D preparations (Table 11.2)

One of the most difficult issues in the management of chronic hypocalcemia is the choice of an

Table 11.2 Drugs useful in treating hypocalcemia and osteopenia resistant to physiological doses of vitamin D.

Nonproprietary name	Abbreviation	Commercial name	Effective daily dose	Time for reversal of toxic effects (days)
Ergocalciferol	Vitamin D_2	Calciferol	1–10 mg	17–60
Calcifediol	$25(OH)D_3$	Calderol	0.05–0.5 mg	7–30
Dihydrotachysterol	—	Dihydrotachysterol	0.1–1 mg	3–14
Alfacalcidol	$1\alpha(OH)D_3$	One-alpha*	1–2 µg	5–10
Calcitriol	$1,25(OH)_2D_3$	Rocaltrol	0.5–1 µg	2–10

*Not available in the USA; available in Canada, the UK, and Japan.

effective form of vitamin D to increase calcium absorption from the gut. Many preparations are available, and the final choice will depend on cost and suitability in the individual patient. An unexplained finding by most clinicians treating patients with hypoparathyroidism is the variation in sensitivity between vitamin D preparations and from time to time in an individual patient. Changes in vitamin D requirements may occur as a consequence of concomitant administrations of drug such as anti-convulsants, thiazide diuretics, and antacids, or the development of electrolyte disturbances such as hypomagnesemia or hyperphosphatemia. In the author's view the wise course is to choose a preparation and stick with it in most patients. If the clinician must change to a new preparation in an individual patient, then a useful rule is to use one-third less to lessen the risks of toxicity. Idiopathic hypoparathyroidism may be much harder to control than postoperative hypoparathyroidism. 1,25-Dihydroxyvitamin D or a similar 1-hydroxylated analog are very convenient to administer, they restore normocalcemia within 3–6 days, and should they cause toxicity, these effects dissipate within several days of their removal. However, although less convenient to use, the older vitamin D preparations such as ergocalciferol (vitamin D_3) are much cheaper. The author advises starting with a dose of ergocalciferol 0.5 mg/day, and raising this by 0.25 mg/day every 2 weeks until the serum calcium reaches 8 mg/dl (2 mmol/l). Table 11.2 shows the various properties of the forms of vitamin D which are important in considering the treatment of patients with hypocalcemia. The major side-effects associated with vitamin D therapy are hypercalcemia, hypercalciuria, soft-tissue calcification, and nephrocalcinosis. Hypercalcemia should be rare in patients who are properly monitored (see below), particularly when the vitamin D metabolites with the longest half-lives are used. When these patients develop hypercalcemia, they have typical clinical features including anorexia, nausea, vomiting, weight loss, polyuria, and polydypsia, and alterations in mental status. Infants and small children may show listlessness and hypotonic muscles. However, hypercalcemia complicating therapy may be difficult to anticipate. Unfortunately, there is a very small window between therapeutic levels and toxic levels with all of the vitamin D preparations. Moreover, with the older preparations with the longest half-lives, toxicity can persist for some time after the drug is withdrawn. To avoid intoxication the patient should be followed at regular intervals, possibly every 1 to 2 months, with monitoring of the serum calcium and the urine calcium. The urine calcium should be maintained at <300 mg/g creatinine and the serum calcium at approximately 8 mg/dl (2 mmol/l) to avoid symptoms of hypercalcemia and hypercalciuria. If higher doses are used so that the serum calcium is in the normal range, then renal calcium excretion may be unacceptably

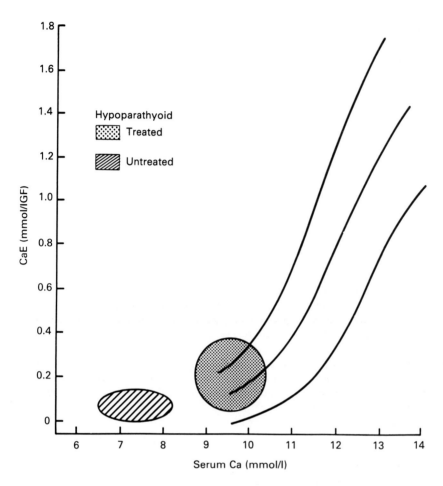

Figure 11.1

Effects of treatment with vitamin D metabolite and oral calcium on renal calcium excretion in hypoparathyroidism.

high (Figure 11.1). When patients become intoxicated with a vitamin D metabolite, the metabolites should be discontinued and fluids should be forced. If hypercalcemia is severe, patients respond well to a short course of glucocorticoids or salmon calcitonin. For reasons which are unexplained, some patients who may be very difficult to treat with vitamin D_2 may be managed more easily with the metabolites with shorter half-lives.

Use of thiazide therapy in the treatment of hypocalcemia

Since it has long been known that thiazides can enhance renal tubular calcium reabsorption and may increase the serum calcium, they have been used as supplemental forms of therapy in patients with longstanding hypoparathyroidism to circumvent the need for vitamin D or to decrease urinary calcium excretion. Hydrochlorothiazide or chlorthalidone may be used.[4] Hydrochlorothiazide in doses of 25–50 mg orally twice daily is an appropriate therapy. This form of therapy can decrease the doses of calcium and calcitriol (or other vitamin D preparations) required to maintain patients asymptomatic. It has not been carefully documented how effective this form of therapy is.

Treatment of hyperphosphatemia

In most patients with hypoparathyroidism the serum calcium is markedly reduced and the serum phosphorus is increased, since lack of PTH leads to enhanced renal tubular phosphate reabsorption. A major risk for these patients is that an increase in the serum calcium in a patient who is hyperphosphatemic may cause soft-tissue calcium deposition. This risk may be enhanced in a patient also treated with vitamin D, because vitamin D therapy will increase phosphate absorption from the gut in addition to increasing the serum calcium. In fact, Eisenberg[5] has found that correction of the serum calcium in patients with hypoparathyroidism increases renal phosphate wasting, so that increasing the calcium–phosphate product is usually not a practical problem. There are other treatments which can limit this risk, but these are rarely needed. They include the use of amphogel (aluminum hydroxide) or acetazolamide either to impair phosphate absorption from the gut or to increase renal phosphate wasting. Theoretically, if vitamin D is used in the treatment of hypoparathyroidism, then a high serum phosphorus would impair conversion of vitamin D to biologically active metabolites, particularly in a patient with PTH deficiency. For this reason 1,25-dihydroxyvitamin D is probably the best (albeit most expensive) form of therapy for patients with hypocalcemia due to hypoparathyroidism.

Parathyroid autotransplantation

Since the current management of hypoparathyroidism with calcium and the vitamin D metabolites is still not entirely satisfactory, it is not surprising that in the past 10 years there has been

a fashion for either autotransplanting parathyroid tissue at the time of parathyroidectomy into the muscles of the forearm or cryo-preserving parathyroid tissue for the later insertion of viable and functioning PTH-secreting cells into the forearm musculature. The new era of parathyroid transplantation began in the mid-1970s following reports by Wells et al[6] and Hickey and Samaan[7] on the successful autotransplantation of functioning parathyroid tissue into the muscles of the forearm, documented by demonstration of a gradient of immunoreactive PTH across the graft. The autotransplantation of parathyroid tissue is used principally in three clinical settings: (a) after total thyroidectomy for cancer, where parathyroid glands are almost certain to be removed; (b) after surgery for secondary hyperparathyroidism in patients with chronic renal disease, where it is extremely difficult to leave sufficient parathyroid tissue both to maintain normal calcium homeostasis and to avoid parathyroid bone disease; and

Table 11.3   Treatment of hypocalcemia.

Chronic
  Vitamin D metabolite
  Oral calcium
  Parathyroid gland autotransplantation (if due to primary hypoparathyroidism)

Urgent
  Intravenous calcium
  Short-acting vitamin D metabolite

(c) after surgery for primary parathyroid gland hyperplasia, where again it is difficult to achieve normocalcemia following the surgical procedure. It is important to consider the advantages and disadvantages of this approach alongside standard medical therapy for the individual patient with hypoparathyroidism, discussed above. In some patients, particularly those in whom hypoparathyroidism after surgery seems certain, autotransplantation should be given serious consideration.

The topic of parathyroid autotransplantation has been extensively reviewed by Niederle et al.[8]

The treatment of hypocalcemia is summarized in Table 11.3.

APPENDIX

Conversion tables for SI units and traditional units

Normal adult ranges for measurements used in the investigation of patients with disorders of calcium homeostasis.

Measurement		SI units	Traditional units
Plasma			
Total calcium		2.20–2.60 mmol/l	9–10.5 mg/dl
Ionized calcium		1.10–1.35 mmol/l	2.20–2.70 mg/dl
Fasting inorganic phosphate		1.0–1.5 mmol/l	3–4.5 mg/dl
Magnesium		0.8–1.3 mmol/l	1.5–2.5 mEq/l
Urine			
Calcium	M	2.5–10 mmol/24h	<300 mg/24h
	F	2.5–9.0 mmol/24h	<250 mg/24h
Phosphate		16–32 mmol/24h	500–1000 mg/24h
Total hydroxyproline	M	55–250 μmol/24h	
	F	75–430 μmol/24h	
Fasting urine			
Calcium/creatinine ratio		0.10–0.32 mmol/mmol	100–400 mg/g
Total hydroxyproline/creatinine		<40 μmol/mmol	<50 mg/g
cAMP		<10 μmol	<5 μmol

M or F denotes sex

Conversion table for SI units and traditional units

Serum total calcium	4 mg/dl	= 1 mmol/l
Serum albumin	4.0 g/dl	= 40 g/l
Serum creatinine	1.0 mg/dl	= 113 μmol/l
Serum phosphorus	3.1 mg/dl	= 1 mmol/l
Serum magnesium	2.4 mg/dl	= 1 mmol/l

REFERENCES

Chapter 1

1 Rasmussen H, Bordier P, *The physiological and cellular basis of metabolic bone disease* (Williams and Wilkins: Baltimore 1974).

2 Rasmussen H, Barrett PQ, Calcium messenger system: an integrated view, *Physiol Rev* (1984) **64**: 938–84.

3 Somlyo AP, Cellular site of calcium regulation, *Nature* (1984) **309**: 516–17.

4 Parfitt AM, The actions of parathyroid hormone on bone. Relation to bone remodelling and turnover, calcium homeostasis and metabolic bone disease. II. PTH and bone cells: bone turnover and plasma calcium regulation, *Metabolism* (1976) **25**: 909–55.

5 Parfitt AM, Equilibrium and disequilibrium hypercalcemia: new light on an old concept, *Metab Bone Dis Relat Res* (1979) **1**: 279–93.

6 Parfitt AM, Bone and plasma calcium homeostasis, *Bone* (1987) **1**: 51–8.

7 Nordin BEC, Peacock M, Role of kidney in regulation of plasma-calcium, *Lancet* (1969) **ii**: 1280–3.

8 Jones SJ, Boyde A, Scanning electron microscopy of bone cells in culture. In: Copp H, Talmage RV, eds, *Endocrinology of calcium metabolism* (Excerpta Medica: Amsterdam 1978) 97–114.

9 Harinck HIJ, Bijvoet OLM, Plantingh AST, Role of bone and kidney in tumor-induced hypercalcemia and its treatment with bisphosphonate and sodium chloride, *Am J Med* (1987) **82**: 1133–42.

10 Parfitt AM, Plasma calcium control at quiescent bone surfaces: a new approach to the homeostatic function of bone lining cells, *Bone* (1989) **10**: 87–8.

11 Staub JF, Tracqui P, Lausson S et al, A physiological view of in vivo calcium dynamics: the regulation of a nonlinear self-organized system, *Bone* (1989) **10**: 77–86.

12 Staub JF, Tracqui P, Brezillon P et al, Calcium metabolism in the rat: a temporal self-organized model, *Am J Physiol* (1988) **254**: R134–49.

13 Moore-Ede MC, Physiology of the circadian timing system: predictive versus reactive homeostasis, *Am J Physiol* (1986) **250**: R735–52.

14 Harms HM, Captaina U, Kulpmann WR et al, Pulse amplitude and frequency modulation of parathyroid hormone in plasma, *J Clin Endocrinol Metab* (1989) **69**: 843–51.

15 Payne RB, Jones DP, Walker AP et al, Clustering of serum calcium and magnesium concentrations in siblings, *Clin Chem* (1986) **32**: 349–50.

16 Copp DH, Moghadan H, Mensen ED et al, The parathyroids and calcium homeostasis. In: Greep RO, Talmage RV, eds, *The parathyroids* (Thomas: Springfield, Illinois 1961) 203–23.

17 Staub JF, Tracqui P, Brezillon P et al, Calcium metabolism in the rat: a temporal self-organized model, *Am J Physiol* (1988) **254**: R134–49.

18 Nordin BEC, Plasma calcium and plasma magnesium homeostasis. In: Nordin BEC, ed., *Phosphate and magnesium metabolism* (Churchill-
Livingstone: Edinburgh 1976) 186–216.

19 Peacock M, Robertson WG, Nordin BEC, Relation
between serum and urine calcium with particular
reference to parathyroid activity, *Lancet* (1969) **i**:
384–6.

20 Tuttle KR, Kunau RT, Mundy GR, Renal tubular
calcium reabsorption is increased in hypercalcemia
of malignancy, *Clin Res* (1989) **37**: 463A.

21 Pak CYC, Kaplan R, Bone H et al, A simple test for
the diagnosis of absorptive, resorptive and renal
hypercalciurias, *N Engl J Med* (1975) **292**: 497–500.

22 Percival RC, Yates AJP, Gray RES et al, Mechanisms
of malignant hypercalcemia in carcinoma of the
breast, *Br Med J* (1985) **291**: 776–9.

Chapter 2

1 Wasserman RH, Chandler JS, Molecular mechanisms of intestinal calcium absorption. In Peck WA,
ed., *Bone and mineral research, annual 3* (Excerpta
Medica: Amsterdam 1985) 181–212.

2 Bell NH, Vitamin D—endocrine system, *J Clin Invest*
(1985) **76**: 1–6.

3 DeLuca HF, Metabolism and mechanism of action
of vitamin D. In: Peck WA, ed., *Bone and mineral
research, annual 1* (Excerpta Medica: Princeton
1982) 7–73.

4 Bourdeau JE, Burg MB, Effect of PTH on calcium
transport across the cortical thick ascending limb of
Henle's loop, *Am J Physiol* (1980) **239**: F121–6.

5 Klahr S, Hruska K, Effects of parathyroid hormone
on the renal reabsorption of phosphorus and
divalent cations. In: Peck WA, ed., *Bone and
mineral research, annual 2* (Excerpta Medica:
Amsterdam 1983) 65–124.

6 Mundy GR, Roodman GD, Osteoclast ontogeny and
function. In: Peck WA, ed., *Bone and mineral
research V* (Elsevier: Amsterdam 1987) 209–80.

7 Baron R, Vignery A, Horowitz M, Lymphocytes,
macrophages and the regulation of bone remodeling. In: Peck WA, ed., *Bone and mineral research,
annual 2* (Elsevier: Amsterdam 1984) 175–243.

8 Garrett IR, Boyce BF, Mundy GR, Osteoclastic bone
resorption in vitro is dependent on oxygen-derived
free radical production, *J Bone Miner Res* (1987) **2**:
373.

9 Caplan AI, Pechak DG, The cellular and molecular
embryology of bone formation. In: Peck WA, ed.,
Bone and mineral research V (Elsevier: Amsterdam
1987) 117–83.

10 Frost HM, Dynamics of bone remodeling. In: *Bone
biodynamics*, Ch. 18 (Little, Brown and Co.: Boston
1964) 315.

11 Rasmussen H, Bordier P, *The physiological and
cellular basis of metabolic bone disease* (Williams
and Wilkins: Baltimore 1974) 29, 40.

12 Howard GA, Bottemiller BL, Turner RT et al, Parathyroid hormone stimulates bone formation and
resorption in organ culture: evidence for a coupling
mechanism, *Proc Natl Acad Sci USA* (1981) **78**:
3204–8.

13 Jones SJ, Boyde A, Scanning electron microscopy
of bone cells in culture. In: Copp DH, Talmage RV,
eds, *Endocrinology of calcium metabolism*
(Excerpta Medica: Amsterdam 1978) 97–114.

14 Beresford JN, Gallagher JA, Poser JW et al, Production of osteocalcin by human bone cells in vitro.
Effects of 1,25(OH)$_2$D$_3$, 24,25(OH)$_2$D$_3$, parathyroid
hormone, and glucocorticoids, *Metab Bone Dis
Relat Res* (1984) **5**: 229–34.

15 Talmage RV, Cooper CW, Toverud SU, The physiological significance of calcitonin. In: Peck WA,
ed., *Bone and mineral research, annual 1* (Excerpta
Medica: Amsterdam 1983) 74–143.

16 Neuman NW, Blood: bone equilibrium, *Calcif Tissue Int* (1982) **34**: 117–20.

17 Bushinsky DA, Chabala JM, Levi-Setti R, Ion microprobe analysis of mouse calvariae in vitro: evidence for a "bone membrane", *Am J Physiol* (1989)
256: E152–58.

18 Levinskas GJ, Neuman WF, The solubility of bone
mineral. I. Solubility studies of synthetic hydroxyapatite, *J Physiol Chem* (1955) **59**: 164–68.

19 Parfitt AM, Plasma calcium control at quiescent bone surfaces, a new approach to the homeostatic function of bone lining cells, *Bone* (1989) **10**: 87–8.

Chapter 3

1   Kronenberg HM, Igarashi T, Freeman MW et al, Structure and expression of the human parathyroid hormone gene, *Recent Prog Horm Res* (1986) **42**: 641–63.

2   Potts JT Jr, Kronenberg HM, Rosenblatt M, Parathyroid hormone: chemistry, biosynthesis, and mode of action, *Adv Protein Chem* (1982) **35**: 323–96.

3   Rosenblatt M, Pre-proparathyroid hormone, proparathyroid hormone, and parathyroid hormone: the biologic role of hormone structure, *Clin Orthop* (1982) **170**: 260–76.

4   Brown EM, Parathyroid secretion in vivo and in vitro. Regulation by calcium and other secretagogues, *Miner Electrolyte Metab* (1982) **8**: 130–50.

5   Barnicot NA, The local action of the parathyroid and other tissues on the bone in intracerebral grafts, *J Anat* (1948) **82**: 233–48.

6   Gaillard PJ, Parathyroid and bone tissue in culture. In: Greep RO, Talmage RV, eds, *The parathyroids* (Thomas: Springfield, Illinois 1961) 20–45.

7   Raisz LG, Bone resorption in tissue culture. Factors influencing the response to parathyroid hormone, *J Clin Invest* (1965) **44**: 103–16.

8   Mundy GR, Roodman GD, Osteoclast ontogeny and function. In: Peck W, ed., *Bone and mineral research V* (Elsevier: Amsterdam 1987) 209–80.

9   McSheehy PMJ, Chambers TJ, Osteoblastic cells mediate osteoclastic responsiveness to parathyroid hormone, *Endocrinology* (1986) **118**: 824–8.

10 Perry HM, Chappel JC, Bellorin-Font E et al, Parathyroid receptors on circulating human mononuclear leukocytes, *J Biol Chem* (1984) **259**: 5531–6.

11 Ibbotson KJ, Roodman GD, McManus LM et al, Identification and characterization of osteoclast-like cells and their progenitors in cultures of feline marrow mononuclear cells, *J Cell Biol* (1984) **99**: 471–80.

12 Kurihara N, Civin C, Roodman GD, Identification of a pure population of committed osteoclast precursors, *J Bone Min Res* (1989) **4**: Abstract 324.

13 Hakeda Y, Hiura K, Sato T et al, Existence of parathyroid hormone binding sites on murine hemopoietic blast cells, *Biochem Biophys Res Commun* (1989) **163**: 1481–6.

14 Lorenzo JA, Raisz LG, Hock JM, DNA synthesis is not necessary for osteoclastic responses to parathyroid hormone in cultured fetal rat long bones, *J Clin Invest* (1983) **72**: 1924–9.

15 Peck WA, Cyclic AMP as a second messenger in the skeletal actions of parathyroid hormone: a decade-old hypothesis, *Calcif Tissue Int* (1979) **29**: 1–4.

16 Jones SJ, Boyde A, Scanning electron microscopy of bone cells in culture. In: Copp DH, Talmage RV, eds, *Endocrinology of calcium metabolism* (Excerpta Medica: Amsterdam 1978) 97–114.

17 Rouleau MF, Mitchell J, Goltzman D, In vivo distribution of parathyroid hormone receptors in bone: evidence that a predominant osseous target cell is not the mature osteoblast, *Endocrinology* (1988) **123**: 187–91.

18 Dietrich JW, Canalis EM, Maina DM et al, Hormonal control of bone collagen synthesis in vitro: effects of parathyroid hormone and calcitonin, *Endocrinology* (1976) **98**: 943–9.

19 Garabedian M, Tanaka Y, Holick MF et al, Response of intestinal calcium transport and bone calcium mobilization to 1,25-dihydroxyvitamin D_3 in thyroparathyroidectomised rats, *Endocrinology* (1974) **94**: 1022–7.

20 Klahr S, Hruska K, Effects of parathyroid hormone on the renal reabsorption of phosphorus and divalent cations. In: Peck WA, ed., *Bone and mineral research, annual 2* (Excerpta Medica: Amsterdam 1983) 65–124.

21 Yamamoto M, Kawanobe Y, Takahashi H et al, Vitamin D deficiency and renal calcium transport in the rat, *J Clin Invest* (1984) **74**: 507–13.

22 Bell NH, Vitamin D—endocrine system, *J Clin Invest* (1985) **76**: 1–6.

23 DeLuca HF, Metabolism and mechanism of action of vitamin D. In: Peck WA, ed., *Bone and mineral research, annual 1* (Excerpta Medica: Princeton 1982) 7–73.

24 Norman AW, Roth J, Orci L, The vitamin D endocrine system: steroid metabolism, hormone, receptors, and biological response (calcium binding proteins), *Endocr Rev* (1982) **3**: 331–66.

25 Haddad JG, Traffic, binding and cellular access of vitamin D sterols. In: Peck WA, ed., *Bone and mineral research V* (Elsevier: Amsterdam 1987) 281–308.

26 Cooke NE, David EV, Serum vitamin D-binding protein is a third member of the albumin and alpha fetoprotein gene family, *J Clin Invest* (1985) **76**: 2420–4.

27 Cooke NE, Rat vitamin D binding protein: determination of the full-length primary structure from cloned cDNA, *J Biol Chem* (1986) **261**: 3441–50.

28 Yang F, Brune JL, Naylor SL et al, Human group-specific component (Gc) demonstrates striking homology with the albumin family of genes, *Proc Natl Acad Sci USA* (1985) **82**: 7994–8.

29 Bikle DD, Gee E, Halloran B, et al, Free 1,25-dihydroxyvitamin D levels in serum from normal subjects, pregnant subjects, and subjects with liver disease, *J Clin Invest* (1984) **74**: 1966–71.

30 Raisz LG, Trummel CL, Wener JA et al, Effect of glucocorticoids on bone resorption in tissue culture, *Endocrinology* (1972) **90**: 961–7.

31 Adams JS, Sharma OP, Gacad MA et al, Metabolism of 25-hydroxyvitamin D_3 by cultured pulmonary alveolar macrophages in sarcoidosis, *J Clin Invest* (1983) **72**: 1856–60.

32 Breslau NA, McGuire JL, Zerwekh JE et al, Hypercalcemia associated with increased serum calcitriol levels in three patients with lymphoma, *Ann Intern Med* (1984) **100**: 1–7.

33 Weisman Y, Harell A, Edelstein S et al, 1α,25-dihydroxyvitamin D_3 and 24,25-dihydroxyvitamin D_3 in vivo synthesis by human decidua and placenta, *Nature* (1979) **281**: 317–19.

34 Gray TK, Lester GE, Lorenc RS, Evidence for extra renal 1α-hydroxylation of 25-hydroxyvitamin D_3 in pregnancy, *Science* (1979) **204**: 1311–13.

35 Haussler MR, Mangelsdorf DJ, Komm BS et al, Molecular biology of the vitamin D hormone. In: Cohn DV, Martin TJ, Meunier PJ, eds, *Calcium regulation and bone metabolism*, Vol. 9 (Excerpta Medica: Amsterdam 1987) 465–74.

36 Evans RM, The steroid and thyroid hormone receptor superfamily, *Science* (1988) **240**: 889–95.

37 Haussler MR, Mangelsdorf DJ, Komm BS et al, Molecular biology of the vitamin D hormone, *Recent Prog Horm Res* (1988) **44**: 263–305.

38 McDonnell DP, Pike JW, O'Malley BW, The vitamin D receptor: a primitive steroid receptor related to thyroid hormone receptor, *J Steroid Biochem* (1988) **30**: 41–6.

39 Marx SJ, Liberman UA, Eil C et al, Hereditary resistance to 1,25-dihydroxyvitamin D, *Recent Prog Horm Res* (1984) **40**: 589–620.

40 Hughes MR, Malloy PJ, Kieback DG et al, Broken zinc fingernails cause inherited rickets, *J NIH Res* (1989) July/August: 94–6.

41 Roodman GD, Ibbotson KJ, MacDonald BR et al, 1,25 dihydroxyvitamin D_3 causes formation of multinucleated cells with several osteoclast characteristics in cultures of primate marrow, *Proc Natl Acad Sci USA* (1985) **82**: 8213–17.

42 Underwood JL, DeLuca HF, Vitamin D is not directly necessary for bone growth and mineralization, *Am J Physiol* (1984) **246**: E493–8.

43 Abe E, Miyaura C, Sakagami H, Differentiation of mouse myeloid leukemia cells induced by 1 alpha,25-dihydroxyvitamin D_3, *Proc Natl Acad Sci USA* (1981) **78**: 4990–4.

44 Reitsma PH, Rothberg PG, Astria SM et al, Regulation of mycogene expression in HL-60 leukaemia cells by a vitamin D metabolite, *Nature* (1983) **306**: 492–4.

45 Key L, Carnes D, Cole S et al, Treatment of congenital osteopetrosis with high dose calcitriol, *N Engl J Med* (1984) **310**: 410–15.

46 Tsoukas CD, Provvedini DM, Manolagas SC, 1,25 dihydroxyvitamin D_3: a novel immunoregulatory hormone, *Science* (1984) **224**: 1438–40.

47 Amento EP, Bhalla AK, Kurnick JT et al, 1 alpha 25-dihydroxyvitamin D_3 induces maturation of the human monocyte cell line U937, and, in association with a factor from human T lymphocytes, augments production of the monokine, mononuclear cell factor, *J Clin Invest* (1984) **73**: 731–9.

48 Fetchick DA, Bertolini DR, Sarin PS et al, Production of 1,25-dihydroxyvitamin D by human T-cell lymphotrophic virus-I transformed lymphocytes, *J Clin Invest* (1986) **78**: 592–6.

49 Provvedini DM, Tsoukas CD, Deftos LJ et al, 1,25 dihydroxyvitamin D_3 receptors in human leukocytes, *Science* (1983) **221**: 1181–3.

50 Ishizuka S, Ishimoto S, Norman AW, Biological activity assessment of 1α, 25-dihydroxyvitamin D_3-26, 23-lactone in the rat, *J Steroid Biochem* (1984) **20**: 611–16.

51 Ishizuka S, Oshid J, Tsuruta H et al, The steriochemical configuration of the natural 1α, 25-dihydroxyvitamin D_3-26, 23-lactone, *Arch Biochem Biophys* (1985a) **242**: 82–9.

52 Ishizuka S, Kiyoki M, Orimo H et al, Biological activity and characteristics of 1α, 25(OH)$_2$D$_3$-26, 23-lactone. In: Norman AW, Schaefer K, Grigoleit HG, Herrath DV, eds, *Vitamin D: Chemical, Biochemical and Clinical Update* (de Gruyter: Berlin 1985) 402–3.

53 Ishizuka S, Norman AW, The difference of biological activity among four diastereoisomers of 1α, 25-dihydroxyvitamin D_3-26, 23-lactone, *J Steroid Biochem* (1986) **25**: 505–10.

54 Ishizuka S, Norman AW, Metabolic pathways from 1α, 25-dihydroxyvitamin D_3 to 1α, 25-dihydroxyvitamin D_3-26, 23-lactone, *J Biol Chem* (1987) **262**: 7165–70.

55 Ishizuka S, Kurihara N, Hakeda Y et al, 1α, 25-dihydroxyvitamin D_3 [1α, 25-(OH)$_2$D$_3$]-26,23-lactone inhibits 1,25-(OH)$_2$D$_3$-mediated fusion of mouse bone marrow mononuclear cells, *Endocrinology* (1988) **123**: 781–6.

56 Kiyoki M, Kurihara N, Ishizuka S et al, The unique action for bone metabolism of 1α, 26-23-lactone, *Biochem Biophys Res Commun* (1985) **127**: 693–8.

57 Kurihara N, Ishizuka S, Kumegawa M et al, 23(S) 25(R)-1,25(OH)$_2$D$_3$-26,23-Lactone, a naturally occurring metabolite of 1,25(OH)$_2$ vitamin D_3,

inhibits osteoclast-like cell formation in human bone marrow cultures (submitted).

58 Hirsch PF, Munson PL, Thyrocalcitonin, *Physiol Rev* (1969) **49**: 548–622.

59 Rosenfeld MG, Mermod JJ, Amara SG et al, Production of a novel neuropeptide encoded by the calcitonin gene via tissue-specific RNA processing, *Nature* (1983) **304**: 129–35.

60 Hendy GN, O'Riordan JLH, The genes that control calcium homeostasis, *Clin Endocrinol (Oxf)* (1984) **21**: 465–70.

61 Friedman J, Au WYW, Raisz LG, Responses of fetal rat bone to thyrocalcitonin in tissue culture, *Endocrinology* (1968) **82**: 149–56.

62 Heersche JNM, Marcus R, Aurbach GD, Calcitonin and the formation of 3',5'-AMP in bone and kidney, *Endocrinology* (1974) **94**: 241–7.

63 Chambers TJ, Magnus CJ, Calcitonin alters the behavior of isolated osteoclasts, *J Pathol* (1982) **136**: 27–40.

64 Baron R, Vignery A, Changes in the osteoclastic pools and the osteoclasts nuclei balance after a single injection of salmon calcitonin in the adult rat. In: Meunier PJ, ed., *Bone histomorphometry* (Lab Armour Montagu: Paris 1976) 147–56.

65 Wener JA, Gorton SJ, Raisz LG, Escape from inhibition of resorption in cultures of fetal bone treated with calcitonin and parathyroid hormone, *Endocrinology* (1972) **90**: 752–9.

66 Binstock ML, Mundy GR, Effect of calcitonin and glucocorticoids in combination in malignant hypercalcemia, *Ann Intern Med* (1980) **93**: 269–72.

67 Farley JR, Tarbaux NM, Hall SL et al, The anti-bone-resorptive agent calcitonin also acts in vitro to directly increase bone formation and bone cell proliferation, *Endocrinology* (1988) **123**: 159–67.

68 Burns DM, Forstrom JM, Friday KE et al, Procalcitonin's amino-terminal cleavage peptide is a bone-cell mitogen, *Proc Natl Acad Sci* (1989) **86**: 9519–23.

69 D'Souza SM, MacIntyre I, Girgis SI et al, Human synthetic calcitonin-gene related peptide inhibits bone resorption in vitro, *Endocrinology* (1986) **119**: 58–61.

70 Gowen M, Nedwin G, Mundy GR, Preferential inhibition of cytokine stimulated bone resorption by recombinant interferon gamma, *J Bone Miner Res* (1986) **1**: 469–74.

71 Heath JK, Saklatvala J, Meikle MC et al, Pig interleukin 1 (catabolin) is a potent stimulator of bone resorption in vitro, *Calcif Tissue Int* (1985) **37**: 95–7.

72 Gowen M, Mundy GR, Actions of recombinant interleukin-1, interleukin-2 and interferon gamma on bone resorption in vitro, *J Immunol* (1986) **136**: 2478–82.

73 Dewhirst, FE, Stashenko PP, Mole JE et al, Purification and partial sequence of human osteoclast-activating factor: identity with interleukin-1 beta, *J Immunol* (1985) **135**: 2562–8.

74 Boyce BF, Aufdemorte TB, Garrett IR et al, Effects of interleukin-1 on bone turnover in normal mice, *Endocrinology* (1989) **125**: 1142–50.

75 Boyce BF, Yates AJP, Mundy GR, Bolus injections of recombinant human interleukin-1 cause transient hypocalcemia in normal mice, *Endocrinology* (1989) **125**: 2780–3.

76 Sabatini M, Boyce B, Aufdemorte T et al, Infusions of recombinant human interleukin-1 alpha and beta cause hypercalcemia in normal mice, *Proc Natl Acad Sci* (1988) **85**: 5235–9.

77 Smith DD, Gowen M, Mundy GR, Effects of interferon-γ and other cytokines on collagen synthesis in fetal rat bone cultures, *Endocrinology* (1987) **120**: 2494–9.

78 Pennica D, Nedwin GE, Hayflick JS et al, Human tumour necrosis factor: precursor structure, expression and homology to lymphotoxin, *Nature* (1984) **312**: 724–9.

79 Bertolini DR, Nedwin GE, Bringman TS et al, Stimulation of bone resorption and inhibition of bone formation in vitro by human tumour necrosis factors, *Nature* (1986) **319**: 516–18.

80 Torti FM, Dieckmann B, Beutler B et al, A macrophage factor inhibits adipocyte gene expression: an in vitro model of cachexia, *Science* (1985) **229**: 867–9.

81 Beutler BA, Milsark IW, Cerami A, Cachectin tumor necrosis factor: production, distribution and metabolic fate in vivo, *J Immunol* (1985) **135**: 3972–7.

82 Johnson RA, Boyce BF, Mundy GR et al, Tumors producing human TNF induce hypercalcemia and osteoclastic bone resorption in nude mice, *Endocrinology* (1989) **124**: 1424–7.

83 Yoneda T, Alsina MM, Chavez JB et al, Hypercalcemia in a human tumor is due to tumor necrosis factor production by host immune cells, *J Bone Min Res* (1989) **4**: Abstract 826.

84 Evans CH, Lymphotoxin—an immunologic hormone with anticarcinogenic and antitumor activity, *Cancer Immunol Immunother* (1982) **12**: 181–90.

85 Gray PW, Aggarwal BB, Benton CV, Cloning and expression of cDNA for human lymphotoxin, a lymphokine with tumour necrosis activity, *Nature* (1984) **312**: 721–4.

86 Higgins PG, Interferons, *J Clin Pathol* (1984) **37**: 109–16.

87 Pestka S, The human interferons—from protein purification and sequence to cloning and expression in bacteria: before, between and beyond, *Arch Biochem Biophys* (1983) **211**: 1–37.

88 Vilcek J, Gresser I, Merigan TC, Regulatory functions of interferons, *Ann NY Acad Sci* (1980) **350**: 1–641.

89 Kirchner H, *Springer Seminars Immunopathol* (1984) **7**: 347–74.

90 Nedwin GE, Svedersky LP, Bringman TS et al, Effect of interleukin-2, interferon-gamma, and mitogens on the production of tumor necrosis factors alpha and beta, *J Immunol* (1985) **135**: 2492–7.

91 Abe E, Tanaka H, Ishimi Y et al, Differentiation-inducing factor purified from conditioned medium of mitogen-treated spleen cell cultures stimulates bone resorption, *Proc Natl Acad Sci USA* (1986) **83**: 5958–62.

92 Suda T, Miyaura C, Ishimi Y et al, The role of differentiation-inducing factors (DIFs) in bone metabolism. In: *X International Conference on Calcium Regulating Hormones*, (Elsevier: Amsterdam, 1989), in press.

93 Tomida M, Yamamoto-Yamaguchi Y, Hozumi M, Purification of a factor inducing differentiation of mouse myeloid leukemic M1 cells from conditioned medium of mouse fibroblast L929 cells, *J Biol Chem* (1984) **259**: 10978–82.

94 Ichikawa Y, Further studies on the differentiation of a cell line of myeloid leukemia, *J Cell Physiol* (1970) **76**: 175–84.

95 Abe ER, Tanaka H, Ishimi Y et al, Differentiation-inducing factor purified from conditioned medium of mitogen-treated spleen cell cultures stimulates bone resorption, *Proc Natl Acad Sci* (1986) **83**: 5958–62.

96 Abe E, Ishimi Y, Takahashi N et al, A differentiation-inducing factor produced by the osteoblastic cell line MC3T3-E1 stimulates bone resorption by promoting osteoclast formation, *J Bone Min Res* (1988) **3**: 635–45.

97 Shiina-Ishimi Y, Abe E, Tanaka H et al, Synthesis of colony-stimulating factor (CSF) and differentiation-inducing factor (D-factor) by osteoblastic cells, clone MC3T3-E1, *Biochem Biophys Res Commun* (1986) **134**: 400–6.

98 Lowe DG, Nunes W, Bombara M et al, Genomic cloning and heterologous expression of human differentiation-stimulating factor, *DNA* (1989) **8**: 351–9.

99 Gearing DP, Gough NM, King JA et al, Molecular cloning and expression of cDNA encoding a murine myeloid leukaemia inhibitory factor (LIF), *Embo J* (1987) **6**: 3995–4002.

100 Reid I, Lowe C, Cornish J et al, Leukemia inhibitory factor: a novel cytokine which stimulates bone resorption in vitro, *J Bone Min Res* (1989) **4**: Abstract 129.

101 Metcalf D, Gearing DP, Fatal syndrome in mice engrafted with cells producing high levels of the leukemia inhibitory factor, *Proc Natl Acad Sci* (1989) **86**: 5948–52.

102 Muraguchi A, Kishimoto T, Miki Y et al, T-cell-replacing factor- (TRF) induced IgG secretion in a human b blastoid cell line and demonstration of acceptors for TRF, *J Immunol* (1981) **127**: 412–16.

103 Hirano T, Yasukawa K, Harada H et al, Complementary DNA for a novel human interleukin (BSF-2) that induces B-lymphocytes to produce immunoglobulin, *Nature* (1986) **324**: 73–6.

104 Bataille R, Jourdan M, Zhang Xue-Guang et al, Serum levels of interleukin-6, a potent myeloma cell growth factor, as a reflect of disease severity in plasma cell dyscrasias, *J Clin Invest* (1989) **84**: 2008–11.

105 Kawano M, Hirano T, Matsuda T et al, Autocrine generation and essential requirement of BSF-2/IL-6 for human multiple myeloma, *Nature* (1988) **322**: 83–6.

106 Kurihara N, Suda T, Akiyama Y et al, Interleukin-6 (IL-6) stimulates osteoclast formation in vitro by inducing IL-1 production, *Exp Hematol* (1989) **17**: 577A (199).

107 Lowik C, van der Plujim G, Hoekman K et al, IL-6 produced by PTH-stimulated osteogenic cells can be a mediator in osteoclast recruitment, *J Bone Min Res* (1989) **4**: Abstract 558.

108 Feyen JHM, Elford P, Dipadova FE et al, Interleukin-6 is produced by bone and modulated by parathyroid hormone, *J Bone Min Res* (1989) **4**: 633–8.

109 Garrett IR, Black KS, Mundy GR, Interactions between interleukin-6 and interleukin-1 in osteoclastic bone resorption in neonatal mouse calvariae, *Calcif Tissue Int* (1990) **46**: A37.

110 Klein DC, Raisz LG, Prostaglandins: stimulation of bone resorption in tissue culture, *Endocrinology* (1970) **86**: 1436–40.

111 Tashjian AH, Voelkel EF, Levine L et al, Evidence that the bone resorption-stimulating factor produced by mouse fibrosarcoma cells is protaglandin E_2: a new model for the hypercalcemia of cancer, *J Exp Med* (1972) **136**: 1329–43.

112 Chambers TJ, Ali NN, Inhibition of osteoclastic motility by protaglandins I_2, E_1, E_2 and 6-oxoE_1, *J Pathol* (1983) **139**: 383–97.

113 Mundy GR, Martin TJ, Hypercalcemia of malignancy—pathogenesis and treatment, *Metabolism* (1982) **31**: 1247–77.

114 Ibbotson KJ, Twardzik DR, D'Souza SM et al, Stimulation of bone resorption in vitro by synthetic transforming growth factor-alpha, *Science* (1985) **228**: 1007–9.

115 Ibbotson KJ, Harrod J, Gowen M et al, Effects of human transforming growth factor (TGF) alpha on bone resorption and formation in vitro, *Proc Natl Acad Sci USA* (1986) **83**: 2228–32.

116 Tashjian AH, Voelkel EF, Lazzaro M et al, Alpha and beta transforming growth factors stimulate prostaglandin production and bone resorption in cultured mouse calvaria, *Proc Natl Acad Sci USA* (1985) **82**: 4535–8.

117   Stern PH, Krieger NS, Nissenson RA et al, Human transforming growth factor alpha stimulates bone resorption in vitro, *J Clin Invest* (1985) **76**: 2016–20.

118   Ibbotson KJ, D'Souza SM, Smith DD et al, EGF receptor antiserum inhibits bone resorbing activity produced by a rat Leydig cell tumor associated with the humoral hypercalcemia of malignancy, *Endocrinology* (1985) **116**: 469–71.

119   Tashjian AH, Voelkel EF, Lloyd W et al, Actions of growth factors on plasma calcium, *J Clin Invest* (1986) **78**: 1405–9.

120   Sporn MB, Roberts AB, Shull JH et al, Polypeptide transforming growth factors isolated from bovine sources and used for wound healing in vivo, *Science* (1983) **219**: 1329–31.

121   Derynck R, Jarrett JA, Chen EY et al, Human transforming growth factor β complementary DNA sequence and expression in normal and transformed cells, *Nature* (1985) **316**: 701–5.

122   Frolik CA, Wakefield LM, Smith DM et al, Characterization of a membrane receptor for transforming growth factor β in normal rat kidney fibroblasts, *J Biol Chem* (1984) **259**: 10 995–11 000.

123   Seyedin SM, Thompson AY, Bentz H et al, Cartilage-inducing factor A: apparent identity to transforming growth factor β, *J Biol Chem* (1986) **261**: 5693–5.

124   Tucker RF, Shipley GD, Moses HL et al, Growth inhibitor from BSC-1 cells closely related to platelet type B transforming growth factor, *Science* (1984) **226**: 705–7.

125   Ling N, Ying SY, Ueno N et al, A homodimer of the β-subunits of inhibin A stimulates the secretion of pituitary follicle stimulating hormone, *Biochem Biophys Res Commun* (1986) **138**: 1129–37.

126   Chenu C, Pfeilschifter J, Mundy GR et al, Transforming growth factor β inhibits formation of osteoclast-like cells in long-term human marrow cultures, *J Bone Miner Res* (1987) **2**: 253.

127   Pfeilschifter JP, Seyedin S, Mundy GR, Transforming growth factor β inhibits bone resorption in fetal rat long bone cultures, *J Clin Invest* (1988) **82**: 680–5.

128   Chenu C, Pfeilschifter J, Mundy GR et al, Transforming growth factor β inhibits formation of osteoclast-like cells in long-term human marrow cultures, *Proc Natl Acad Sci* (1988) **85**: 5683–7.

129   Bolander ME, Joyce ME, Boden SD et al, Estrogen receptor mRNA expression during fracture healing in the rat detected by polymerase chain reaction amplification, *J Bone Min Res* (1989) **4**: Abstract 567.

130   Seyedin SM, Thomas TC, Thompson AY et al, Purification and characterization of two cartilage-inducing factors from bovine demineralized bone, *Proc Natl Acad Sci* (1985) **82**: 2267–71.

131   Pfeilschifter J, Mundy GR, Modulation of transforming growth factor β activity in bone cultures by osteotropic hormones, *Proc Natl Acad Sci* (1987) **84**: 2024–8.

132   Pfeilschifter J, D'Souza SM, Mundy GR, Effects of transforming growth factor β on osteoblastic osteosarcoma cells, *Endocrinology* (1987) **121**: 212–218.

133   Pfeilschifter J, Wolf O, Naumann A et al, Chemotactic response of osteoclast-like cells to transforming growth factor β, *J Bone Min Res* (1990) in press.

134   Oreffo ROC, Mundy GR, Seyedin S et al, Activation of the bone derived latent TGFβ complex by isolated osteoclasts, *Biochem Biophys Res Commun* (1989) **158**: 817–23.

135   Cochran BH, The molecular action of platelet-derived growth factor. In: *Advances in cancer research*, Vol. 45 (Academic Press: New York 1985), 183–216.

136   Stiles CD, The molecular biology of platelet-derived growth factor, *Cell* (1983) **33**: 653–5.

137   Mornex JF, Martinet Y, Yamauchi K, Spontaneous expression of the c-sis gene and release of a platelet-derived growth factor-like molecule by human alveolar macrophages, *J Clin Invest* (1986) **78**: 61–6.

138   Shimokado K, Raines EW, Madtes DK et al, A significant part of macrophage-derived growth factor consists of at least two forms of PDGF, *Cell* (1985) **43**: 277–86.

139 Deuel TF, Huang JS, Huang SS, Expression of a platelet-derived growth factor-like protein in Simian sarcoma virus transformed cells, *Science* (1983) **221**: 1348–50.

140 Doolittle RF, Hunkapiller MW, Hood LE, Simian sarcoma virus oncogene, v-sis, is derived from the gene (or genes) encoding a platelet-derived growth factor, *Science* (1983) **221**: 275–7.

141 Heldin CH, Westermark B, Wasteson A, Specific receptors for platelet-derived growth factor on cells derived from connective tissue and glia, *Proc Natl Acad Sci USA* (1981) **78**: 3664–8.

142 Graves DT, Antoniades HN, Williams SR et al, Evidence for functional platelet-derived growth factor receptors on MG-63 human osteosarcoma cells, *Cancer Res* (1984) **44**: 2966–70.

143 Hauschka PV, Mavrakos AE, Iafrati MD et al, Growth factors in bone matrix, *J Biol Chem* (1986) **261**: 12 665–74.

144 Tashjian AH, Hohmann EL, Antoniades HN et al, Platelet derived growth factor stimulates bone resorption via a prostaglandin mediated mechanism, *Endocrinology* (1982) **111**: 118–24.

145 Canalis E, Platelet derived growth factor stimulates DNA and protein synthesis in cultured fetal rat calvaria, *Metabolism* (1980) **30**: 970–5.

146 Graves DT, Owen AJ, Barth RK, Detection of c-sis transcripts and synthesis of PDGF-like proteins by human osteosarcoma cells, *Science* (1984) **226**: 972–4.

147 Graves DT, Owen AJ, Antoniades HN, Evidence that a human osteosarcoma cell line which secretes a mitogen similar to platelet-derived growth factor requires growth factors present in platelet-poor plasma, *Cancer Res* (1983) **43**: 83–7.

148 Mundy GR, Shapiro JL, Bandelin JG et al, Direct stimulation of bone resorption by thyroid hormones, *J Clin Invest* (1976) **58**: 529–34.

149 Mundy GR, Raisz LG, Drugs for disorders of bones, *New Ethic Med Prog* (1974) **11**: 165–99.

150 Lindsay R, Hart DM, Aitken JM et al, Long-term prevention of postmenopausal osteoporosis by oestrogen: evidence for an increased bone mass after delayed onset of oestrogen treatment, *Lancet* (1976) **i**: 1038–41.

151 Lindsay R, Hart DM, Forrest C et al, Prevention of spinal osteoporosis in oophorectomised women, *Lancet* (1980) **ii**: 1151–3.

152 Caputo CB, Meadows D, Raisz LG, Failure of estrogens and androgens to inhibit bone resorption in tissue culture, *Endocrinology* (1975) **98**: 1065–8.

153 Komm BS, Terpening CM, Benz DJ et al, Estrogen binding receptor MRNA, and biologic response in osteoblast-like osteosarcoma cells, *Science* (1988) **214**: 81–4.

154 Eriksen EF, Colvard DS, Berg NJ, et al, Evidence of estrogen receptors in normal human osteoblast-like cells, *Science* (1988) **241**: 84–6.

155 Keeting PE, Bonewald LF, Colvard TC et al, Estrogen-mediated release of transforming growth factor-beta by normal human osteoblast-like cells, *J Bone Min Res* (1989) **4**: Abstract 655.

156 Canalis EM, Dietrich JW, Maina DM et al, Hormonal control of bone collagen synthesis in vitro. Effects of insulin and glucagon, *Endocrinology* (1977) **100**: 668.

157 Bradley TR, Metcalf D, The growth of mouse bone marrow cells in vitro, *Aust J Exp Biol Med Sci* (1966) **44**: 287–300.

158 Ichikawa Y, Pluznik DH, Sachs L, In vitro control of the development of macrophage and granulocyte colonies, *Proc Natl Acad Sci USA* (1966) **56**: 488–95.

159 Metcalf D, The granulocyte macrophage colony stimulating factors, *Science* (1985) **229**: 16–22.

160 Jubinsky PT, Stanley ER, Purification of hemopoietin I: a multilineage hemopoietic growth factor, *Proc Natl Acad Sci USA* (1985) **82**: 2764–8.

161 Leary A, Strauss LC, Civin CI et al, Disparate differentiation in hemopoietic colonies derived from human paired progenitors, *Blood* (1985) **66**: 327–32.

162 Fung MC, Hapel AJ, Ymer S et al, Molecular cloning of cDNA for murine interleukin-3, *Nature* (1984) **307**: 233–6.

163 Gabrilove JL, Welte K, Lu L et al, Constitutive production of leukemia differentiation, colony-stimulating, erythroid burst-promoting, and pluripoietic factors by a human hepatoma cell line:

characterization of the leukemia differentiation factor, *Blood* (1985) **66**: 407–15.

164 Gough NM, Gough J, Metcalf D et al, Molecular cloning of cDNA encoding a murine haemato-poietic growth regulator, granulocyte-macrophage colony stimulating factor, *Nature* (1984) **309**: 763–7.

165 Wong GG, Witek JS, Temple PA et al, Human GM-CSF: molecular cloning of complementary DNA and purification of the natural and recombinant proteins, *Science* (1985) **228**: 810–15.

166 Gasson JC, Weisbart RH, Kaufman SE et al, Purified human granulocyte-macrophage colony-stimulating factor: direct action on neutrophils, *Science* (1984) **226**: 1339–42.

167 Stanley ER, Heard PM, Factors regulating macrophage production and growth, *J Biol Chem* (1977) **252**: 4305–431.

168 Waheed A, Shadduck RK, Purification and properties of L-cell derived colony stimulating factor, *J Lab Clin Med* (1979) **94**: 180–94.

169 Ben Avran CM, Shively JE, Shadduck RK et al, Amino-terminal amino acid sequence of murine colony stimulating factor-1, *Proc Natl Acad Sci USA* (1985) **82**: 4486–9.

170 Kawasaki ES, Ladner MB, Wang AM et al, Molecular cloning of a complementary DNA encoding human macrophage-specific colony stimulating factor (CSF-1), *Science* (1985) **230**: 291–6.

171 Sherr CJ, Rettenmier CW, Sacca R et al, The c-fms proto-oncogene product is related to the receptor for the mononuclear phagocyte growth factor, CSF-1, *Cell* (1985) **41**: 665–76.

172 Welte K, Platzer E, Lu L et al, Purification and biochemical characterization of human pluri-potent hematopoietic colony-stimulating factor, *Proc Natl Acad Sci USA* (1985) **82**: 1526–30.

173 Souza LM, Boone TC, Gabrilove J et al, Recom-binant human granulocyte colony stimulating fac-tor: effects on normal and leukemic myeloid cells, *Science* (1986) **232**: 61–5.

174 Nagata S, Tsuchiya M, Asano S et al, Molecular cloning and expression of cDNA for human granu-locyte colony stimulating factor, *Nature* (1986) **319**: 415–18.

175 Lee MY, Baylink DJ, Hypercalcemia, excessive bone resorption, and neutrophilia in mice bearing a mammary carcinoma, *Proc Soc Exp Biol Med* (1983) **172**: 424–9.

176 Oshawa NY, Ueyama Y, Morita K, Kondo Y, Heterotransplantation of human functioning tumors to nude mice. In: Nomura T, Oshawa N, Tamaoki N, Fujiwara F, eds, *Proc Second Int Workshop Nude Mice* (University of Tokyo Press: Tokyo 1977), 395–405.

177 Kondo Y, Sato K, Ohkawa H et al, Association of hypercalcemia with tumors producing colony-stimulating factor(s), *Cancer Res* (1983) **43**: 2368–74.

178 Asano S, Urabe A, Okabe T et al, Demonstration of granulopoietic factor(s) in the plasma of nude mice transplanted with a human lung cancer and in the tumor tissue, *Blood* (1977) **49**: 845–52.

179 Sato K, Kuratomi Y, Yamamoto K et al, Primary squamous cell carcinoma of the thyroid associ-ated with marked leukocytosis and hypercalcemia, *Cancer* (1981) **48**: 2080–3.

180 Sato K, Mimura H, Ueyama Y et al, Pathogenesis of hypercalcemia in nude mice transplanted with a human colony-stimulating factor (CSF)-producing tumor. A new paraneoplastic syndrome of granulocytosis and hypercalcemia. In: Cohn DV, Fujita T, Potts TJ Jr, Talmage RV, eds, *Endocrine control of bone and metabolism*, Vol. 8A (Excerpta Medica: Amsterdam 1984) 288–291. (Int Congr Ser 619.)

181 Sato S, Hisako M, Han DC et al, Production of bone-resorbing activity and colony-stimulating activity in vivo and in vitro by a human squamous cell carcinoma associated with hypercalcemia and leukocytosis, *J Clin Invest* (1986) **78**: 145–54.

182 Burger EH, Van der Meer JWM, Van de Gevel JS et al, In vitro formation of osteoclasts from long term bone cultures of bone marrow mononuclear phagocytes, *J Exp Med* (1982) **156**: 1604–14.

183 Wiktor-Jedrzejczak W, Ahmed A, Szczylik C et al, Hematological characterisation of congenital osteopetrosis in op/op mouse, *J Exp Med* (1982) **156**: 1516–27.

184 Takahashi N, Mundy GR, Roodman GD, Recom-binant human gamma interferon inhibits forma-tion of human osteoclast like cells, *J Immunol* (1986) **137**: 3541–9.

185 MacDonald BR, Mundy GR, Clark S et al, Effects of human recombinant CSF-GM and highly purified CSF-1 on the formation of multinucleated cells with osteoclast characteristics in long-term bone marrow cultures, *J Bone Miner Res* (1986) **1**: 227–33.

186 Horton MA, Rimmer ET, Lewis D et al, Cell surface characterization of the human osteoclast: phenotype relationship to other bone marrow-derived cell types, *J Pathol* (1984) **144**: 281–94.

Chapter 4

1 Delay J, Lemperiere T, Feline A et al, Une observation d'hyperparathyroidisme primitif révéle par des troubles psychiatriques aigus, *Ann Med Interne (Paris)* (1970) **121**: 457–62.

2 Fitz TE, Hallman BL, Mental changes associated with hyperparathyroidism, *Arch Intern Med* (1952) **89**: 547–51.

3 Petersen P, Psychiatric disorders in primary hyperparathyroidism, *J Clin Endocrinol* (1968) **28**: 1491–5.

4 Simpsom JA, Aphonia and deafness in hyperparathyroidism, *Br Med J* (1954) **1**: 494–6.

5 Streeto JM, Acute hypercalcemia simulating basilar-artery insufficiency, *N Engl J Med* (1969) **280**: 427–9.

6 Edwards GA, Daum SM, Increased spinal fluid protein in hyperparathyroidism and other hypercalcemia states, *Arch Intern Med* (1959) **104**: 29–36.

7 Evaldsson U, Ertekin C, Ingvar DH et al, Encephalopathia hypercalcemia, *J Chronic Dis* (1969) **22**: 431–49.

8 Smith JP, Boronow RC, Moure JM, Hypercalcemia accompanying ovarian mesonephroma without skeletal metastasis, *South Med J* (1968) **61**: 375–8.

9 Strickland NJ, Bold AM, Medd WE, Bronchial carcinoma with hypercalcaemia simulating cerebral metastases, *Br Med J* (1967) **3**: 590–2.

10 Mavligit G, Corticosteroid therapy in hypercalcemia of malignant disease, *Lancet* (1970) **ii**: 1188–9.

11 Donegan WL, Spiro HM, Parathyroids and gastric secretion, *Gastroenterology* (1960) **38**: 750–9.

12 Spiro HM, Hyperparathyroidism, parathyroid 'adenomas', and peptic ulcer, *Gastroenterology* (1960) **39**: 544–52.

13 Spiro HM, *Clinical gastroenterology* (Macmillan: New York 1983) 430–1.

14 Barreras RF, Donaldson RM, Role of calcium in gastric hypersecretion, parathyroid adenoma and peptic ulcer, *N Engl J Med* (1967) **276**: 1122–4.

15 Barreras RF, Donaldson RM, Effects of induced hypercalcemia on human gastric secretion, *Gastroenterology* (1960) **52**: 670–5.

16 Weidmann P, Massry SG, Coburn JW et al, Blood pressure effects of acute hypercalcemia, *Ann Intern Med* (1972) **76**: 741–5.

17 Moore WT, Smith LH, Experience with a calcium infusion test in parathyroid disease, *Metabolism* (1963) **12**: 447–51.

18 Earl JM, Kurtzman NA, Moser RH, Hypercalcemia and hypertension, *Ann Intern Med* (1966) **64**: 378–81.

19 Shackney S, Hasson J, Precipitous fall in serum calcium, hypotension, and acute renal failure after intravenous therapy for hypercalcemia, *Ann Intern Med* (1967) **66**: 906–16.

20 Soffer A, Toribara T, Moore-Jones D et al, Clinical applications and untoward reactions of chelation in cardiac arrhythmias, *Arch Intern Med* (1960) **106**: 824–34.

21 Bronsky D, Dubin A, Kushner DS et al, Calcium and the electrocardiogram. III. The relationship of the intervals of the electrocardiogram to the level of serum calcium, *Am J Cardiol* (1961) **7**: 840–3.

22 Muggia FM, Heinemann HO, Hypercalcemia associated with neoplastic disease, *Ann Intern Med* (1970) **73**: 281–90.

23 Crum WB, Till HJ, Hyperparathyroidism with Wenckebach's Phenomenon, *Am J Cardiol* (1960) **6**: 838–40.

24 Lown B, Black H, Moore FD, Digitalis, electrolytes and the surgical patient, *Am J Cardiol* (1960) **6**: 309–37.

25 Bower JO, Mengle HAK, The additive effect of calcium and digitalis, *JAMA* (1936) **106**: 1151–3.

26 Petersen MJ, Edelman IS, Calcium inhibition of the action of vasopressin on the urinary bladder of the toad, *J Clin Invest* (1964) **43**: 583–94.

27 Zeffren JL, Heinemann HO, Reversible defect in renal concentrating mechanism in patients with hypercalcemia, *Am J Med* (1962) **33**: 54–63.

28 Suki WN, Eknoyan G, Rector FC et al, The renal diluting and concentration mechanism in hypercalcemia, *Nephron* (1969) **6**: 50–61.

29 Heinemann HO, Reversible defect in renal ammonium excretion in patients with hypercalcemia, *Metabolism* (1963) **12**: 792–803.

30 Manitius A, Levitin H, Beck D et al, On the mechanism of impairment of renal concentrating ability in hypercalcemia, *J Clin Invest* (1960) **39**: 693–7.

31 Steck IE, Deutsch H, Reed CI et al, Further studies on intoxication with vitamin D, *Ann Intern Med* (1937) **10**: 951–64.

32 Epstein FH, Calcium nephropathy. In: Strauss MB, Welt LG, eds, *Diseases of the kidney* (Little, Brown: Boston 1971) 903–31.

33 McLean FC, Hastings AB, A biological method for the estimation of calcium ion concentration, *J Biol Chem* (1934) **107**: 337–50.

34 Wills MR, McGowan GK, Plasma-chloride levels in hyperparathyroidism and other hypercalcaemic states, *Br Med J* (1964) **1**: 1153–6.

35 Lufkin EG, Kao PC, Heath H III, Parathyroid hormone radioimmunoassays in the differential diagnosis of hypercalcemia due to primary hyperparathyroidism or malignancy, *Ann Intern Med* (1987) **106**: 559–60.

36 Blind E, Schmidt-Gayk H, Armbruster FP et al, Measurement of intact human parathyrin by an extracting two-site immunoradiometric assay, *Clin Chem* (1987) **33**: 1376–81.

37 Blind E, Schmidt-Gayk H, Scharla S et al, Two-site assay of intact parathyroid hormone in the investigation of primary hyperparathyroidism and other disorders of calcium metabolism compared with a midregion assay, *J Clin Endocrinol Metab* (1988) **67**: 353–60.

38 Nussbaum SR, Zahradnik RJ, Lavigne JR et al, Highly sensitive two-site immunoradiometric assay of parathyrin, and its clinical utility in evaluating patients with hypercalcemia, *Clin Chem* (1987) **33**: 1364–7.

39 Brown RC, Aston JP, Weeks I et al, Circulating intact parathyroid hormone measured by a two-site immunochemiluminometric asay, *J Clin Endocrinol Metab* (1987) **65**: 407–14.

40 Dent CE, Some problems of hyperparathyroidism, *Br Med J* (1962) **2**: 1419–25.

41 Watson L, Moxham J, Fraser P, Hydrocortisone suppression test and discriminant analysis in differential diagnosis of hypercalcaemia, *Lancet* (1980) **i**: 1320–5.

Chapter 5

1 Mundy GR, Cove DH, Fisken R, Primary hyperparathyroidism: changes in the pattern of clinical presentation, *Lancet* (1980) **i**: 1317–20.

2 Mundy GR, Martin TJ, The hypercalcemia of malignancy: pathogenesis and management, *Metabolism* (1982) **31**: 1247–77.

3 Mundy GR, Ibbotson KJ, D'Souza SM et al, The hypercalcemia of cancer: clinical implications and pathogenic mechanisms, *N Engl J Med* (1984) **310**: 1718–27.

4 Mundy GR, Hypercalcemia of malignancy, *Kidney Int* (1987) **31**: 142–55.

5 Mundy GR, Hypercalcemia of malignancy revisited, *J Clin Invest* (1988) **82**: 1–6.

6 Valentin-Opran A, Charhon SA, Meunier PJ et al, Quantitative histology of myeloma induced bone changes, *Br J Haematol* (1982) **52**: 601–10.

7 Stewart AF, Vignery A, Silvergate A et al, Quantitative bone histomorphometry in humoral hypercalcemia of malignancy—uncoupling of bone cell activity, *J Clin Endocrinol Metab* (1982) **55**: 219–27.

8 Bringhurst FR, Bierer BE, Godeau F et al, Humoral hypercalcemia of malignancy. Release of prostaglandin-stimulating bone-resorbing factor in vitro by human transitional-cell carcinoma cells, *J Clin Invest* (1986) **77**: 456–64.

9   Albright F, Case records of the Massachusetts General Hospital—Case 39061, *N Engl J Med* (1941) **225**: 789–96.

10   Lafferty FW, Pseudohyperparathyroidism, *Med Balt* (1966) **45**: 247–60.

11   Tashjian AH, Levine L, Munson PL, Immunoassay of parathyroid hormone by quantitative complement fixation, *Endocrinology* (1964) **74**: 244–54.

12   Sherwood LM, O'Riordan JLH, Aurbach GD et al, Production of parathyroid hormone by nonparathyroid tumors, *J Clin Endocrinol Metab* (1967) **27**: 140–6.

13   Buckle RM, McMillan M, Mallison C, Ectopic secretion of parathyroid hormone by a renal adenocarcinoma in a patient with hypercalcemia, *Br Med J* (1970) **4**: 724–6.

14   Benson RC, Jr, Riggs BL, Pickard BM et al, Radioimmunoassay of parathyroid hormone in hypercalcemic patients with malignant disease, *Am J Med* (1974) **56**: 821–6.

15   Powell D, Singer FR, Murray TM et al, Nonparathyroid humoral hypercalcemia in patients with neoplastic disease, *N Engl J Med* (1973) **289**: 176–81.

16   Simpson EL, Mundy GR, D'Souza SM et al, Absence of parathyroid hormone messenger RNA in nonparathyroid tumors associated with hypercalcemia, *N Engl J Med* (1983) **309**: 325–30.

17   Moseley JM, Kubota M, Diefenbach-Jagger H et al, Parathyroid hormone-related protein purified from a human lung cancer cell line, *Proc Natl Acad Sci USA* (1987) **84**: 5048–52.

18   Yoshimoto K, Yamasaki R, Sakai H et al, Ectopic production of parathyroid hormone by small cell lung cancer in a patient with hypercalcemia, *J Clin Endocrinol Metab* (1989) **68**: 976–81.

19   Stewart AF, Horst R, Deftos LJ et al, Biochemical evaluation of patients with cancer-associated hypercalcemia: evidence for humoral and nonhumoral groups, *N Engl J Med* (1980) **303**: 1377–83.

20   Stewart AF, Insogna KL, Goltzman D et al, Identification of adenylate cyclase-stimulating activity and cytochemical glucose-6-phosphate dehydrogenase-stimulating activity in extracts of tumors from patients with humoral hypercalcemia

of malignancy, *Proc Natl Acad Sci USA* (1981) **53**: 941–7.

21   Goltzman D, Stewart AF, Broadus AE, Malignancy associated hypercalcemia: evaluation with a cytochemical bioassay for parathyroid hormone, *J Clin Endocrinol Metab* (1981) **53**: 899–904.

22   Rodan SB, Insogna KL, Vignery AMC et al, Factors associated with humoral hypercalcemia of malignancy stimulate adenylate cyclase in osteoblastic cells, *J Clin Invest* (1983) **72**: 1511–15.

23   Strewler GJ, Williams RD, Nissenson RA, Human renal carcinoma cells produce hypercalcemia in the nude mouse and a novel protein recognized by parathyroid hormone receptors, *J Clin Invest* (1983) **71**: 769–74.

24   Suva LJ, Winslow GA, Wettenhall REH et al, A parathyroid hormone-related protein implicated in malignant hypercalcemia: cloning and expression, *Science* (1987) **237**: 893–6.

25   Stewart AF, Wu T, Goumas D et al, N-terminal amino acid sequence of two novel tumor-derived adenylate cyclase-stimulating proteins: identification of parathyroid hormone-like and parathyroid hormone-unlike domains, *Biochem Biophys Res Commun* (1987) **146**: 672–8.

26   Strewler GJ, Stern PH, Jacobs JW et al, Parathyroid hormone-like protein from human renal carcinoma cells: structural and functional homology with parathyroid hormone, *J Clin Invest* (1987) **80**: 1803–7.

27   D'Souza SM, Ibbotson KJ, Mundy GR, Failure of PTH antagonists to inhibit in vitro bone resorbing activity produced by two animal models of the humoral hypercalcemia of malignancy, *J Clin Invest* (1984) **74**: 1104–7.

28   Merendino JJ, Insogna KL, Milstone LM et al, A parathyroid hormone-like protein from cultured human keratinocytes, *Science* (1986) **231**: 388–90.

29   Seshadri MS, Cornish CJ, Mason RS et al, Parathyroid hormone-like bioactivity in tumours from patients with oncogenic osteomalacia, *Clin Endocrinol (Oxf)* (1985) **23**: 689–97.

30   Kemp BE, Moseley JM, Rodda CP et al, Parathyroid hormone-related protein of malignancy: active synthetic fragments, *Science* (1987) **238**: 1568–70.

31   Horiuchi N, Caulfield MP, Fisher JE et al, Similarity of synthetic peptide from human tumor to parathyroid hormone in vivo and in vitro, *Science* (1987) **238**: 1566–7.

32   Yates AJP, Gutierrez GE, Smolens P et al, Effects of a synthetic peptide of a parathyroid hormone-related protein on calcium homeostasis, renal tubular calcium reabsorption and bone metabolism, *J Clin Invest* (1988) **81**: 932–8.

33   Stewart AF, Mangin M, Wu T et al, Synthetic human parathyroid hormone-like protein stimulates bone resorption and causes hypercalcemia in rats, *J Clin Invest* (1988) **81**: 596–600.

34   Stewart AF, Horst R, Deftos LJ et al, Biochemical evaluation of patients with cancer-associated hypercalcemia: evidence for humoral and nonhumoral groups, *New Engl J Med* (1980) **303**: 1377-83.

35   Wills MR, McGowan GK, Plasma-chloride levels in hyperparathyroidism and other hypercalcaemic states, *Br Med J* (1964) **1**: 1153–6.

36   Gutierrez G, Poser JW, Katz MS et al, Mechanisms of hypercalcemia of malignancy. In: *Bailliere's Clinics in Endocrinology and Metabolism* (1990) **4**: 119–38.

37   Mundy GR, Ibbotson KJ, D'Souza SM, Tumor products and the hypercalcemia of malignancy, *J Clin Invest* (1985) **76**: 391–5.

38   Sabatini M, Boyce B, Aufdemorte T et al, Infusions of recombinant human interleukin-1α and β cause hypercalcemia in normal mice, *Proc Natl Acad Sci USA* (1988) **85**: 5235–9.

39   Tashjian AH, Voelkel EF, Lloyd W et al, Actions of growth factors on plasma calcium, *J Clin Invest* (1986) **78**: 1405–9.

40   Mundy GR, The hypercalcemia of malignancy, *Kidney Int* (1987) **31**: 142–55.

41   Sabatini M, Carrett IR, Mundy GR, TNF potentiates the effects of interleukin-1 on bone resorption in vitro, *J Bone Min Res* (1987) **2** (suppl 1): 34.

42   Takahashi N, McDonald BR, Hon J et al, Recombinant human transforming growth factor alpha stimulates the formation of osteoclast-like cells in long term human marrow cultures, *J Clin Invest* (1986) **78**: 894–8.

43   Yates AJP, Mundy GR, (unpublished observations).

44   Sato K, Fujii Y, Kasono K et al, Parathyroid hormone-related protein and interleukin-1α synergistically stimulate bone resorption in vitro and increase the serum calcium concentration in mice in vivo, *Endocrinology* (1989) **124**: 2172–8.

45   Gutierrez GE, Mundy GR, Derynck R et al, Inhibition of parathyroid hormone-responsive adenylate cyclase in clonal osteoblast-like cells by transforming growth factor α and epidermal growth factor, *J Biol Chem* (1987) **262**: 15845–50.

46   Moseley JM, Kemp BE, Rodda CP et al, Parathyroid hormone-related protein of malignancy: biological actions of synthetic peptides and immunocytochemical localisation, *Calcif Tissue Int* (1987) Suppl **42**: 174.

47   Ikeda K, Arnold A, Mangin M et al, Expression of transcripts encoding a parathyroid hormone-related peptide in abnormal human parathyroid tissues, *J Clin Endocrinol Metab* (1989) **69**: 1240–8.

48   Ernst M, Rodan GA, Thiede MA, Rapid induction of parathyroid hormone-like peptide (PTH-LP) in keratinocytes: implication of PTH-LP in the proliferative response to epidermal growth factor, *J Bone Min Res* (1989) **4**: Abstract 309.

49   Thiede MA, In vivo regulation of a calcium mobilizing parathyroid hormone-like peptide by prolactin, *J Bone Min Res* (1989) **4**: Abstract 308.

50   Goltzman D, Hendy GN, Banville D, Parathyroid hormone-like peptide: molecular characterization and biological properties, *Trends Endo Metab* (1989) Sept/Oct, 39–44.

51   Rizzoli R, Bonjour JP, High extracellular calcium increases the production of a parathyroid hormone-like activity by cultured Leydig tumor cells associated with humoral hypercalcemia, *J Bone Min Res* (1989) **4**: 839–44.

52   Mangin M, Wegg AC, Dreyer BE et al, Identification of a cDNA encoding a parathyroid hormone-like peptide from a human tumor associated with humoral hypercalcemia of malignancy, *Proc Natl Acad Sci USA* (1988) **85**: 597–601.

53   Mangin M, Ikeda K, Dreyer BE et al, Two distinct tumor-derived, parathyroid hormone-like peptides result from alternative ribonucleic acid splicing, *Mol Endocrinol* (1988) **2**: 1049–55.

54 Broadus AE, Mangin M, Ikeda K et al, Humoral hypercalcemia of malignancy: identification of a novel parathyroid hormone-like peptide, *N Engl J Med* (1988) **319**: 556–63.

55 Orloff JJ, Wu TL, Stewart AF, Parathyroid hormone-like proteins: biochemical responses and receptor interactions, *Endocr Rev* (1989) **10**: 476–95.

56 Goltzman D, Hendy GN, Banville D, Parathyroid hormone-like peptide: Molecular characterization and biological properties, *Trends Endo Metab* (1989) Sept/Oct, 39–44.

57 Rodda CP, Kubota M, Heath JA et al, Evidence for a novel parathyroid hormone-related protein in fetal lamb parathyroid glands and sheep placenta: Comparisons with a similar protein implicated in humoral hypercalcemia of malignancy, *J Endocrinol* (1988) **117**: 261–71.

58 Thiede MA, Rodan GA, Expression of a calcium-mobilizing parathyroid hormone-like peptide in lactating mammary tissue, *Science* (1988) **242**: 278–80.

59 Budayr AA, Halloran BP, King JC et al, High levels of a parathyroid hormone-like protein in milk, *J Bone Min Res* (1989) **4**: Abstract 82.

60 Insogna KL, Stewart AF, Morris CA et al, Native and a synthetic analogue of the malignancy-associated parathyroid hormone-like protein have in vitro transforming growth factor-like properties, *J Clin Invest* (1989) **83**: 1057–60.

61 Gutierrez GE, Mundy GR, Manning DR et al, Transforming growth factor β enhances parathyroid hormone stimulation of adenylate cyclase in clonal osteoblast-like cells, *J Cell Physiol* (in press).

62 Kukreja SC, Shevrin DH, Wimbiscus SA et al, Antibodies to parathyroid hormone-related protein lower serum calcium in athymic mouse models of malignancy-associated hypercalcemia due to human tumors, *J Clin Invest* (1988) **82**: 1798–1802.

63 Mehdizadeh S, Alaghband-Zadeh J, Gusterson B et al, Bone resorption and circulating PTH-like bioactivity in an animal model of hypercalcemia of malignancy, *Biochem Biophys Res Commun* (1989) **161**: 1166–71.

64 Goltzman D, Hendy GN, Banville D, Parathyroid hormone-like peptide: Molecular characterization and biological properties, *Trends Endo Metab* (1989) Sept/Oct, 39–44.

65 Goltzman D, Henderson B, Loveridge N, Cytochemical bioassay of parathyroid hormones: characteristics of the assay and analysis of circulating hormonal forms, *J Clin Invest* (1988) **65**: 1309.

66 Sporn MB, Todaro GJ, Autocrine secretion and malignant transformation of cells, *N Engl J Med* (1980) **303**: 878–80.

67 Todaro GJ, Fryling D, DeLarco JE, Transforming growth factors produced by certain human tumor cells: polypeptides that interact with epidermal growth factor receptors, *Proc Natl Acad Sci USA* (1980) **77**: 5258–62.

68 Roberts AB, Anzano MA, Lamb LC et al, Isolation from murine sarcoma cells of novel transforming growth factors potentiated by EGF, *Nature* (1982) **295**: 417–19.

69 Anzano MA, Roberts AB, Smith LM et al, Purification by reverse-phase high-performance liquid chromatography of an epidermal growth factor-dependent transforming growth factor, *Anal Biochem* (1982) **125**: 217–24.

70 Marquardt H, Todaro GJ, Human transforming growth factor: production by a melanoma cell line, purification, and initial characterization, *J Biol Chem* (1982) **257**: 5220–5.

71 Marquardt H, Hunkapiller MW, Hood LE et al, Rat transforming growth factor type I: structure and relation to epidermal growth factor, *Science* (1984) **223**: 1079–82.

72 Lee DC, Rose TM, Webb NR et al, Cloning and sequence analysis of a cDNA for rat transforming growth factor α, *Nature* (1985) **313**: 489–91.

73 Derynck R, Roberts AB, Winkler ME et al, Human transforming growth factor α: precursor structure and expression in E. coli, *Cell* (1984) **38**: 287–97.

74 Bringman TS, Lindquist PB, Derynck R, Different transforming growth factor-alpha species are derived from a glycosylated and palmitoylated transmembrane precursor, *Cell* (1987) **48**: 429–40.

75 Ibbotson KJ, Twardzik DR, D'Souza SM et al, Stimulation of bone resorption in vitro by synthetic transforming growth factor-alpha, *Science* (1985) **228**: 1007–9.

76 Ibbotson KJ, Harrod J, Gowen M et al, Effects of human transforming growth factor (TGF) alpha on bone resorption and formation in vitro, *Proc Natl Acad Sci USA* (1986) **83**: 2228–32.

77 Tashjian AH, Voelkel EF, Lazzaro M et al, Alpha and beta transforming growth factors stimulate prostaglandin production and bone resorption in cultured mouse calvaria, *Proc Natl Acad Sci USA* (1985) **82**: 4535–8.

78 Stern PH, Krieger NS, Nissenson RA et al, Human transforming growth factor-alpha stimulates bone resorption in vitro, *J Clin Invest* (1985) **76**: 2016–20.

79 Ibbotson KJ, D'Souza SM, Ng KW et al, Tumor-derived growth factor increases bone resorption in a tumor associated with the humoral hypercalcemia of malignancy, *Science* (1983) **221**: 1292–4.

80 Ibbotson KJ, D'Souza SM, Smith DD et al, EGF receptor antiserum inhibits bone resorbing activity produced by a rat Leydig cell tumor associated with the humoral hypercalcemia of malignancy, *Endocrinology* (1985) **116**: 469–71.

81 Mundy GR, Ibbotson KJ, D'Souza SM et al, Evidence that transforming growth factor alpha production causes bone resorption and hypercalcemia in squamous cell carcinoma of the lung, *Clin Res* (1985) **33**: 5743.

82 Tashjian AH, Voelkel EF, Lloyd W et al, Actions of growth factors on plasma calcium, *J Clin Invest* (1986) **78**: 1405–9.

83 Shreiber AB, Winkler ME, Derynck R, Transforming growth factor-alpha: a more potent angiogenic mediator than epidermal growth factor, *Science* (1986) **232**: 1250–3.

84 Brown JP, Twardzik DR, Marquardt H et al, Vaccinia virus encodes a polypeptide homologous to epidermal growth factor and transforming growth factor, *Nature* (1985) **313**: 491–2.

85 Oshawa NY, Ueyama Y, Morita K, Kondo Y, Heterotransplantation of human functioning tumors to nude mice. In: Nomura T, Oshawa N, Tamaoki N, Fujiwara F, eds, *Proc Second Int Workshop Nude Mice* (University of Tokyo Press: Tokyo 1977), 395–405.

86 Kondo Y, Sato K, Ohkawa H et al, Association of hypercalcemia with tumors producing colony-stimulating factor(s), *Cancer Res* (1983) **43**: 2368–74.

87 Sato K, Kuratomi Y, Yamamoto K et al, Primary squamous cell carcinoma of the thyroid associated with marked leukocytosis and hypercalcemia, *Cancer* (1981) **48**: 2080–3.

88 Asano S, Urabe A, Okabe T et al, Demonstration of granulopoietic factor(s) in the plasma of nude mice transplanted with a human lung cancer and in the tumor tissue, *Blood* (1977) **49**: 845–52.

89 Sato K, Mimura H, Ueyama Y et al, Pathogenesis of hypercalcemia in nude mice transplanted with a human colony-stimulating factor (CSF)-producing tumor. A new paraneoplastic syndrome of granulocytosis and hypercalcemia. In: Cohn DV, Fujita T, Potts TJ Jr, Talmage RV, eds, *Endocrine control of bone and metabolism*, Vol 8A (Excerpta Medica: Amsterdam 1984) 288–291. (Int Congr Ser 619).

90 Sato S, Hisako M, Han DC et al, Production of bone-resorbing activity and colony-stimulating activity in vivo and in vitro by a human squamous cell carcinoma associated with hypercalcemia and leukocytosis, *J Clin Invest* (1986) **78**: 145–54.

91 Yoneda T, Nishikawa N, Nishimura R et al, Three cases of oral squamous cancer associated with leukocytosis, hypercalcemia, or both, *Oral Surgery* (1989) **68**: 604–11.

92 Block NL, Whitmore WF, Leukemoid reaction, thrombocytosis and hypercalcemia associated with bladder cancer, *J Urol* (1973) **110**: 660–3.

93 Hocking W, Goodman J, Golde D, Granulocytosis associated with tumor cell production of colony-stimulating activity, *Blood* (1983) **61**: 600–3.

94 Lee MY, Baylink DJ, Hypercalcemia, excessive bone resorption, and neutrophilia in mice bearing a mammary carcinoma, *Proc Soc Exp Biol Med* (1983) **172**: 424–9.

95 Sato K, Miumra H, Han DC et al, Production of bone resorbing activity and colony stimulating activity in vivo and in vitro by a human squamous cell carcinoma associated with hypercalcemia and leukocytosis, *J Clin Invest* (1986) **78**: 145–54.

96 MacDonald BR, Mundy GR, Clark S et al, Effects of human recombinant CSF-GM and highly purified CSF-1 on the formation of multinucleated cells with osteoclast characteristics in long term bone marrow cultures, *J Bone Miner Res* (1986) **1**: 227–33.

97 Yoneda T, Alsina MM, Chavez JB et al, Hypercalcemia in a human tumor is due to tumor necrosis factor production by host immune cells, *J Bone Min Res* (1989) **4**: Abstract 826.

98 Sabatini M, Bonewald L, Chavez J et al, Production of GM-CSF by a human tumor associated with leukocytosis and hypercalcemia induces cytokine production by host cells, *J Bone Min Res* (1989) **4**: Abstract 149.

99 D'Souza SM, Ibbotson KJ, Smith DD et al, Production of a macromolecular bone resorbing factor by the hypercalcemic variant of the Walker rat carcinosarcoma, *Endocrinology* (1984) **115**: 1746–52.

100 Sammon PJ, Wronski TJ, Flueck JA et al, Humoral hypercalcemia of malignancy: evidence for interleukin-1 activity as a bone resorbing factor released by human transitional-cell carcinoma cells. In: Cohn DV, Martin TJ, Meunier PJ, eds, *Calcium regulation and bone metabolism* (Elsevier Science: Amsterdam 1987) 383–96.

101 Sato K, Fujii Y, Kasono K et al, Interleukin-1α and PTH-like factor are responsible for humoral hypercalcemia associated with esophageal carcinoma cells (EC-GI), *J Bone Miner Res* (1987) **2**(Suppl 1): 387.

102 Sabatini M, Boyce B, Aufdemorte T et al, Infusions of recombinant human interleukin-1α and β cause hypercalcemia in normal mice, *Proc Natl Acad Sci USA* (1988) (in press).

103 Fried RM, Voelkel EF, Rice RH et al, Two squamous cell carcinomas not associated with humoral hypercalcemia produce a potent bone resorption-stimulating factor which is interleukin-1α, *Endocrinology* (1989) **125**: 742–51.

104 Klein DC, Raisz LG, Prostaglandins: stimulation of bone resorption in tissue culture, *Endocrinology* (1970) **86**: 1436–40.

105 Franklin RB, Tashjian AH, Intravenous infusion of prostaglandin E$_2$ raises plasma calcium concentration in the rat, *Endocrinology* (1975) **97**: 240–3.

106 Ito H, Sanada T, Katamaya et al, Letter: iodomethacin-responsive hypercalcemia, *N Engl J Med* (1975) **293**: 558–9.

107 Tashjian AH, Voelkel EF, Levine L et al, Evidence that the bone resorption-stimulating factor produced by mouse fibrosarcoma cells is prostaglandin E$_2$: a new model for the hypercalcemia of cancer, *J Exp Med* (1972) **136**: 1329–43.

108 Thomson BM, Mundy GR, Chambers TJ, Tumor necrosis factors α and β induce osteoblastic cells to stimulate osteoclastic bone resorption, *J Immunol* (1987) **138**: 775–9.

109 Seyberth HW, Segre GV, Morgan JL et al, Prostaglandins as mediators of hypercalcemia associated with certain types of cancer, *N Engl J Med* (1975) **293**: 1278–83.

110 Sabatini M, Yates AJ, Garrett R et al, Increased production of tumor necrosis factor by normal immune cells in a model of the humoral hypercalcemia of malignancy. Lab Investigation (in press).

111 Peacock M, Robertson WG, Nordin BEC, Relation between serum and urine calcium with particular reference to parathyroid activity, *Lancet* (1969) **i**: 384–6.

112 Ralston SH, Gardner MD, Dryburgh FJ et al, Influence of urinary sodium excretion on the clinical assessment of renal tubular calcium reabsorption in hypercalcemic man, *J Clin Pathol* (1986) **39**: 641–6.

113 Tuttle KR, Mundy GR, Kunau RT, Renal tubular calcium reabsorption is increased in hypercalcemia of malignancy, *Clin Res* (1989) **37**: Abstract 463A.

114 Harinck HIJ, Bijvoet OLM, Plantingh AST, Role of bone and kidney in tumor-induced hypercalcemia and its treatment with bisphosphonate and sodium chloride, *Am J Med* (1987) **82**: 1133–42.

115 Martodam RR, Thornton KS, Sica DA et al, The effects of dichloromethylene diphosphonate on hypercalcemia and other parameters of the humoral hypercalcemia of malignancy in the rat Leydig cell tumor, *Calcif Tissue Int* (1983) **35**: 512–19.

116 Rizzoli R, Caverzasio J, Fleisch H et al, Parathyroid hormone-like changes in renal calcium and phosphate reabsorption induced by Leydig cell tumor in thyroparathyroidectomized rats, *Endocrinology* (1986) **119**: 1004–9.

117 Hirschel-Scholtz S, Caverzasio J, Rizzoli R et al, Normalization of hypercalcemia associated with a decrease in renal calcium reabsorption in Leydig cell tumor-bearing rats treated with WR-2721, *J Clin Invest* (1986) **78**: 319–22.

118 Percival RC, Yates AJP, Gray RES et al, Mechanisms of malignant hypercalcemia in carcinoma of the breast, *Br Med J* (1985) **291**: 776–9.

119 Kanis JA, Cundy T, Heynen G et al, The patho-physiology of hypercalcaemia, *Metab Bone Dis Relat Res* (1980) **2**: 151–9.

120 Yamamoto I, Kitamura N, Aoka J et al, Circulating 1,25-dihydroxyvitamin D concentrations in patients with renal cell carcinoma-associated hypercalcemia are rarely suppressed, *J Clin Endocrinol Metab* (1987) **64**: 175–9.

121 Kukreja SC, York PA, Nalbantian-Brandt C et al, Effect of hypercalcemia-producing tumor on $1,25(OH)_2D_3$ biosynthesis in athymic mice, *Am J Physiol* (1989) **256**: E309–14.

122 Fukumoto S, Matsumoto T, Yamoto H et al, Suppression of serum 1,25-dihydroxyvitamin D in humoral hypercalcemia of malignancy is caused by elaboration of a factor that inhibits renal 1,25-dihydroxyvitamin D_3 production, *Endocrinology* (1989) **124**: 2057–62.

123 Gordan GS, Cantino TJ, Erhardt L et al, Osteolytic steroid in human breast cancer, *Science* (1966) **151**: 1226–8.

124 Haddad JG, Cowranz SJ, Avioli LV, Circulating phytosterols in normal females, lactating mothers, and breast cancer patients, *J Clin Endocrinol Metab* (1970) **30**: 174–80.

125 Shigeno C, Yamamoto I, Dokoh S et al, Identification of 1,24(R)-dihydroxyvitamin D_3-like bone-resorbing lipid in a patient with cancer-associated hypercalcemia, *J Clin Endocrinol Metab* (1985) **61**: 761–8.

126 Galasko CSB, Burn JI, Hypercalcemia in patients with advanced mammary cancer, *Br Med J* (1971) **3**: 573–7.

127 Graham WP, Gardner B, Thomas AN et al, Hypercalcemia in carcinoma of the female breast, *Surg Gynecol Obstet* (1963) **117**: 709–14.

128 Jessiman AG, Emerson K Jr, Shah RC et al, Hypercalcemia in carcinoma of the breast, *Ann Surg* (1963) **157**: 377–93.

129 Sklaroff DM, Charles ND, Bone metastases from breast cancer at the time of the radical mastectomy, *Surg Gynecol Obstet* (1968) **127**: 763–8.

130 Galasko CSB, The detection of skeletal metastases from mammary cancer by gamma camera scintigraphy, *Br J Surg* (1969) **56**: 757–64.

131 Hoffman HC, Marty R, Bone scanning: its value in the preoperative evaluation of patients with suspicious breast masses, *Am J Surg* (1972) **124**: 194–9.

132 Coleman RE, Rubens RD, Bone metastases and breast cancer, *Cancer Treat Rev* (1985) **12**: 251–70.

133 Stewart J, King R, Hayward J et al, Estrogen and progesterone receptors: correlation of response rates, site and timing of receptor analysis, *Breast Cancer Res Treat* (1982) **2**: 243–50.

134 Hickey RC, Samaan NA, Jackson GL, Hypercalcaemia in patients with breast cancer, *Arch Surg* (1981) **116**: 545–51.

135 Bloom HJG, Richardson WW, Histological grading and prognosis in breast cancer: a study of 1409 cases of which 359 have been followed for 15 years, *Br J Cancer* (1957) **II**: 359–77.

136 Percival RL, Yates AJP, Gray RES et al, Mechanism of malignant hypercalcaemia in carcinoma of the breast, *Br Med J* (1988) **291**: 776–9.

137 Galasko CSB, Mechanisms of bone destruction in the development of skeletal metastases, *Nature* (1976) **263**: 507–8.

138 Galasko CSB, Bennett A, Relationship of bone destruction in skeletal metastases to osteoclast activation and prostaglandins, *Nature* (1976) **263**: 508–10.

139 Eilon G, Mundy GR, Direct resorption of bone by human breast cancer cells in vitro, *Nature* (1978) **276**: 726–8.

140 Eilon G, Mundy GR, Effects of inhibition of microtubule assembly on bone mineral release and enzyme release by human breast cancer cells, *J Clin Invest* (1981) **67**: 69–76.

141 Boyde A, Maconnachie E, Reid SA et al, Scanning electron microscopy in bone pathology: review of methods. Potential and applications, *Scan Electron Microsc* (1986) **IV**: 1537–54.

142 Salomon DS, Zweibel JA, Bano M et al, Presence of transforming growth factors in human breast cancer cells, *Cancer Res* (1984) **44**: 4069–77.

143 Dickson RB, Bates SE, McManaway ME et al, Characterization of estrogen responsive transforming activity in human breast cancer cell lines, *Cancer Res* (1986) **46**: 1707–13.

144 Perroteau I, Salomon D, DeBortoli M et al, Immunological detection and quantitation of alpha transforming growth factors in human breast carcinoma cells, *Breast Cancer Res Treat* (1986) **7**: 201–10.

145 Lippman ME, Dickson RB, Bates S et al, Autocrine and paracrine growth regulation of human breast cancer, *Breast Cancer Res Treat* (1986) **7**: 59–70.

146 Salomon DS, Kidwell WR, Liu S et al, Presence of alpha TGF mRNA in human breast cancer cell lines and in human breast carcinomas, *Breast Cancer Res Treat* (1986) **8**: 106A.

147 Derynck R, Goeddel DV, Ullrich A et al, Synthesis of messenger RNAs for transforming growth factors alpha and beta and the epidermal growth factor receptor by human tumors, *Cancer Res* (1987) **47**: 707–12.

148 Knabbe C, Lippman ME, Wakefield LM et al, Evidence that transforming growth factor β is a hormonally regulated negative growth factor in human breast cancer cells, *Cell* (1987) **48**: 417–28.

149 Powles TJ, Clark SA, Easty BN et al, The inhibition by aspirin and indomethacin of osteolytic tumor deposits and hypercalcemia in rats with Walker tumour, and its possible application to human breast cancer, *Br J Cancer* (1973) **28**: 316–21.

150 Powles TJ, Dowsett M, Easty BN et al, Breast cancer osteolysis, bone metastases, and anti-osteolytic effect of aspirin, *Lancet* (1976) **i**: 608–10.

151 Bennett A, McDonald AM, Simpson JS et al, Breast cancer, prostaglandins, and bone metastases, *Lancet* (1975) **i**: 1218–20.

152 Greaves M, Ibbotson KJ, Atkins D et al, Prostaglandin as mediators of bone resorption in renal and breast tumors, *Clin Sci* (1980) **58**: 201–10.

153 Hong SL, Polsky-Cynkin R, Levine L, Stimulation of prostaglandin biosynthesis by vasoactive substances in methylcholanthrene-transformed mouse BALB-3T3, *J Biol Chem* (1976) **251**: 776–80.

154 Powles TJ, Muindi J, Coombes RC, Mechanisms for development of bone metastases and effects of anti-inflammatory drugs. In: Powles TJ, Bockman RS, Honn KV, Ramwell P, eds, *Prostaglandins and related lipids*, Vol. 2 (Alan R. Liss: New York 1982) 541–53.

155 Valentin A, Eilon G, Saez S et al, Estrogens and anti-estrogens stimulate release of bone resorbing activity by cultured human breast cancer cells, *J Clin Invest* (1985) **75**: 726–31.

156 Garcia M, Augereau P, Briozzo P et al, The estrogen-induced 52K cathepsin D secreted by human breast cancer: structure and biological functions. In: Bresciani F, King RJB, Lippman ME, Raynaud JP, eds, *Progress in Cancer Research and Therapy*, (Raven Press: New York 1988) 263–9.

157 Wo Z, Bonewald L, Oreffo ROC et al, The potential role of procathepsin D secreted by breast cancer cells in bone resorption. In: *X International Conference on Calcium Regulating Hormones* (Elsevier: Amsterdam 1989) in press.

158 Ralston S, Fogelman I, Gardner MD et al, Hypercalcemia and metastatic bone disease: is there a causal link? *Lancet* (1982) **ii**: 903–5.

159 Isales C, Carcangui ML, Stewart AF, Hypercalcemia in breast cancer, *Am J Med* (1987) **82**: 1143–7.

160 Hermann JB, Kirsten E, Krakauer JS, Hypercalcemic syndrome associated with androgenic and estrogenic therapy, *J Clin Endocrinol* (1949) **9**: 1–12.

161 Hall TC, Dederick MM, Nevinny HB, Prognostic value of hormonally induced hypercalcemia in breast cancer, *Cancer Chemother Rep* (1963) **30**: 21–3.

162 Kennedy BJ, Tibbetts DM, Nathanson IT et al, Hypercalcemia, a complication of hormone therapy of advanced breast cancer, *Cancer Res* (1953) **13**: 445–9.

163 Patterson JS, Baum M, Safety of tamoxifen, *Lancet* (1978) **i**: 105.

164 Spooner D, Evans BD, Tamoxifen and life-threatening hypercalcemia, *Lancet* (1979) **ii**: 413–4.

165 Cornbleet M, Bondy PK, Powles TJ, Fatal irreversible hypercalcemia in breast cancer, *Br Med J* (1977) **1**: 145.

166 Singhakowinta A, Saunders DW, Brooks SC et al, Clinical application of estrogen receptor in breast cancer, *Cancer* (1980) **46**: 2932–8.

167 Campbell FC, Blamey RW, Elston CW et al, Oestrogen receptor status and sites of metastasis in breast cancer, *Br J Cancer* (1981) **44**: 456–9.

168 Samaan NA, Buzdar AU, Aldinger KA et al, Estrogen receptor: a prognostic factor in breast cancer, *Cancer* (1981) **47**: 554–60.

169 Stewart JF, King RJB, Sexton SA et al, Oestrogen receptors, sites of metastatic disease and survival in recurrent breast cancer, *Eur J Cancer* (1981) **17**: 449–53.

170 Sanchez J, Baker V, Miller DM, Basic mechanisms of metastasis, *Am J Med Sci* (1986) **292**: 376–85.

171 Dresden MH, Heilman SA, Schmidt JD, Collagenolytic enzymes in human neoplasms, *Cancer Res* (1972) **32**: 993–6.

172 Hashimoto K, Yamanishi Y, Maeyens E et al, Collagenolytic activities of squamous cell carcinoma of the skin, *Cancer Res* (1973) **33**: 2790–801.

173 Sylven B, Biochemical and enzymatic factors involved in cellular detachment. In: Garattini S, Franchi F, eds, *Chemotherapy of cancer dissemination and metastasis* (Raven Press: New York 1973).

174 Weiss L, *Fundamental aspects of metastasis* (North-Holland: Amsterdam 1976).

175 Coman DR, Mechanisms responsible for the origin and distribution of blood bone tumor metastases: a review, *Cancer Res* (1952) **13**: 397–404.

176 Fidler IJ, Mechanisms of cancer invasion and metastasis. In: Becker FF, ed., *Cancer: a comprehensive treatise. Biology of tumors: surfaces, immunology and comparative pathology*, Vol. 4 (Plenum Press: New York 1975).

177 Malmgren RA, Studies of circulating tumor cells in cancer patients. In: Denoix P, ed., *Mechanisms of invasion of cancer* (Springer-Verlag: New York 1967).

178 Salisbury AJ, The significance of the circulating cancer cell, *Cancer Treat Rev* (1975) **2**: 55–72.

179 Clifton EE, Agostino D, Factors affecting the development of metastatic cancer. Effect of alterations in the clotting mechanism, *Cancer* (1962) **15**: 276–83.

180 Gasic GJ, Gasic TB, Galanti N et al, Platelet–tumor cell interaction in mice. The role of platelets in the spread of malignant disease, *Int J Cancer* (1973) **11**: 704–18.

181 Warren BA, Environment of the blood–bone–tumor embolus adherent to the vessel wall, *J Med* (1973) **4**: 150–77.

182 Fidler IJ, The relationship of embolic homogeneity, number, size and viability to the evidence of experimental metastasis, *Eur J Cancer* (1973) **9**: 223–7.

183 Smith RT, Landy M, *Immune surveillance* (Academic Press: New York 1971).

184 Weiss L, *The cell periphery, metastasis, and other contact phenomena* (North-Holland: Amsterdam 1967).

185 Weiss L, Contact between cancer and other cells. A biophysical approach. In: Garattini S, Franchi G, eds, *Chemotherapy of cancer dissemination and metastases* (Raven Press: New York 1973).

186 Folkman J, Tumor angiogenesis, *Adv Cancer Res* (1974) **19**: 331–58.

187 Folkman J, Tumor angiogenesis factor, *Cancer Res* (1975) **34**: 2109–13.

188 Folkman J, Klagsburn M, Angiogenic factors, *Science* (1987) **235**: 442–7.

189 Paget J, The distribution of secondary growths in cancer of the breast, *Lancet* (1889) **i**: 571–3.

190 Hart IR, 'Seed and soil' revisited: mechanisms of site specific metastasis, *Cancer Metastasis Rev* (1982) **1**: 5–16.

191 Kalinowski DT, Goodwin CA, Metastatic disease developing in Paget's disease of bone in the distal end of the femur, *Skeletal Radiol* (1981) **7**: 229–31.

192 Powell N, Metastatic carcinoma in association with Paget's disease of bone, *Br J Radiol* (1983) **56**: 582–5.

193 Mundy GR, Varani J, Orr W et al, Resorbing bone is chemotactic for monocytes, *Nature* (1978) **275**: 132–5.

194 Orr W, Varani J, Gondek MD et al, Chemotactic responses of tumor cells to products of resorbing bone, *Science* (1979) **203**: 176–9.

195 Orr FW, Varani J, Gondek MD et al, Partial characterization of a bone derived chemotactic factor for tumor cells, *Am J Pathol* (1980) **99**: 43–52.

196 Mundy GR, DeMartino S, Rowe DW, Collagen and collagen derived fragments are chemotactic for tumor cells, *J Clin Invest* (1981) **68**: 1102–5.

197 Spiro TP, Mundy GR, In vitro chemotaxis of Walker 256 carcinosarcoma cells is dependent on microtubule and microfilament function, *Clin Res* (1979) **27**: 392A.

198 Manishen WJ, Sivanathan K, Orr FW, Resorbing bone stimulates tumor cell growth. A role for the host microenvironment in bone metastasis, *Am J Pathol* (1986) **123**: 39–45.

199 Fisher B, Fisher ER, Feduska N, Trauma and the localization of tumor cells, *Cancer* (1967) **20**: 23–30.

200 Hayashi H, Yoshida K, Ozaki T et al, Chemotactic factor associated with invasion of cancer cells, *Nature* (1970) **226**: 174–5.

201 Ozaki T, Yoshida K, Ushijima K et al, Studies on the mechanisms of invasion in cancer. II. In vivo effects of a factor chemotactic for cancer cells, *Int J Cancer* (1971) **7**: 93–100.

202 Ushijima K, Nishi H, Ishikura A et al, Characterization of two different factors chemotactic for cancer cells from tumor tissue, *Virchows Arch* (1976) **21**: 119–31.

Chapter 6

1 Snapper I, Kahn A, *Myelomatosis* (Karger: Basel 1971).

2 Durie BGM, Salmon SE, Mundy GR, Relation of osteoclast activating factor production to the extent of bone disease in multiple myeloma, *Br J Haematol* (1981) **47**: 21–30.

3 Lazor MZ, Rosenberg LE, Mechanisms of adrenal-steroid reversal of hypercalcemia in multiple myeloma, *N Engl J Med* (1964) **270**: 749–55.

4 Mundy GR, Raisz LG, Cooper RA et al, Evidence for the secretion of an osteoclast stimulating factor in myeloma, *N Engl J Med* (1974) **291**: 1041–6.

5 Valentin-Opran A, Charhon SA, Meunier PJ et al, Quantitative histology of myeloma induced bone changes, *Br J Haematol* (1982) **52**: 601–10.

6 Mundy GR, Bertolini DR, Bone destruction and hypercalcemia in plasma cell myeloma, *Semin Oncol* (1986) **13**: 291–9.

7 Woolfenden JM, Pitt MJ, Durie BGM et al, Comparison of bone scintigraphy and radiography in multiple myeloma, *Radiology* (1980) **134**: 723–8.

8 Wahner HW, Kyle RA, Beabout JW, Scintigraphic evaluation of the skeleton in multiple myeloma, *Mayo Clin Proc* (1980) **55**: 739–46.

9 Lindstrom E, Lindstrom FD, Skeletal scintigraphy with technetium diphosphonate in multiple myeloma—a comparison with skeletal x-ray, *Acta Med Scand* (1980) **208**: 289–91.

10 Leonard RCF, Owen JP, Proctor SJ et al, Multiple myeloma: radiology or bone scanning, *Clin Radiol* (1981) **32**: 291–5.

11 Deding A, Ingeberg S, Jorgensen PS, Whole body bone scintigraphy compared to conventional radiography in detecting bone lesions in multiple myeloma, *Dan Med Bull* (1981) **28**: 207–9.

12 Ludwig H, Kumpan W, Sinzinger H, Radiography and bone scintigraphy in multiple myeloma: a comparative analysis, *Br J Radiol* (1982) **55**: 173–81.

13 Bataille R, Magub M, Grevier J et al, Serum β_2-M in multiple myeloma: relation to presenting features and clinical status, *Eur J Cancer Clin Oncol* (1982) **18**: 59–66.

14 Schechter GP, Wahl LM, Horton JE, In vitro bone resorption by human myeloma cells. In: Potter M, ed., *Progress in myeloma* (Elsevier/North-Holland: New York 1980) 67–80.

15 McDonald DF, Schofield BH, Prezioso EM et al, Direct bone resorbing activity of murine myeloma cells, *Cancer Lett* (1983) **19**: 119–24.

16 Eilon G, Mundy GR, Direct resorption of bone by human breast cancer cells in vitro, *Nature* (1978) **276**: 726–8.

17 Mundy GR, Bertolini DB, Bone destruction and hypercalcemia in plasma cell myeloma, *Sem Oncol* (1986) **13**: 291–9.

18 Bardwick PA, Zvaifler NJ, Gill GN et al, Plasma cell dyscrasia with polyneuropathy, organomegaly, endocrinopathy, M protein, and skin changes: the POEMS syndrome, *Medicine* (1980) **59**: 311–22.

19 Canalis E, McCarthy T, Centrella M, Identity of bone-derived growth factor β_2 microglobulin, *J Bone Miner Res* (1987) **2**: 251.

20 Bataille R, Chevalier J, Rossi M et al, Bone scintigraphy in plasma-cell myeloma, *Radiology* (1982) **145**: 801–4.

21 Morell A, Riesen W, Serum β_2-M, serum creatinine and bone marrow plasma cells in benign and malignant monoclonal gammopathy, *Acta Haematol* (1980) **65**: 87–93.

22 Norfolk D, Child JA, Cooper EH et al, Serum β_2-microglobulin in myelomatosis: potential value in stratification and monitoring, *Br J Cancer* (1980) **42**: 510–14.

23 Mundy GR, Luben RA, Raisz LG et al, Bone resorbing activity in supernatants from lymphoid cell lines, *N Engl J Med* (1974) **290**: 867–71.

24 Gailani S, McLimans WF, Mundy GR et al, Controlled environment culture of bone marrow explants from human myeloma, *Cancer Res* (1976) **36**: 1299–304.

25 Josse RG, Murray TM, Mundy GR et al, Observations on the mechanism of bone resorption induced by multiple myeloma marrow culture fluids and partially purified osteoclast-activating factor, *J Clin Invest* (1981) **67**: 1472–81.

26 Luben RA, Mundy GR, Trummel CL et al, Partial purification of osteoclast activating factor from phytohemagglutinin-stimulated human leukocytes, *J Clin Invest* (1974) **53**: 1473–80.

27 Raisz LG, Luben RA, Mundy GR, Effects of osteoclast activating factor from human leukocytes on bone metabolism, *J Clin Invest* (1975) **56**: 408–13.

28 Yoneda T, Mundy GR, Prostaglandins are necessary for osteoclast activating factor (OAF) production by activated peripheral blood leukocytes, *J Exp Med* (1979) **149**: 279–83.

29 Yoneda T, Mundy GR, Monocytes regulate osteoclast activating factor production by releasing prostaglandins, *J Exp Med* (1979) **150**: 338–50.

30 Mundy GR, Rick ME, Turcotte R et al, Pathogenesis of hypercalcemia in lymphosarcoma cell leukemia—role of an osteoclast activating factor like substance and mechanism of action for glucocorticoid therapy, *Am J Med* (1978) **654**: 600–6.

31 Fetchick DA, Bertolini DR, Sarin P et al, Production of 1,25-dihydroxyvitamin D by human T cell lymphotropic virus-I tranformed lymphocytes, *J Clin Invest* (1986) **78**: 592–6.

32 Dewhirst FE, Stashenko PP, Mole JE et al, Purification of partial sequences of human osteoclast activating factor: identity with interleukin 1 beta, *J Immunol* (1985) **135**: 2562–8.

33 Gowen M, Wood OD, Ihrie EJ et al, An interleukin 1 like factor stimulates bone resorption in vitro, *Nature* (1983) **306**: 378–80.

34 Gowen M, Mundy GR, Actions of recombinant interleukin-1, interleukin-2 and interferon gamma on bone resorption in vitro, *J Immunol* (1986) **136**: 2478–82.

35 Bertolini DR, Nedwin GE, Bringman TS, Stimulation of bone resorption and inhibition of bone formation in vitro by human tumour necrosis factors, *Nature* (1986) **319**: 516–18.

36 Garrett IR, Durie BGM, Nedwin GE et al, Production of the bone resorbing cytokine lymphotoxin by cultured human myeloma cells, *N Engl J Med* (1987) **317**: 526–32.

37 Bataille R, Jourdan M, Zhang Xue-Guang et al, Serum levels of interleukin-6, a potent myeloma cell growth factor, as a reflect of disease severity in plasma cell dyscrasias, *J Clin Invest* (1989) **84**: 2008–11.

38 Garrett IR, Black KS, Mundy GR, Interactions between interleukin-6 and interleukin-1 in osteoclastic bone resorption in neonatal mouse calvariae, *Calcif Tissue Int* (1990) in press.

39 Kawano M, Yamamoto I, Iwato K et al, Interleukin-1 beta rather than lymphotoxin as the major bone resorbing activity in human multiple myeloma, *Blood* (1989) **73**: 1646–9.

40 Cozzolino F, Torcia M, Aldinucci D et al, Production of interleukin-1 by bone marrow myeloma cells, *Blood* (1989) **74**: 387–80.

41 Merlini G, Fitzpatrick LA, Siris ES et al, A human myeloma immunoglobulin G binding four moles of calcium associated with asymptomatic hypercalcemia, *J Clin Immunol* (1984) **4**: 185–96.

42 Binstock ML, Mundy GR, Effects of calcitonin and glucocorticoids in combination in hypercalcemia of malignancy, *Ann Intern Med* (1980) **93**: 269–72.

43 Mundy GR, Martin TJ, The hypercalcemia of malignancy: pathogenesis and management, *Metabolism* (1982) **31**: 1247–77.

44 Van Breukelen FJM, Bijvoet OLM, Van Oosterom AT, Inhibition of osteolytic bone lesions by (3-amino-1-hydroxypropylidene)-1,1-bisphosphonate (A.P.D.), *Lancet* (1979) **i**: 803–5.

45 Siris ES, Sherman WH, Baquiran DC et al, Effects of dichloromethylene diphosphonate on skeletal mobilization of calcium in multiple myeloma, *N Engl J Med* (1980) **302**: 310–15.

46 Jacobs TP, Gordon AC, Silverberg SJ et al, Neoplastic hypercalcemia: physiologic response to intravenous etidronate disodium, *Am J Med* (1987) **82**(2A): 42–50.

47 Singer FR, Fernandez M, Therapy of hypercalcemia of malignancy, *Am J Med* (1987) **82**(2A): 34–41.

48 Cohen P, Fluoride and calcium therapy for myeloma bone lesions, *JAMA* (1966) **198**: 583–6.

49 Carbone PP, Zipkin I, Sokoloff L et al, Fluoride effect on bone in plasma cell myeloma, *Arch Intern Med* (1968) **121**: 130–40.

50 Cohen P, Gardner FH, Induction of subacute skeletal fluorosis in a case of multiple myeloma, *N Engl J Med* (1964) **271**: 1129–33.

51 Harly JB, Schilling A, Glidewell O, Ineffectiveness of fluoride therapy in multiple myeloma, *N Engl J Med* (1972) **286**: 1283–8.

52 Kyle RA, Jowsey J, Phil D et al, Multiple myeloma bone disease. The comparative effect of sodium fluoride and calcium carbonate or placebo, *N Engl J Med* (1975) **293**: 1334–8.

53 Adatia AK, Dental tissues and Burkitt's tumor, *Oral Surg* (1968) **25**: 221–34.

54 Spiegel A, Greene M, Magrath I et al, Hypercalcemia with suppressed parathyroid hormone in Burkitt's lymphoma, *Am J Med* (1978) **64**: 691–702.

55 Mudde AH, van den Berg H, Boshuis PG et al, Ectopic production of 1,25-dihydroxyvitamin D by B-cell lymphoma as a cause of hypercalcemia, *Cancer* (1987) **59**: 1543–6.

56 Broder S, Bunn PA, Jaffe ES et al, T-cell lymphoproliferative syndrome associated with human T-cell leukemia/lymphoma virus, *Ann Intern Med* (1984) **100**: 543–57.

57 Grossman B, Schechter GP, Horton JE et al, Hypercalcemia associated with T-cell lymphoma-leukemia, *Am J Clin Pathol* (1981) **75**: 149–55.

58 Bunn PA, Schechter TP, Jaffe E et al, Clinical course of retrovirus-associated adult T-cell lymphoma in the United States, *N Engl J Med* (1983) **309**: 257–64.

59 Blayney DW, Jaffe ES, Fisher RI et al, The human T-cell leukemia/lymphoma virus, lymphoma, lytic bone lesions, and hypercalcemia, *Ann Intern Med* (1983) **98**: 144–51.

60 Canellos GP, Hypercalcemia in malignant lymphoma and leukemia, *Ann NY Acad Sci* (1974) **230**: 240–6.

61 Kinoshita K, Kamihira S, Ikeda S et al, Clinical, hematologic, and pathologic features of leukemic T-cell lymphoma, *Cancer* (1982) **50**: 1554–62.

62 Brigham BA, Bunn PA, Horton JE et al, Skeletal manifestations in cutaneous T-cell lymphomas, *Arch Dermatol* (1982) **118**: 461–7.

63 Singer FR, Powell D, Minkin C et al, Hypercalcemia in reticulum cell sarcoma without hyperparathyroidism or skeletal metastases, *Ann Intern Med* (1973) **78**: 365–9.

64 Bertolini DR, Sarin P, Mundy GR, Production of macromolecular bone resorbing activity by human T cell leukemia virus (HTLV) transformed cell lines, *Calcif Tissue Int* (1984) **36**: 284A.

65 Breslau NA, McGuire JL, Zerwekh JE et al, Hypercalcemia associated with increased serum calcitriol levels in three patients with lymphoma, *Ann Intern Med* (1984) **100**: 1–7.

66 Rosenthal N, Insogna KL, Godsall JW et al, Elevations in circulating 1,25 dihydroxyvitamin D in three patients with lymphoma-associated hypercalcemia, *J Clin Endocrinol Metab* (1985) **60**: 29–33.

67 Zaloga GP, Eil C, Medbery CA, Humoral hypercalcemia in Hodgkin's disease, *Arch Intern Med* (1985) **145**: 155–7.

68 Davies M, Hayes ME, Mawer EB et al, Abnormal vitamin D metabolism in Hodgkin's lymphoma, *Lancet* (1985) **i**: 1186–8.

69 Dodd CD, Winkler CF, Williams ME et al, Calcitriol levels in hypercalcemic patients with adult T-cell lymphoma, *Arch Intern Med* (1986) **146**: 1971–3.

70 Fukumoto S, Matsumoto T, Ikeda K et al, Clinical evaluation of calcium metabolism in adult T-cell leukemia/lymphoma, *Arch Intern Med* (1988) **148**: 921–5.

71 Burt ME, Brennan MF, Incidence of hypercalcemia and malignant neoplasm, *Arch Surg* (1980) **115**: 704–7.

72 Adams JS, Fernandez M, Gacad MA et al, Vitamin D metabolite-mediated hypercalcemia and hypercalciuria patients with AIDS- and non-AIDS-associated lymphoma, *Blood* (1989) **73**: 235–9.

73 Motokura T, Fukumoto S, Matsumoto T et al, Parathyroid hormone related protein in adult T-cell leukemia-lymphoma, *Ann Intern Med* (1989) **111**: 484–8.

74 Walker IR, Lymphoma with hypercalcemia, *Can Med Assoc J* (1974) **111**: 928–30.

75 Kabakow B, Mines MF, King FH, Hypercalcemia in Hodgkin's disease, *N Engl J Med* (1957) **256**: 59–62.

76 Korek-Amorosa J, Scheinman HZ, Clemett AR et al, Hypercalcemia and extensive lymph-node calcification in a patient with Hodgkin's disease prior to therapy, *Br J Radiol* (1974) **47**: 905–7.

77 Eldar M, Shoenfeld Y, Douer D et al, Hypercalcemia in Hodgkin's disease, *Haemotologica* (1980) **65**: 459.

78 Needle MA, Chandra B, Hypercalcemia, Hodgkin's disease and calcitriol (Letter), *Ann Intern Med* (1984) **100**: 916.

79 Schaefer K, Saupe J, Pauls A et al, Hypercalcemia and elevated serum 1,25 dihydroxyvitamin D_3 in a patient with Hodgkin's lymphoma, *Klin Wochenschr* (1986) **64**: 89–91.

80 Adams ND, Gray RW, Lemann J et al, Effects of calcitriol administration on calcium metabolism in healthy men, *Kidney Int* (1982) **21**: 90–7.

81 Maierhofer WJ, Gray RW, Cheung HS et al, Bone resorption stimulated by elevated serum 1,25-$(OH)_2$-vitamin D concentrations in healthy men, *Kidney Int* (1983) **24**: 555–60.

82 Moses AM, Spencer H, Hypercalcemia in patients with malignant lymphoma, *Ann Intern Med* (1963) **59**: 531–6.

83 Laugen RH, Carey RM, Wills MR, Hess CE, Hypercalcaemia associated with chronic lymphocytic leukemia, *Arch Intern Med* (1979) **139**: 1307–9.

84 McMillan P, Mundy G, Mayer P, Hypercalcaemia and osteolytic bone lesions in chronic lymphocytic leukaemia, *Br Med J* (1980) **281**: 1107.

85 Heath H III, Weller RE, Mundy GR, A model for study of the hypercalcemia of cancer, *Calif Tissue Int* (1980) **30**: 127–33.

Chapter 7

1 Parfitt AM, Equilibrium and disequilibrium hypercalcemia : new light on an old concept, *Metab Bone Dis Relat Res* (1979) **1**: 279–93.

2 Fleisch H, Russell RGG, Francis MD, Diphosphonates inhibit hydroxyapatite dissolution in vitro and bone resorption in tissue culture and in vivo, *Science* (1969) **165**: 1262–4.

3 Francis MD, Russell RGG, Fleisch H, Diphosphonates inhibit formation of calcium phosphate crystals in vitro and pathological calcification in vivo, *Science* (1969) **165**: 1264–6.

4 Russell RGG, Muhlbauer RC, Bisaz S et al, The influence of pyrophosphate, condensed phosphates phosphonates, and other phosphate compounds on the dissolution of hydroxyapatite in vitro and on bone resorption induced by parathyroid hormone in tissue culture and in thyroparathyroidectomized rats, *Calcif Tissue Res* (1970) **6**: 183–96.

5 Russell RGG, Smith R, Diphosphonates—experimental and clinical aspects, *J Bone Joint Surg [Br]* (1973) **55**: 66–86.

6 Frijlink WB, Bijvoet OLM, te Velde J et al, Treatment of Paget's disease with (3'-amino-1-hydroxy propylidene)-1.1-bisphosphonate (A.P.D.), *Lancet* (1979) **i**, 799–803.

7 Meunier PJ, Chapuy MC, Alexandre C, Effects of disodium dichloromethylene disphosphonate on Paget's disease of bone, *Lancet* (1979) **ii**: 489–92.

8   Hughes DE, MacDonald BR, Russell RGG et al, Inhibition of osteoclast-like cell formation by bisphosphonates in long-term cultures of human bone marrow, *J Clin Invest* (1989) **83**: 1930–5.

9   Reitsma PH, Teitelbaum SL, Bijvoet OLM et al, Differential action of the bisphosphonates (3-amino-1-hydroxypropylidene)-1,1-bisphosphonate (APD) and disodium dichloromethylidene disphosphonate (Cl₂MDP) on rat macrophage-mediated bone resorption in vitro, *J Clin Invest* (1982) **70**: 927–33.

10 Lowik CW, Boonekamp PM, van de Ploym G et al, Bisphosphonates can reduce osteoclastic bone resorption by two different mechanisms, *Adv Exp Med Biol* (1986) **208**: 275–81.

11 Flanagan AM, Chambers TJ, Dichloromethylene bisphosphonate (Cl₂MBP) inhibits bone resorption through injury to osteoclasts that resorb Cl₂MBP-coated bone, *Bone Miner* (1989) **6**: 33–43.

12 Mundy GR, Wilkinson R, Heath DA, Comparative study of available medical therapy for hypercalcemia of malignancy, *Am J Med* (1983) **74**: 421–32.

13 Ryzen E, Rude RK, Elbaum N et al, Use of intravenous etidronate disodium in the treatment of hypercalcemia. In: Garattini S, ed., *Bone resorption, metastasis, and diphosphonates* (Raven Press: New York 1985) 99–108.

14 Sleeboom HP, Bijvoet OLM, van Oosterom AT et al, Comparison of intravenous (3-amino-1-hydroxypropylidene)-1, 1-bisphosphonate and volume repletion in tumour-induced hypercalcemia, *Lancet* (1983) **ii**: 239–43.

15 Yates AJP, Jerums GJ, Murray RML et al, A comparison of single and multiple intravenous infusions of 3-amino-hydroxypropylidene-1, 1-bisphosphonate (APD) in the treatment of hypercalcemia of malignancy, *Aust NZ J Med* (1987) **17**: 387–91.

16 Morton AR, Cantrill JA, Craid AE et al, Single dose versus daily intravenous aminohydroxypropylidene bisphosphonate (APD) for hypercalcemia of malignancy, *Brit Med J* (1988) **296**: 811–14.

17 Thiebaud D, Jaeger Ph, Jacquet AF et al, Dose response in the treatment of hypercalcaemia of malignancy by a single infusion of the bisphosphonate AHPrBP (APD), *J Clin Oncol* (1988) **6**: 762–8.

18 Portmann L, Haefliger JM, Bill G et al, Un traitment simple de l'hypercalcemie tumorale: l'amino-hydroxypropylidene bisphosphonate (APD) iv, *Schweiz Med Wochenschr* (1983) 113, **51**: 1960–3.

19 Thiebaud D, Jaeger P, Jacquet AF et al, A single-day treatment of tumor-induced hypercalcemia by intravenous amino-hydroxypropylidene bisphosphonate, *J Bone Min Res* (1986) **6**: 555–62.

20 Cantwell B, Harris AL, Single high dose aminohydroxypropylidene diphosphonate infusions to treat cancer-associated hypercalcemia, *Lancet* (1986) **i**: 165–6.

21 Kaplan RA, Geno WB, Poindexter C et al, Metabolic effects of diphosphonate in primary hyperparathyroidism, *J Clin Pharmacol* (1977) **17**: 410–19.

22 Chapuy MC, Meunier PJ, Alexandre CM et al, Effects of disodium dichloromethylene diphosphonate on the hypercalcemia produced by bone metastases, *J Clin Invest* (1980) **65**: 1243–7.

23 Jacobs TP, Siris ES, Bilezikian JP et al, Hypercalcemia of malignancy. Treatment with intravenous dichloromethylene diphosphonate, *Ann Intern Med* (1981) **94**: 312–16.

24 Siris ES, Sherman WH, Baquiran DC et al, Effects of dichloromethylene diphosphonate on skeletal mobilization of multiple myeloma, *N Engl J Med* (1980) **302**: 310–15.

25 Siris ES, Hyman GA, Canfield RE, Effects of dichloromethylene diphosphonate in women with breast carcinoma metastatic to the skeleton, *Am J Med* (1983) **74**: 401–6.

26 Van Breukelen FJM, Bijvoet OLM, Van Oosterom AT, Inhibition of osteolytic bone lesions by (3-aminio-1-hydroxypropylidene)-1.1.bisphosphonate (A.P.D.), *Lancet* (1979) **i**: 803–5.

27 Sleeboom HP, Bijvoet OLM, Treatment of tumour-induced hypercalcaemia. In: Garattini S, ed., *Bone resorption, metastasis, and diphosphonates* (Raven Press: New York 1985) 59–78.

28 Ralston SH, Patel U, Fraser WD et al, Comparison of three intravenous bisphosphonates in cancer-associated hypercalcaemia, *Lancet* (1989) **2**: 1180–2.

29 Body JJ, Borkowski A, Cleeren A et al, Treatment of malignancy-associated hypercalcaemia with intravenous aminohydroxypropylidene diphosphonate, *J Clin Oncol* (1986) **8**: 1177–83.

30 Body JJ, Pot M, Borkowski A et al, Dose-response study of aminohydroxypropylidene bisphosphonate in tumor-associated hypercalcemia, *Am J Medicine* (1987) **82**: 957–63.

31 Portmann L, Haefliger JM, Bill G et al, Un traitement simple de l'hypercalcemie tumorale: l'aminohydroxypropylidene bisphosphonate (APD) iv, *Schweiz Med Wochenschr* (1983) 113, **51**: 1960–3.

32 Cappelli R, Adami S, Bnalla AK et al, Bisphosphonates and the acute phase response. In: Cohn DV, Martin TJ, Meunier PJ, eds, *Calcium Regulation and Bone Metabolism. Basic and Clinical Aspects*, vol 9 (Excerpta Medica: Amsterdam 1987) 699.

33 Coleman RE, Rubens RD, 3 (amino-1, 1-hydroxypropylidene) bisphosphonate (APD) for hypercalcemia of breast cancer, *Brit J Cancer* (1987) **56**: 465–9.

34 Ralston SH, Alzaid AA, Gardner MD et al, Treatment of cancer associated hypercalcemia with combined aminohydroxypropylidene diphosphonate and calcitonin, *Brit Med J* (1986) **292**: 1549–50.

35 Singer FR, Fernandez M, Therapy of hypercalcemia of malignancy, *Am J Med* (1987) **82**(Suppl 2A): 34–41.

36 Burckhardt P, ed., *Disodium pamidronate (APD) in the treatment of malignancy-related disorders*, (Hans Huber Publishers: Toronto 1989).

37 Rubens RD, ed., *The management of bone metastases and hypercalcaemia by osteoclast inhibition*, (Hogrefe and Huber Publishers: Toronto–Lewiston NY–Bern–Stuttgart).

38 Kanis JA, ed., *Bone Supplement: Clodronate – A new perspective in the treatment of neoplastic bone disease*, (Pergamon Press: New York 1987).

39 Garattini S, ed., *Bone Resorption, Metastasis, and Diphosphonates*, (Raven Press: New York 1984).

40 Kofman S, Medrek TJ, Alexander RW, Mithramycin in the treatment of embryonal cancer, *Cancer* (1964) **17**: 938–48.

41 Brown JH, Kennedy BJ, Mithramycin in the treatment of disseminated testicular neoplasms, *N Engl J Med* (1965) **272**: 111–18.

42 Parsons V, Baum M, Seif M, Effect of mithramycin on calcium and hydroxyproline metabolism in patients with malignant disease, *Br Med J* (1967) **1**: 474–7.

43 Singer FR, Neer RM, Murray TM et al, Mithramycin therapy of intractable hypercalcemia due to parathyroid carcinoma, *N Engl J Med* (1970) **283**: 634–6.

44 Perlia CP, Gubisch NJ, Wolter J et al, Mithramycin treatment of hypercalcemia, *Cancer* (1970) **25**: 389–94.

45 Ryan WG, Schwartz TB, Northrop G, Mithramycin in Paget's disease of bone, *JAMA* (1970) **213**: 1153–7.

46 Coombes RC, Dady P, Parsons C et al, Mithramycin therapy: an adjunct to conventional treatment of hypercalcemia and bone metastases in breast cancer, *Metab Bone Dis Relat Res* (1980) **2**: 199–202.

47 Yarbro JW, Kennedy BJ, Barnum CP, Mithramycin inhibition of ribonucleic acid synthesis, *Cancer Res* (1966) **26**: 36–9.

48 Wollheim MS, Yarbro JW, Kennedy BJ, Effect of mithramycin on HeLa cells, *Cancer* (1968) **21**: 22–5.

49 Northrop G, Taylor SG III, Northrup RL, Biochemical effects of mithramycin on cultured cells, *Cancer Res* (1969) **29**: 1916–19.

50 Minkin C, Inhibition of parathyroid hormone stimulated bone resorption in vitro by the antibiotic mithramycin, *Calcif Tissue Res* (1973) **13**: 249–57.

51 Cortes EP, Holland JF, Moskowitz R et al, Effects of mithramycin on bone resorption in vitro, *Cancer Res* (1972) **32**: 74–6.

52 Mundy GR, Martin TJ, The hypercalcemia of malignancy: pathogenesis and management, *Metabolism* (1982) **31**: 1247–77.

53 Copp DH, Cameron EC, Cheney EC et al, Evidence for calcitonin—a new hormone from the parathyroid that lowers blood calcium, *Endocrinology* (1962) **70**: 638–49.

54 Binstock ML, Mundy GR, Effect of calcitonin and glucocorticoids in combination on the hypercalcemia of malignancy, *Ann Intern Med* (1980) **93**: 269–72.

55 Au WYW, Calcitonin treatment of hypercalcemia due to parathyroid carcinoma—synergistic effect of prednisone on long-term treatment of hypercalcemia, *Arch Intern Med* (1975) **135**: 1594–7.

56 Ralston SH, Gardner MD, Jenkins AS et al, Comparison of aminohydroxypropylidene diphosphon-

ate, mithramycin and corticosteroids/calcitonin in treatment of cancer associated hypercalcaemia, *Lancet* (1985) **ii**: 207–10.

57   Hosking DJ, Gilson D, Comparison of the renal and skeletal actions of calcitonin in the treatment of severe hypercalcaemia of malignancy, *Q J Med* (1984) **211**: 359–68.

58   Thiebaud D, Burckhardt P, Jaeger Ph et al, Effectiveness of salmon calcitonin administered as suppositories in tumor-induced hypercalcemia, *Am J Med* (1987) **82**: 745–50.

59   Hosking DJ, Bijvoet OLM, Therapeutic uses of calcitonin. In Parsons JA, ed., *Endocrinology of calcium metabolism* (Raven Press: New York 1982) 485–535.

60   Raisz LG, Niemann I, Effect of phosphate, calcium and magnesium on bone resorption and hormonal responses in tissue culture, *Endocrinology* (1969) **85**: 446–52.

61   Lorenzo JA, Holtrop ME, Raisz LG, Effects of phosphate on calcium release, lysosomal enzyme activity in the medium, and osteoclast morphometry in cultured fetal rat bones, *Metab Bone Dis Relat Res* (1984) **5**: 187–90.

62   Ayala G, Chertow BS, Shah JH et al, Acute hyperphosphatemia and acute persistent renal insufficiency induced by oral phosphate therapy, *Ann Intern Med* (1975) **83**: 520–1.

63   Dudley FJ, Blackburn CRB, Extraskeletal calcification complicating oral neurophosphate therapy, *Lancet* (1970) **ii**: 628–30.

64   Carey RW, Schmitt GW, Kopald HH, Massive extraskeletal calcification during phosphate treatment of hypercalcemia, *Arch Intern Med* (1968) **122**: 150–5.

65   Dent CE, Some problems of hyperparathyroidism, *Br Med J* (1962) **2**: 1419–25.

66   Mundy GR, Rick ME, Turcotte R et al, Pathogenesis of hypercalcemia in lymphosarcoma cell leukemia—role of an osteoclast activating factor-like substance and mechanism of action for glucocorticoid therapy, *Am J Med* (1978) **65**: 600–6.

67   Ito H, Sanada T, Katamaya et al, Letter: indomethacin-responsive hypercalcemia, *N Engl J Med* (1975) **293**: 558–9.

68   Seyberth HW, Segre GV, Morgan JL et al, Prostaglandins as mediators of hypercalcemia associated with certain types of cancer, *N Engl J Med* (1975) **293**: 1278–83.

69   Foster BJ, Clagett-Carr K, Hoth D et al, Gallium nitrate: the second metal with clinical activity, *Cancer Treat Rep* (1986) **70**: 1311–19.

70   Warrell RP Jr, Bockman RS, Coonley CJ et al, Gallium nitrate inhibits calcium resorption from bone and is effective treatment for cancer-related hypercalcemia, *J Clin Invest* (1984) **73**: 1487–90.

71   Warrell RP Jr, Israel R, Frisone M et al, Gallium nitrate for acute treatment of cancer-related hypercalcemia. A randomized, double-blind comparison to calcitonin, *Ann Intern Med* (1988) **108**: 669–74.

72   Bonjour J-P, Philippe J, Guelpa G et al, Bone and renal components in hypercalcemia of malignancy and responses to a single infusion of clodronate, *Bone* (1988) **9**: 123–30.

73   Kukla LJ, Abramson EC, McGuire WP et al, Cisplatinum treatment for malignancy-associated humoral hypercalcemia in an athymic mouse model, *Calcif Tissue Int* (1984) **36**: 559–62.

74   Lad TE, Mishoulam HM, Shevrin DH et al, Treatment of cancer-associated hypercalcemia with cisplatin, *Arch Intern Med* (1987) **147**: 329–32.

75   Glover D, Riley L, Carmichael K et al, Hypocalcemia and inhibition of parathyroid hormone secretion after administration of WR-2721 (a radioprotective and chemoprotective agent), *N Engl J Med* (1983) **309**: 1137–41.

76   Weaver ME, Morrissey J, McConkey C Jr et al, WR-2721 inhibits parathyroid adenylate cyclase, *Am J Physiol* (1987) **252**: E197–201.

77   Larsson R, Nygren P, Wallfelt C et al, Dual effects of a new hypocalcemic agent, WR-2721, on cytoplasmic Ca^{2+} and parathyroid hormone release of dispersed parathyroid cells from patients with hyperparathyroidism, *Biochem Pharmacol* (1986) **35**: 4237–41.

78   Hirschel-Scholz S, Caverzasio J, Bonjour J-P, Inhibition of parathyroid hormone secretion and parathyroid hormone-independent diminution of tubular calcium reabsorption by WR-2721, a unique hypocalcemic agent, *J Clin Invest* (1985) **76**: 1851–6.

79 Attie MF, Fallon D, Spar B et al, Bone and parathyroid inhibitory effects of S-2(3-aminopropylamino) ethylphosphorothioic acid. Studies in experimental animals and cultured bone cells, *J Clin Invest* (1985) **75**: 1191–7.

80 Glover DJ, Shaw L, Glick JH et al, Treatment of hypercalcemia in parathyroid cancer with WR-2721, S-2-(3-aminopropylamino) ethyl-phosphorthioic acid, *Ann Intern Med* (1985) **103**: 55–7.

81 Hirschel-Scholz S, Jung A, Fischer JA et al, Suppression of parathyroid secretion after administration of WR-2721 in a patient with parathyroid carcinoma, *Clin Endocrinol* (1985) **23**: 313–18.

82 Hirschel-Scholz S, Caverzasio J, Rizzoli R et al, Normalization of hypercalcemia associated with a decrease in renal calcium absorption in Leydig cell tumor-bearing rats treated with WR-2721, *J Clin Invest* (1986) **78**: 319–22.

83 Weiss J, Walker ST, Fallon M et al, In vivo and in vitro effects of WR-2721 in experimental hypercalcemia in the rat, *J Pharmacol Exp Ther* (1986) **238**: 969–73.

84 Turrisi AT, Glover DJ, Hurwitz S et al, Final report of the phase I trial of single-dose WR-2721 [S-2-(3-aminopropylamino) ethylphosphorothioic acid], *Cancer Treat Rep* (1986) **70**: 1389–93.

85 Glover D, Glick JN, Weiler C et al, Phase I/II trials of WR-2721 and cis-platinum, *J Radiat Oncol Biol Phys* (1986) **12**: 1509–12.

86 Hosking DJ, Cowley A, Bucknall CA, Rehydration in the treatment of severe hypercalcaemia, *Q J Med* (1981) Ser. L: 473–81.

87 Bounameux HM, Schifferli J, Montani J-P et al, Renal failure associated with intravenous diphosphonates, *Lancet* (1983) i: 471.

88 Kanis JA, Preston CJ, Yates AJP et al, Effects of intravenous diphosphonates on renal function, *Lancet* (1983) i: 1328.

89 Suki WN, Yium JJ, von Minden M et al, Acute treatment of hypercalcemia with furosemide, *N Engl J Med* (1970) **283**: 836–40.

90 Albright F, Bauer W, Claflin D et al, Studies in parathyroid physiology, *J Clin Invest* (1932) **11**: 411–35.

91 Goldsmith RS, Ingbar SH, Inorganic phosphate treatment of hypercalcemia of diverse etiologies, *N Engl J Med* (1966) **274**: 1–7.

92 Fulmer DH, Dimich AB, Rothschild EO et al, Treatment of hypercalcemia, comparison of intravenously administered phosphate sulfate and hydrocortisone, *Arch Intern Med* (1972) **129**: 923–30.

93 Shackney S, Hasson J, Precipitous fall in serum calcium, hypotension, and acute renal failure after intravenous phosphate therapy for hypercalcemia, *Ann Intern Med* (1967) **66**: 906–16.

94 Breur RI, LeBauer J, Caution in the use of phosphates in the treatment of severe hypercalcemia, *J Clin Endocrinol Metab* (1967) **27**: 695–8.

95 Fierer JA, Wagner BM, Strebel RF, Metastatic calcification of the myocardium, *Am J Cardiol* (1970) **26**: 423–6.

96 Carey RW, Schmitt GW, Kopald HH, Massive extra skeletal calcification during phosphate treatment of hypercalcemia, *Arch Intern Med* (1968) **122**: 150–5.

97 Miach PJ, Dawborn JK, Martin TJ et al, Management of the hypercalcemia of malignancy by peritoneal dialysis, *Med J Aust* (1975) **1**: 782–4.

Chapter 8

1 Sandstrom I, Om en ny kortel nos menniska och artskolliga daggijur, *Ups Lakarefor Forh* (1880) **15**: 441. (Cited by Seipel CM, An English translation of Sandstrom's *Glandulae Parathyreoideae*, *Bull Hist Med* (1938) **6**: 179–22.)

2 Albright F, Reifenstein EC Jr, *The parathyroid glands and metabolic bone disease: selected studies* (Williams and Wilkins: Baltimore 1948).

3 Von Recklinghausen FD, Die Fibrose oder Deformirende Osteitis, die Osteomalacie, und die Osteoplastische Carcinose in Ihren Gegenseitigen Beziehungen Festschr Rudolph Virchow zu seinem 71 Geburtstage Gewidmet, Berlin (1891) 1–89.

4 Askanazy M, Ueber Osteitis Deformans ohne Osteides, *Gew Arb Path Anat Bakt Tuebingen* (1904) **4**: 398–422.

5 Erdheim J, Tetania Parathyreopriva, *Mitt Granzgeb Med Chir* (1906) **16**: 632–744.

6   Mandl F, Therapeutischer Versuch bei einem Falle von Ostitis Fibrosa Generalisata mittels Exstirpation eines Epithelkorperchen Tumors, *Zentralbl Chir* (1926) **53**: 260–4.

7   Cope O, Keynes WM, Roth SI et al, Primary chief-cell hyperplasia of the parathyroid glands. A new entity in the surgery of hyperparathyroidism, *Ann Surg* (1958) **148**: 375–88.

8   Cope O, Hyperparathyroidism—too little, too much surgery? *N Engl J Med* (1976) **295**: 100–1.

9   Hannon RR, Shorr E, McClellan WS et al, A case of osteitis fibrosa cystica (osteomalacia?) with evidence of hyperactivity of the parathyroid bodies, *J Clin Invest* (1930) **8**: 215–77.

10 Bauer W, Albright F, Aub JC, A case of osteitis fibrosa cystica (osteomalacia?) with evidence of hyperactivity of the parathyroid bodies, *J Clin Invest* (1930) **8**: 229–48.

11 Bauer W, Hyperparathyroidism: a distinct disease entity, *J Bone Joint Surg* (1933) **15**: 135–41.

12 Bauer W, Federman DD, Hyperparathyroidism epitomized: the case of Captain Charles E. Martell, *Metabolism* (1962) **11**: 21–9.

13 Boonstra CE, Jackson CE, Serum calcium survey for hyperparathyroidism, *Am J Clin Pathol* (1971) **55**: 523–6.

14 Mundy GR, Cove DH, Fisken R, Primary hyperparathyroidism: changes in the pattern of clinical presentation, *Lancet* (1980) **i**: 1317–20.

15 Heath H III, Hodgson SF, Kennedy MA, Primary hyperparathyroidism: incidence, morbidity, and potential economic impact in a community, *N Engl J Med* (1980) **302**: 189–93.

16 Paloyan E, Lawrence AM, Primary hyperparathyroidism: pathology and therapy, *JAMA* (1981) **246**: 1344.

17 Shane E, Bilezikian JP, Parathyroid carcinoma: a review of 62 patients, *Endocr Rev* (1982) **3**: 218–26.

18 Jackson CE, Cerny JC, Block MA et al, Probably clonal origin of parathyroid 'adenomas', *Surgery* (1982) **92**: 875–9.

19 Arnold A, Kronenberg HM, Approaches to the clonality of human tumors and their application to parathyroid adenomas. In: Kleerekoper M, Krane SM, eds, *Clinical Disorders of Bone and Mineral Metabolism* (Mary Ann Liebert: New York 1989) 311–16.

20 Arnold A, Staunton CE, Kim HG et al, Monoclonality and abnormal parathyroid hormone genes in parathyroid adenomas, *N Engl J Med* (1988) **318**: 658–62.

21 Brandi ML, Aurbach GD, Fitzpatrick LA et al, Parathyroid mitogenic activity in plasma from patients with familial multiple endocrine neoplasia Type 1, *N Engl J Med* (1986) **314**: 1287–93.

22 Peters G, Brookes S, Smith R et al, Tumorigenesis by mouse mammary tumor virus: evidence for a common region for provirus integration in mammary tumors, *Cell* (1983) **33**: 369–77.

23 Katz A, Braunstein GD, Clinical, biochemical, and pathologic features of radiation-associated hyperparathyroidism, *Arch Intern Med* (1983) **143**: 79–82.

24 Brown EM, Broadus AE, Brennan MF et al, Direct comparison in vivo and in vitro of suppressibility of parathyroid function of calcium in primary hyperparathyroidism, *J Clin Endocrinol Metab* (1979) **48**: 604–10.

25 Brown EM, Four-parameter model of the sigmoidal relationship between parathyroid hormone release and extracellular calcium concentration in normal and abnormal parathyroid tissue, *J Clin Endocrinol Metab* (1983) **56**: 572–81.

26 Brown EM, Chen CJ, Leboff MS, Physiology and pathophysiology of Ca^{++}-regulated PTH release. In: Kleerekoper M, Krane SM, eds, *Clinical Disorders of Bone and Mineral Metabolism* (Mary Ann Liebert: New York 1989) 317–22.

27 Fitzpatrick LA, Brandi ML, Aurbach GD, Calcium-controlled secretion is effected through a guanine nucleotide regulatory protein in parathyroid cells, *Endocrinology* (1986) **119**: 2700–3.

28 Marx SJ, Brandi ML, In Peck WA, ed., *Familial primary hyperparathyroidism* (Elsevier: Amsterdam 1987) 375–407.

29 Fourman P, Royer P, *Calcium metabolism and the bone* (FA Davis Co.: Philadelphia 1968).

30 Habener JF, Potts JT, Jr, Clinical features of primary hyperparathyroidism. In: DeGroot LJ, et

al., *Endocrinology*, Vol. 2, (Grune and Stratton: New York 1979) 693–701.

31   Mallette LE, Bilezikian JP, Heath DA et al, Primary hyperparathyroidism: clinical and biochemical features, *Medicine* (1974) **53**: 127–46.

32   Hellstrom J, Ivemark BI, Primary hyperparathyroidism, *Acta Chir Scand* (1962) **294**(Suppl): 1–133.

33   Melick RA, Henneman PH, Clinical and laboratory studies of 207 consecutive patients in a kidney stone clinic, *N Engl J Med* (1958) **259**: 307–14.

34   Dauphine RT, Riggs BL, Scholz DA, Back pain and vertebral crush fractures: an unemphasized mode of presentation for primary hyperparathyroidism, *Ann Intern Med* (1975) **83**: 365–7.

35   Genant HK, Baron JM, Straus FH et al, Osteosclerosis in primary hyperparathyroidism, *Am J Med* (1975) **59**: 104–13.

36   Pak CYC, Stewart A, Kaplan R et al, Photon absorptiometric analysis of bone density in primary hyperparathyroidism, *Lancet* (1975) **ii**: 7–8.

37   Jowsey J, Riggs BL, Bone changes in a patient with hypervitaminosis A, *J Clin Endocrinol Metab* (1968) **28**: 1833–5.

38   Belanger LF, In: Gaillard PJ, Talmage RV, Budy AM, eds, *The parathyroid glands: ultrastructure, secretion and function* (University of Chicago Press: Chicago 1965) 137–43.

39   Meunier P, Vignon G, Bernard J et al, Clinical aspects of metabolism and bone disease, Proceedings of an International Symposium, *Int Congr Ser* (1972) **270**: 215.

40   Bordier PJ, Arnaud CD, Hawker C et al, Clinical aspects of metabolism and bone disease, Proceedings of an International Symposium, *Int Congr Ser* (1973) **270**: 222–8 (Excerpta Medica: Amsterdam).

41   Jones SJ, Boyde A, Scanning electron microscopy of bone cells in culture. In: Copp DH, Talmage RV, eds, *Endocrinology of calcium metabolism* (Excerpta Medica: Amsterdam 1978) 97–114.

42   Steinbach L, Gordan GS, Eisenberg E et al, Primary hyperparathyroidism: a correlation of roentgen, clinical, and pathologic features, *Am J Roentgenol, Radium Ther, Nucl Med* (1961) **86**: 329–43.

43   Doyle FH, Some quantitative radiological observations in primary and secondary hyperparathyroidism, *Br J Radiol* (1966) **39**: 161–7.

44   Patten BM, Bilezikian JP, Mallette LE et al, Neuromuscular disease in primary hyperparathyroidism, *Ann Intern Med* (1974) **80**: 182–93.

45   Cope O, Culver PJ, Mixter CG et al, Pancreatis, diagnostic clue to hyperparathyroidism, *Ann Surg* (1957) **145**: 857–63.

46   Bywaters EGL, Dixon ASJ, Scott JT, Joint lesions of hyperparathyroidism, *Ann Rheum Dis* (1963) **22**: 171–87.

47   Wang CA, Miller LM, Weber AL et al, Pseudogout, *Am J Surg* (1969) **117**: 558–65.

48   Grahame R, Sutor DJ, Mitchener MB, Crystal deposition in hyperparathyroidism, *Ann Rheum Dis* (1971) **30**: 597–604.

49   McCarty DJ, Diagnostic mimicry in arthritis—patterns of joint involvement associated with calcium pyrophosphate dihydrate crystal deposits, *Bull Rheum Dis* (1975) **25**: 804–9.

50   Bunch TW, Hunder GG, Ankylosing spondylitis and primary hyperparathyroidism, *JAMA* (1973) **225**: 1108–9.

51   Preston FS, Adicoff A, Hyperparathyroidism with avulsion of three major tendons, *N Engl J Med* (1962) **266**: 968–71.

52   Preston ET, Avulsion of both quadriceps tendons in hyperparathyroidism, *JAMA* (1972) **221**: 406–7.

53   Hellstrom J, Clinical experiences of twenty one cases of hyperparathyroidism with special reference to the prognosis following parathyroidectomy, *Acta Chir Scand* (1950) **100**: 391–421.

54   Hellstrom J, Primary hyperparathyroidism observations in a series of 50 cases, *Acta Endocrinol (Copenh)* (1954) **16**: 30–58.

55   Hellstrom J, Hyperparathyroidism and gastroduodenal ulcer, *Acta Chir Scand* (1959) **116**: 207–21.

56   Britton DC, Thompson MH, Johnston IDA et al, Renal function following parathyroid surgery in primary hyperparathyroidism, *Lancet* (1971) **ii**: 74–5.

57 Moore WT, Smith LH, Experience with a calcium infusion test in parathyroid disease, *Metabolism* (1963) **12**: 447–51.

58 Anderson DC, Stewart WK, Piercy DM, Calcifying panniculitis with fat and skin necrosis in a case of uraemia with autonomous hyperparathyroidism, *Lancet* (1968) **ii**: 323–5.

59 Hartenstein H, Gardner L, Tetany of the newborn associated with maternal parathyroid adenoma, *New Eng J Med* (1966) **274**: 266–8.

60 Shangold MM, Dor N, Welt S et al, Hyperparathyroidism and pregnancy: A review, *Obstet Gynecol Surv* (1982) **37**: 217–28.

61 Ludwig GD, Hyperparathyroidism in relation to pregnancy, *N Engl J Med* (1962) **267**: 637–42.

62 Higgins RV, Hisley JC, Primary hyperparathyroidism in pregnancy. A report of 2 cases, *J Reprod Med* (1988) **33**: 726–30.

63 Lowe DK, Orwall ES, McClung MR et al, Hyperparathyroidism and pregnancy, *Am J Surg* (1983) **145**: 611–14.

64 Croom RD, Thomas CG, Primary hyperparathyroidism during pregnancy, *Surgery* (1984) **96**: 1109–18.

65 Kristofferson A, Duhlgren S, Lithner F et al, Primary hyperparathyroidism in pregnancy, *Surgery* (1984) **97**: 326–30.

66 Lueg M, Dawkins W, Primary hyperparathyroidism and pregnancy, *S Med J* (1983) **76**: 1389–92.

67 Patterson R, Hyperparathyroidism in pregnancy, *Obstet Gynecol* (1987) **70**: 457–62.

68 Gaeke RF, Kaplan EL, Lindheimer M et al, Maternal primary hyperparathyroidism of pregnancy. Successful treatment by parathyroidectomy, *JAMA* (1977) **238**: 508–9.

69 Gertner JM, Constan DR, Kliger AS et al, Pregnancy as a state of physiologic absorptive hypercalciuria, *Am J Med* (1986) **81**: 451–6.

70 Gray T, Lowe W, Lester G, Vitamin D and pregnancy: The maternal-fetal metabolism of vitamin D, *Endocr Rev* (1981) **2**: 264–74.

71 Rodda CP, Kubota M, Heath JA et al, Evidence for a novel parathyroid hormone-related protein in fetal lamb parathyroid glands and sheep placenta: Comparisons with a similar protein implicated in humoral hypercalcemia of malignancy, *J Endocrinol* (1988) **117**: 261–71.

72 Loveridge N, Caple IW, Rodda C et al, Further evidence for a parathyroid hormone related protein in fetal parathyroid gland of sheep, *Q J Exp Physiol* (1988) **73**: 781–4.

73 Nolan RB, Hayles AB, Woolner LB, Adenoma of the parathyroid gland in children, *Am J Dis Child* (1960) **99**: 622–8.

74 Reinfrank RF, Edwards TL Jr, Parathyroid crisis in a child, *JAMA* (1961) **178**: 468–71.

75 Rajasuriya K, Peiris OA, Ratnaike VT et al, Parathyroid adenomas in childhood, *Am J Dis Child* (1964) **107**: 442–9.

76 Chaves-Carballo E, Hayles AB, Parathyroid adenoma in children, *Am J Dis Child* (1966) **112**: 553–7.

77 Steendijk R, Metabolic bone disease in children, *Clin Orthop* (1971) **77**: 247–75.

78 Thakker RV, Bouloux P, Wooding C et al, Association of parathyroid tumors in multiple endocrine neoplasia type 1 with loss of alleles on chromosome 11, *N Engl J Med* (1989) **321**: 218–24.

79 Friedman E, Sakaguchi K, Bale AE et al, Clonality of parathyroid tumors in familial multiple endocrine neoplasia Type 1, *N Engl J Med* (1989) **321**: 213–18.

80 Knudson AG Jr, Mutation and cancer: statistical study of retinoblastoma, *Proc Natl Acad Sci* (1971) **68**: 820–3.

81 Gagel R, The pathogenesis and clinical course of multiple endocrine neoplasia, type 2a. In: Kleerekoper M, Krane SM, eds, *Clinical Disorders of Bone and Mineral Metabolism* (Mary Ann Liebert: New York 1989) 563–71.

82 Howard JE, Follis RH Jr, Yendt ER et al, Hyperparathyroidism. Case report illustrating spontaneous remission due to necrosis of adenoma, and a study of the incidence of necrosis in parathyroid adenoma, *J Clin Endocrinol Metab* (1953) **13**: 997–1008.

83 DeGroote JW, Acute intermittent hyperparathyroidism with hemorrhage into a parathyroid adenoma, *JAMA* (1969) **208**: 2160–1.

84 Watson L, Moxham J, Fraser P, Hydrocortisone suppression test and discriminant analysis in differential diagnosis of hypercalcaemia, *Lancet* (1980) **i**: 1320–5.

85 Wills MR, McGowan GK, Plasma-chloride levels in hyperparathyroidism and other hypercalcaemic states, *Br Med J* (1964) **1**: 1153–6.

86 Lufkin EG, Kao PC, Heath H III, Parathyroid hormone radioimmunoassays in the differential diagnosis of hypercalcemia due to primary hyperparathyroidism or malignancy, *Ann Intern Med* (1987) **106**: 559–60.

87 Benson RC, Riggs BL, Pickard BM et al, Radioimmunoassay of parathyroid hormone in hypercalcemic patients with malignant disease, *Am J Med* (1974) **56**: 821–6.

88 Powell D, Singer FR, Murray TM et al, Non-parathyroid humoral hypercalcemia in patients with neoplastic disease, *N Engl J Med* (1973) **289**: 176–81.

89 Raisz LG, Yajnik CH, Bockman RS et al, Comparison of commercially available parathyroid hormone immunoassays in the differential diagnosis of hypercalcemia due to primary hyperparathyroidism or malignancy, *Ann Intern Med* (1979) **91**: 739–40.

90 European PTH Study Group (EPSG), Interlaboratory comparison of radioimmunological parathyroid hormone determination, *Eur J Clin Invest* (1978) **8**: 149–54.

91 Mallette LE, Review: primary hyperparathyroidism, an update: incidence, etiology, diagnosis and treatment, *Am J Med Sci* (1987) **293**: 239–49.

92 Silverberg SJ, Shane E, De La Cruz et al, Skeletal disease in primary hyperparathyroidism, *J Bone Min Res* (1989) **4**: 283–91.

93 Nussbaum SR, Zahradnik RJ, Lavigne JR et al, Highly sensitive two-site immunoradiometric assay of parathyrin, and its clinical utility in evaluating patients with hypercalcemia, *Clin Chem* (1987) **33**: 1364–7.

94 Klee GG, Shikegawa, J, Trainer TD, CAP survey of parathyroid hormone assays, *Arch Pathol Lab Med* (1986) **110**: 588–91.

95 Scholz DA, Purnell DC, Asymptomatic primary hyperparathyroidism. 10 year prospective study, *Mayo Clin Proc* (1981) **56**: 473–8.

96 Wilson RJ, Rao DS, Ellis B et al, Mild asymptomatic primary hyperparathyroidism is not a risk factor for vertebral fractures, *Ann Int Med* (1988) **109**: 959–62.

97 Rao DS, Wilson RJ, Kleerekoper M et al, Lack of biochemical progression or continuation of accelerated bone loss in mild asymptomatic primary hyperparathyroidism: evidence for biphasic disease course, *J Clin Endocrinol Metab* (1988) **67**: 1294–8.

98 Bilezikian JP, Primary hyperparathyroidism, *Trends in Endo Metab* (1989) September/October 3–6.

99 Lafferty RW, Hubay CA, Primary hyperparathyroidism. A review of the long-term surgical and nonsurgical morbidities as a basis for a rational approach to treatment, *Arch Intern Med* (1989) **149**: 789–96.

100 Bilezikian JP, The medical management of primary hyperparathyroidism, *Ann Intern Med* (1982) **96**: 198–202.

101 Bilezikian JP, Surgery or no surgery for primary hyperparathyroidism, *Ann Intern Med* (1985) **102**: 402–3.

102 Mundy GR, Wilkinson R, Heath DA, Comparative study of available medical therapy for hypercalcemia of malignancy, *Am J Med* (1983) **74**: 421–32.

103 Shane E, Baquiran DC, Bilezikian JP, Effects of dichloromethylene diphosphonate on serum and urinary calcium in primary hyperparathyroidism, *Ann Intern Med* (1981) **95**: 23–7.

104 Gallagher JC, Nordin BEC, Treatment with oestrogens of primary hyperparathyroidism in postmenopausal women, *Lancet* (1972) **i**: 503–7.

105 Marcus R, Madvig P, Crim M et al, Conjugated estrogens in the treatment of postmenopausal women with hyperparathyroidism, *Ann Intern Med* (1984) **100**: 633–40.

106 Selby PL, Peacock M, Ethinyl estradiol and norethindrone in the treatment of primary hyperparathyroidism in postmenopausal women, *N Engl J Med* (1986) **314**: 1481–5.

107 Au WYW, Calcitonin treatment of hypercalcemia due to parathyroid carcinoma—synergistic effect of prednisone on long-term treatment of hypercalcemia, *Arch Intern Med* (1975) **135**: 1594–7.

108 Binstock ML, Mundy GR, Effects of calcitonin and glucocorticoids in combination in hypercalcemia of malignancy, *Ann Intern Med* (1980) **93**: 269–72.

109 Singer FR, Neer RM, Murray TM et al, Mithramycin therapy of intractable hypercalcemia due to parathyroid carcinoma, *N Engl J Med* (1970) **283**: 634–6.

110 Wang CA, Surgical management of primary hyperparathyroidism. In: DeGroot LJ, ed., *Endocrinology*, Vol. 2 (Grune and Stratton: New York 1979) 735–7.

111 Wang CA, Surgery of the parathyroid glands, *Adv Surg* (1971) **5**: 109–27.

112 Nussbaum SR, Thompson AR, Hutcheson KA et al, Intraoperative measurement of parathyroid hormone in the surgical management of hyperparathyroidism, *Surgery* (1988) **104**: 1121–7.

113 Mallette LE, Blevins T, Jordan PH et al, Autogenous parathyroid grafts for generalized primary parathyroid hyperplasia: Contrasting outcomes in sporadic hyperplasia versus multiple endocrine neoplasia type I, *Surgery* (1987) **101**: 738–45.

114 Rizzoli R, Green J, Marx SJ, Primary hyperparathyroidism in familial multiple endocrine neoplasia type 1, *Am J Med* (1985) **78**: 467–74.

115 Spiegel AM, Eastman ST, Attie MF et al, Intraoperative measurements of urinary cyclic AMP to guide surgery for primary hyperparathyroidism, *N Engl J Med* (1980) **303**: 1457–60.

116 Eisenberg H, Pallota J, Sherwood LM, Selective arteriography, venography and venous hormone assay in diagnosis and localization of parathyroid lesions, *Am J Med* (1974) **56**: 810–20.

117 Bilezikian JP, Doppman JL, Shimkin PM et al, Preoperative localization of abnormal parathyroid tissue. Cumulative experience with venous sampling and arteriography, *Am J Med* (1973) **55**: 505–14.

118 Edis AJ, Sheedy PF, Beahrs OH et al, Results of reoperation for hyperparathyroidism, with evaluation of preoperative localization studies, *Surgery* (1978) **84**: 384–93.

119 Powell D, Murray TM, Pollard JJ et al, Parathyroid localization using venous catheterization and radioimmunoassay, *Arch Intern Med* (1973) **131**: 645–8.

120 Brennan MF, Doppman JL, Krudy AG et al, Assessment of techniques for preoperative parathyroid gland localization in patients undergoing reoperation for hyperparathyroidism, *Surgery* (1981) **91**: 6–11.

121 Krudy AG, Doppman JL, Brennan MF et al, The detection of mediastinal parathyroid glands by computed tomography, selective arteriography, and venous sampling, *Radiology* (1981) **140**: 739–44.

122 Karo JJ, Maas LC, Kaine H et al, Ultrasonography and parathyroid adenoma, *JAMA* (1978) **239**: 2163–4.

123 Arima M, Yokoi H, Sonoda T, Preoperative identification of tumor of the parathyroid by ultrasonotomography, *Surg Gynecol Obstet* (1975) **141**: 242–4.

124 Sample WF, Mitchell SP, Bledsoe RC, Parathyroid ultrasonography, *Radiology* (1978) **127**: 485–90.

125 Mallette LE, Malini S, Review: the role of parathyroid ultrasonography in the management of primary hyperparathyroidism, *Am J Med Sci* (1989) **298**: 51–8.

126 Stark DS, Gooding GAW, Moss AA et al, Parathyroid imaging: of high resolution CT and high-resolution sonography, *Am J Roentgenol* (1983) **141**: 633–8.

127 Ferlin G, Borsato N, Camerani M et al, New perspectives in localizing enlarged parathyroids by technetium–thallium subtraction scan, *J Nucl Med* (1983) **24**: 438–41.

128 Clark OH, Okerlund MD, Moss AA et al, Localization studies in patients with persistent or recurrent hyperparathyroidism, *Surgery* (1985) **98**: 1083–94.

129 Okerlund MD, Sheldon K, Corpuz S et al, A new method with high sensitivity and specificity for localization of abnormal parathyroid glands, *Ann Surg* (1984) **200**: 381–8.

130 Skibber JM, Reynolds JC, Spiegel AM et al, Computerized technetium/thallium scans and parathyroid reoperation, *Surgery* (1985) **98**: 1077–82.

131 Krubsack AJ, Wilson SD, Lawson TL et al, Prospective comparison of radionuclide, computed tomographic and sonographic localization of parathyroid tumors, *World J Surg* (1986) **10**: 579–85.

132   Winzelberg GG, Parathyroid imaging, *Ann Intern Med* (1987) **107**: 64–70.

133   Editorial, Parathyroid gland localization, *Lancet* (1986) **ii**: 726–7.

134   Muller H, Sex, age and hyperparathyroidism, *Lancet* (1969) **i**: 449–50.

135   McGeown MG, Sex, age and hyperparathyroidism, *Lancet* (1969) **i**: 887–8.

136   Watson L, Primary hyperparathyroidism, *Clin Endocrinol Metab* (1974) **3**: 215–35.

Chapter 9

1   Mundy GR, Raisz LG, Thyrotoxicosis and calcium metabolism, *Miner Electrolyte Metab* (1979) **2**: 285–92.

2   Auwerx J, Bouillon R, Mineral and bone metabolism in thyroid disease: a review, *Q J Med* (1986) **232**: 737–52.

3   Baxter JD, Bondy PK, Hypercalcemia of thyrotoxicosis, *Ann Intern Med* (1966) **65**: 429–42.

4   Gordon DL, Swanich S, Ekvili V et al, The serum calcium level and its significance in hyperthyroidism: a prospective study, *Am J Med Sci* (1974) **268**: 31–6.

5   Farnsworth AE, Dobyns BM, Hypercalcaemia and hyperthyreosis, *Med J Aust* (1974) **2**: 782–4.

6   Follis RH, Skeletal changes associated with hyperthyroidism, *Bull Johns Hopkins Hosp* (1953) **92**: 405–21.

7   Meunier PJ, Bianchi GGS, Edouard CM et al, Bony manifestations of thyrotoxicosis, *Orthop Clin North Am* (1972) **3**: 745–74.

8   Parfitt AM, The actions of parathyroid hormone on bone: relation to bone remodeling and turnover, calcium homeostasis, and metabolic bone disease. I, IV. Mechanisms of calcium transfer between blood and bone and their cellular basis: morphological and kinetic approaches to bone turnover, *Metabolism* (1976) **25**: 809–44.

9   Bekier A, Der Nachweis der (thyreogenen Osteopathie) mit Hilfe moderner Photonenabsorptionstechnik, *Schweiz Med Wochenschr* (1975) **105**: 304–7.

10   Cohn SH, Roginsky MS, Aloia JF et al, Alteration in elemental body composition in thyroid disorders, *J Clin Endocrinol Metab* (1973) **36**: 742–9.

11   Mundy GR, Shapiro L, Bandelin JG et al, Direct stimulation of bone resorption by thyroid hormones, *J Clin Invest* (1976) **58**: 529–34.

12   Marcus R, Cyclic nucleotide phosphodiesterase from bone: characterization of the enzyme and studies of inhibition by thyroid hormones, *Endocrinology* (1975) **96**: 400–8.

13   Karlberg BE, Henriksson KG, Andersson RGG, Cyclic adenosine 3′,5′-monophosphate concentration in plasma, adipose tissue and skeletal muscle in normal subjects and in patients with hyper- and hypothyroidism, *J Clin Endocrinol Metab* (1974) **39**: 96–101.

14   Krane SM, Brownell GL, Skanbury JB et al, The effect of thyroid disease on calcium metabolism in man, *J Clin Invest* (1956) **35**: 874–87.

15   Bouillon R, Muls E, deMoor P, Influence of thyroid function on the serum concentration of 1,25-dihydroxyvitamin D_3, *J Clin Endocrinol Metab* (1980) **51**: 793–7.

16   Aub JC, Bauer W, Heath C et al, Studies of calcium and phosphorus metabolism. III. The effects of the thyroid hormone in thyroid disease, *J Clin Invest* (1929) **7**: 97–137.

17   Koutras DA, Pandos PG, Koukoulommat AS et al, Radiological signs of bone loss in hyperthyroidism, *Br J Radiol* (1973) **46**: 695–8.

18   Meema HE, Schatz DL, Simple radiologic demonstration of cortical bone loss in thyrotoxicosis, *Radiology* (1970) **97**: 9–15.

19   Kivirikko KT, Laitinen O, Lamberg BA, Value of urine and serum hydroxyproline in the diagnosis of thyroid disease, *J Clin Endocrinol Metab* (1965) **25**: 1347–52.

20   Rude RK, Oldham SB, Singer FR et al, Treatment of thyrotoxic hypercalcemia with propranolol, *N Engl J Med* (1976) **294**: 431–3.

21   Jones MK, Papapoulos SE, Control of hypercalcaemia in thyrotoxicosis, *Postgrad Med J* (1979) **55**: 891–3.

22 Feely J, Propranolol and the hypercalcaemia of thyrotoxicosis, *Acta Endocrinol (Copenh)* (1981) **98**: 528–32.

23 Plum F, Dunning NF, The effect of therapeutic mobilization on hypercalciuria following acute polyomyolitis, *Arch Intern Med* (1958) **101**: 528–36.

24 Heath H, Earll JM, Schaaf M et al, Serum ionized calcium during bed rest in fracture patients and normal men, *Metabolism* (1972) **21**: 633–40.

25 Wolf A, Chuinard R, Riggens R et al, Immobilization-hypercalcemia: a case report and review of the literature, *Clin Orthop* (1976) **118**: 124–9.

26 Rosen JF, Wolin DA, Finberg L, Immobilization-hypercalcemia after single limb fractures in children and adolescents, *Am J Dis Child* (1978) **132**: 560–4.

27 Minaire P, Meunier P, Edouard C et al, Quantitative histological data on disuse osteoporosis, *Calcif Tissue Res* (1974) **17**: 57–73.

28 Lanyon LE, Rubin CT, Baust G, Modulation of bone loss during calcium insufficiency by controlled dynamic loading, *Calcif Tissue Int* (1986) **38**: 209–16.

29 Lanyon LE, Biomechanical factors in adaptation of bone structure to function, *Curr Concepts Bone Fragility* (1986) 19–33 (Springer-Verlag: Berlin).

30 Smith PH, Robertson WG, Stone formation in the immobilized patient. In: Hodgkinson A, Nordin BEC, eds, *Renal stone research symposium* (Churchill: London 1969).

31 Stewart AF, Adler M, Byers CM et al, Calcium homeostasis in immobilization: an example of resorptive hypercalciuria, *N Engl J Med* (1982) **306**: 1136–40.

32 Singer FR, Adams JS, Abnormal calcium homeostasis in sarcoidosis, *N Engl J Med* (1986) **315**: 755–7.

33 Harrell GT, Fisher S, Blood chemical changes in Boeck's sarcoid with particular reference to protein, calcium and phosphatase values, *J Clin Invest* (1939) **18**: 687–93.

34 Mayock RL, Bertrand P, Morrison CE et al, Manifestations of sarcoidosis, *Am J Med* (1963) **35**: 67–89.

35 Taylor RL, Lynch HR Jr, Wysor WG Jr, Seasonal influence of sunlight on the hypercalcemia of sarcoidosis, *Am J Med* (1963) **34**: 221–7.

36 Goldstein RA, Israel HL, Becker KL et al, The infrequency of hypercalcemia in sarcoidosis, *Am J Med* (1971) **51**: 21–30.

37 Winnacker JL, Becker KL, Katz S, Endocrine aspects of sarcoidosis, *N Engl J Med* (1968) **278**: 483–92.

38 Bell NH, Gill JR Jr, Bartter FC, On the abnormal calcium absorption in sarcoidosis. Evidence for increased sensitivity to vitamin D, *Am J Med* (1964) **36**: 500–13.

39 Avioli LV, Haddad JG, Vitamin D: current concepts, *Metab Clin Exp* (1973) **22**: 507–31.

40 Bell NH, Stern PH, Pantzer E et al, Evidence that increased circulating 1,25-dihydroxyvitamin D is the probable cause for abnormal calcium metabolism in sarcoidosis, *J Clin Invest* (1979) **64**: 218–25.

41 Papapoulos SE, Clemens TL, Fraher LJ et al, 1,25-Dihydroxycholecalciferol in the pathogenesis of the hypercalcemia of sarcoidosis, *Lancet* (1979) **i**: 627–30.

42 Sandler LM, Winearls CG, Fraher LJ et al, Studies of the hypercalcemia of sarcoidosis: effects of steroids and exogenous vitamin D_3 on the circulating concentrations of 1,25-dihydroxyvitamin D_3, *Q J Med* (1984) **53**: 165–80.

43 Meyrier A, Valeyre D, Bouillon R et al, Resorptive versus absorptive hypercalciuria in sarcoidosis: correlations with 25-hydroxyvitamin D_3 and 1,25-dihydroxyvitamin D_3 and parameters of disease activity, *Q J Med* (1985) **54**: 269–81.

44 Barbour GL, Coburn JW, Slatopolsky E et al, Hypercalcemia in an anephric patient with sarcoidosis: evidence for extrarenal generation of 1,25-dihydroxyvitamin D, *N Engl J Med* (1981) **305**: 440–3.

45 Adams JS, Sharma OP, Gacad MA et al, Metabolism of 25-hydroxyvitamin D_3 by cultured pulmonary alveolar macrophages in sarcoidosis, *J Clin Invest* (1983) **72**: 1856–60.

46 Adams JS, Gacad MA, Characterization of 1α-hydroxylation of vitamin D_3 sterols by cultured alveolar macrophages from patients with sarcoidosis, *J Exp Med* (1985) **161**: 755–65.

47 Adams JS, Singer FR, Gacad MA et al, Isolation and structural identification of 1,25-dihydroxy-vitamin D_3 produced by cultured alveolar macrophages in sarcoidosis, *J Clin Endocrinol Metab* (1985) **60**: 960–6.

48 Mason RS, Frankel T, Chan Y-L et al, Vitamin D conversion by sarcoid lymph node homogenate, *Ann Intern Med* (1984) **100**: 59–61.

49 Dent CE, Watson L, Hyperparathyroidism and sarcoidosis, *Br Med J* (1966) **1**: 646–9.

50 Winnacker JL, Becker KL, Friedlander M et al, Sarcoidosis and hyperparathyroidism, *Am J Med* (1969) **46**: 305–12.

51 Dent CE, Flynn FV, Nabarro JDN, Hypercalcaemia and impairment of renal function in generalized sarcoidosis, *Br Med J* (1953) **2**: 808–10.

52 Shulman LE, Schoenrich EH, Harvey AM, The effects of adrenocorticotropic hormone (ACTH) and cortisone on sarcoidosis, *Bull Johns Hopkins Hosp* (1952) **91**: 371–415.

53 Hunt BJ, Yendt ER, The response of hypercalcemia in sarcoidosis to chloroquine, *Ann Intern Med* (1963) **59**: 554–64.

54 O'Leary TJ, Jones G, Yip A et al, The effects of chloroquine on serum 1,25-dihydroxyvitamin D and calcium metabolism in sarcoidosis, *N Engl J Med* (1986) **315**: 727–30.

55 Voipio H, Incidence of chloroquine retinopathy, *Acta Ophthalmol (Copenh)* (1966) **44**: 349–54.

56 Mackenzie AH, Dose refinements in long-term therapy of rheumatoid arthritis with antimalarials, *Am J Med* (1983) **75**(IA): 40–5.

57 Gkonos PJ, London R, Hendler E, Hypercalcemia and elevated 1,25-dihydroxyvitamin D levels in a patient with end-stage renal disease and active tuberculosis, *N Engl J Med* (1984) **311**: 1683–5.

58 Felsenfeld AJ, Drezner MK, Llach F, Hypercalcemia and elevated calcitriol in a maintenance dialysis patient with tuberculosis, *Arch Intern Med* (1986) **146**: 1941–5.

59 Bell NH, Shary J, Shaw S et al, Hypercalcemia associated with increased circulatory 1,25-dihydroxyvitamin D in a patient with pulmonary tuberculosis, *Calcif Tissue Int* (1985) **37**: 588–91.

60 Kantarjian HM, Saad MF, Estey EH et al, Hypercalcemia in disseminated candidiasis, *Am J Med* (1983) **74**: 721–4.

61 Kozeny GA, Barbato AL, Bansal VK et al, Hypercalcemia associated with silicone-induced granulomas, *N Engl J Med* (1984) **311**: 1103–5.

62 Jurney TH, Hypercalcemia in a patient with eosinophilic granuloma, *Am J Med* (1984) **76**: 527–8.

63 Zaloga GP, Chernow B, Eil C, Hypercalcemia and disseminated cytomegalovirus infection in the acquired immunodeficiency syndrome, *Ann Intern Med* (1985) **102**: 331–3.

64 Davies M, Adams PH, The continuing risk of vitamin D intoxication, *Lancet* (1978) **ii**: 621–3.

65 Nordin BEC, *Metabolic bone and stone disease* (Williams and Wilkins Co.: Baltimore 1973).

66 Trummel CL, Raisz LG, Blunt JS et al, 25-Hydroxycholecalciferol, *Science* (1969) **163**: 1450–1.

67 Raisz LG, Trummel CL, Holick MF et al, 1,25-Dihydroxycholecalciferol: a potent stimulator of bone resorption in tissue culture, *Science* (1972) **175**: 768–9.

68 Klein L, Direct measurement of bone resorption and calcium conservation during vitamin D deficiency or hypervitaminosis D, *Proc Natl Acad Sci USA* (1980) **77**: 1818–22.

69 Davies M, Mawer EB, Freemont AJ, The osteodystrophy of hypervitaminosis D: a metabolic study, *Q J Med* (1986) **61**: 911–19.

70 Hess AF, Lewis JM, Clinical experience with irradiated ergosterol, *JAMA* (1928) **91**: 783–8.

71 Anning ST, Dawson J, Dolbe DE et al, The toxic effects of calciferol, *Q J Med* (1948) **17**: 203–28.

72 Mawer EB, Hann JT, Berry JL et al, Vitamin D metabolism in patients intoxicated with ergocalciferol, *Clin Sci* (1985) **68**: 141–5.

73 Harris PWR, An unusual case of calcinosis due to vitamin D intoxication, *Guy's Hosp Rep* (1969) **118**: 533–41.

74 Ham AW, Lewis MD, Hypervitaminosis D rickets: the action of vitamin D, *Br J Exp Pathol* (1934) **15**: 228–34.

75 Nielson HE, Romer FK, Melsen F et al, 1-alpha-hydroxyvitamin D$_3$ treatment of non-dialyzed patients with chronic renal failure. Effects of bone, mineral metabolism and kidney function, *Clin Nephrol* (1980) **13**: 103–8.

76 Kanis JA, Cundy T, Earnshaw M et al, Treatment of renal bone disease with 1-alpha-hydroxylated derivatives of vitamin D$_3$. Clinical, biochemical, radiographic and histological responses, *Q J Med* (1979) **48**: 289–322.

77 Massry SG, Goldstein DA, Is calcitriol (1,25(OH)$_2$D$_3$) harmful to renal function, *JAMA* (1979) **242**: 1875–6.

78 Arnold RM, Bras G, Observations on the morbid anatomy and histology of Manchester wasting disease of cattle in Jamaica and related conditions in other countries of the Americas, *Am J Vet Res* (1956) **17**: 630–9.

79 Worker NA, Carrillo BJ, Enteque Seco calcification and wasting in grazing animals in the Argentine, *Nature* (1967) **215**: 72–4.

80 Haussler MR, Wasserman RH, McCain TA, 1,25 Dihydroxyvitamin D$_3$-glycoside. Identification of a calcinogenic principle of *Solanum malacoxylon*, *Life Sci* (1976) **18**: 1049–56.

81 Shike M, Sturtridge WC, Tam CS et al, A possible role of vitamin D in the genesis of parenteral-nutrition-induced metabolic bone disease, *Ann Intern Med* (1981) **95**: 560–8.

82 Ott SM, Maloney NA, Klein GL et al, Aluminum is associated with low bone formation in patients receiving chronic parenteral nutrition, *Ann Intern Med* (1983) **98**: 910–14.

83 Kimberg DV, Baerg RD, Gershon E et al, Effect of cortisone treatment on the active transport of calcium by the small intestine, *J Clin Invest* (1971) **50**: 1309–21.

84 Streck WF, Waterhouse C, Haddad JG, Glucocorticoid effects in vitamin D intoxication, *Arch Intern Med* (1979) **139**: 974–7.

85 Heyburn PJ, Francis RM, Peacock M, Acute effects of saline, calcitonin and hydrocortisone on plasma calcium in vitamin D intoxication, *Br Med J* (1979) **1**: 232–3.

86 Wolf G, Multiple functions of vitamin A, *Physiol Rev* (1984) **64**: 873–937.

87 Dowling JE, Wald G, The biological function of vitamin A acid, *Proc Natl Acad Sci USA* (1960) **46**: 587–608.

88 Evans RM, The steroid and thyroid hormone receptor superfamily, *Science* (1988) **240**: 889.

89 Wolbach SB, Vitamin A deficiency and excess in relation to skeletal growth, *J Bone Joint Surg* (1947) **29**: 171–92.

90 Frame B, Jackson CE, Reynolds WA et al, Hypercalcemia and skeletal effects in chronic hypervitaminosis A, *Ann Intern Med* (1974) **80**: 44–8.

91 Rodahl K, Moore T, The vitamin A content and toxicity of bear and seal liver, *Biochem J* (1943) **37**: 166–8.

92 Fell HB, Mellanby E, Effect of hypervitaminosis A on foetal mouse bones cultivated in vitro, *Br Med J* (1950) **2**: 535–9.

93 Fell HB, Mellanby E, The effect of hypervitaminosis A on embryonic limb bones cultured in vitro, *J Physiol (Lond)* (1952) **116**: 320–49.

94 Dingle JT, Lucy JA, Fell HB, Studies on the mode of action of excess of vitamin A. I: Effect of excess vitamin A on the metabolism and composition of embryonic chick-limb cartilage grown in organ culture, *Biochem J* (1961) **79**: 497–500.

95 Fell HB, Thomas L, The influence of hydrocortisone on the action of excess vitamin A on limb-bone rudiments in culture, *J Exp Med* (1961) **114**: 343–62.

96 Lucy JA, Dingle JT, Fell HB, Studies on the mode of action of excess of vitamin A. 2: A possible role of intracellular proteases in the degradation of cartilage matrix, *Biochem J* (1961) **79**: 500–8.

97 Raisz LG, Inhibition of actinomycin D of bone resorption induced by parathyroid hormone on vitamin A, *Proc Soc Exp Biol Med* (1965) **119**: 614–17.

98 Barnicot NA, The local action of vitamin A on bone, *J Anat* (1950) **84**: 374–87.

99 Oreffo ROC, Teti A, Triffitt JT et al, Effect of vitamin A on bone resorption: evidence for direct stimulation of isolated chicken osteoclasts by retinol and retinoic acid, *J Bone Min Res* (1988) **3**: 203–10.

100 Ng KW, Gummer PR, Michelangeli VP et al, Regulation of alkaline phosphatase expression in a neonatal rat clonal calvarial cell strain by retinoic acid, *J Bone Min Res* (1988) **3**: 53–61.

101 Ng KW, Hudson PJ, Power BE et al, Retinoic acid and tumour necrosis factor-alpha act in concert to control the level of alkaline phosphatase mRNA, *J Mol Endocrinol* (1989) **3**: 57–64.

102 Taylor AB, Stern PH, Bell NH, Abnormal regulation of circulating 25-hydroxyvitamin D in the Williams Syndrome, *N Engl J Med* (1982) **306**: 972–5.

103 Fanconi G, 'Uber Chronische Storungen des Calcium- und Phosphatstoffwechels im Kinde- salter, *Schweiz med Wochenschr* (1951) **81**: 908– 12.

104 Williams JCP, Barratt-Boyes BG, Lower JB, Supra- valvular aortic stenosis, *Circulation* (1961) **24**: 1311–18.

105 Black JA, Bonham Carter RE, Association between aortic stenosis and facies of severe infantile hyper- calcaemia, *Lancet* (1963) **ii**: 745–9.

106 Garcia RE, Friedman WF, Kabach MM et al, Idiopathic hypercalcemia and supravalvular aortic stenosis: documentation of a new syndrome, *N Engl J Med* (1964) **271**: 117–20.

107 Forbes GV, Bryson MF, Manning J et al, Impaired calcium homeostasis in the infantile hypercal- cemic syndrome, *Acta Paediatr Scand* (1972) **61**: 305–9.

108 Culler FL, Joves KL, Deftos LJ, Impaired calcitonin secretion in patients with Williams syndrome, *J Pediatr* (1985) **107**: 720–3.

109 Parfitt AM, Chlorothiazide-induced hypercalcemia in juvenile osteoporosis and hyperparathyroid- ism, *N Engl J Med* (1969) **281**: 55–9.

110 Parfitt AM, The interactions of thiazide diuretics with parathyroid hormone and vitamin D, *J Clin Invest* (1972) **51**: 1879–88.

111 Duarte CG, Winnacker JL, Becker KL et al, Thiazide-induced hypercalcemia, *N Engl J Med* (1971) **284**: 828–30.

112 Christensson T, Hellstrom K, Wengle B, Hypercal- cemia and primary hyperparathyroidism: preva- lence in patients receiving thiazides as detected in a health screen, *Arch Intern Med* (1977) **137**: 1138– 42.

113 Seitz H, Jaworski ZF, Effect of hydrochlorothiazide on serum and urinary calcium and urinary citrate, *Can Med Assoc J* (1966) **90**: 414–20.

114 Stote RM, Smith H, Wilson DM et al, Hydrochlor- othiazide effects on serum calcium and immuno- reactive parathyroid hormone concentrations, *Ann Intern Med* (1972) **77**: 587–91.

115 Van der Sluys Veer J, Birkenhager JC, Smeenk D, Effect of orally administered diuretics on calcium metabolism: with special emphasis on serum calcium in hyperparathyroidism, *Proc IV Eur Symp Calcified Tissues*, Leiden. Neth Int Congr Ser No. 20, Gaillard PJ, Van der Hooff LA, Steendijk J, eds. (Excerpta Medica: Amsterdam 1966) 96–8.

116 Costanzo LS, Windhager EE, Relationship between clearance of Ca and Na: effect of distal diuretics and PTH, *Am J Physiol* (1976) **230**: 67–73.

117 Koppel MH, Massry SG, Shinaberger JH et al, Thiazide-induced rise in serum calcium and mag- nesium in patients on maintenance hemodialysis, *Ann Intern Med* (1970) **72**: 895–901.

118 Hardt LL, Rivers AB, Toxic manifestations follow- ing the alkaline treatment of peptic ulcer, *Arch Intern Med* (1923) **31**: 171–80.

119 Sippy BW, Gastric and duodenal ulcer: medical cure by an efficient removal of gastric juice corrosion, *JAMA* (1915) **64**: 1625–30.

120 Cope CL, Base changes in the alkalosis produced by the treatment of gastric ulcer with alkalies, *Clin Sci* (1936) **2**: 298–300.

121 Orwoll ES, The milk-alkali syndrome: current concepts, *Ann Intern Med* (1982) **97**: 242–8.

122 Burnett CH, Commons RR, Albright F et al, Hyper- calcemia without hypercalciuria or hypophos- phatemia, calcinosis and renal insufficiency: a syndrome following prolonged intake of milk and alkali, *N Engl J Med* (1949) **240**: 787–94.

123 Clarkson EM, McDonald SJ, de Wardener HE, The effects of a high intake of calcium carbonate in normal subjects and in subjects with chronic renal failure, *Clin Sci* (1966) **30**: 425–38.

124 Dworetzky M, Reversible metastatic calcification (milk drinkers syndrome), *JAMA* (1954) **155**: 830– 2.

125 Dufault FX, Tobias GJ, Potentially reversible renal failure following excessive calcium and alkali intake in peptic ulcer therapy, *Am J Med* (1954) **16**: 231–6.

126 McMillan DE, Freeman RB, The milk-alkali syndrome: a study of the acute disorder with comments on the development of the chronic condition, *Medicine* (1965) **44**: 485–501.

127 Randall RE Jr, Strauss MB, McNeely WF, The milk-alkali syndrome, *Arch Intern Med* (1961) **107**: 163–81.

128 Snapper I, Fradley WG, Wilson VE, Metastatic calcification and nephrocalcinosis from medical treatment of peptic ulcer, *Arch Intern Med* (1954) **93**: 807–17.

129 Stiel JN, Mitchell CA, Radcliff FJ et al, Hypercalcemia in patients with peptic ulceration receiving large doses of calcium carbonate, *Gastroenterology* (1967) **53**: 900–4.

130 Wenger J, Kersner JB, Palmer WL, The milk-alkali syndrome: hypercalcemia, alkalosis and azotemia following calcium carbonate and milk therapy of peptic ulcer, *Gastroenterology* (1957) **33**: 745–69.

131 Werner P, Kuschner M, Riley EA, Reversible metastatic calcification associated with excessive milk and alkali intake, *Am J Med* (1953) **14**: 108–15.

132 Epstein FH, Calcium and the kidney, *Am J Med* (1968) **45**: 700–12.

133 Sutton RAL, Wong NLM, Dirks JH, Effects of metabolic acidosis and alkalosis on sodium and calcium transport in the dog kidney, *Kidney Int* (1979) **15**: 520–33.

134 Seldin DW, Rector FC Jr, The generation and maintenance of metabolic alkalosis, *Kidney Int* (1972) **1**: 306–21.

135 Kapsner P, Langsdorf L, Marcus R et al, Milk-alkali syndrome in patients treated with calcium carbonate after cardiac transplantation, *Arch Intern Med* (1986) **146**: 1965–8.

136 Foley TP, Harrison HC, Arnaud CD et al, Familial benign hypercalcemia, *J Pediatr* (1972) **81**: 1060–7.

137 Marx SJ, Spiegel AM, Brown EM et al, Family studies in patients with primary parathyroid hyperplasia, *Am J Med* (1977) **62**: 698–706.

138 Marx SJ, Spiegel AM, Brown EM et al, Divalent cation metabolism. Familial hypocalciuric hypercalcemia versus typical primary hyperparathyroidism, *Am J Med* (1978) **65**: 235–42.

139 Marx SJ, Spiegel AM, Brown EM et al, Circulating parathyroid hormone activity: familial hypocalciuric hypercalcemia versus typical primary hyperparathyroidism, *J Clin Endocrinol Metab* (1978) **47**: 1190–7.

140 Marx SJ, Familial hypocalciuric hypercalcemia, *N Engl J Med* (1980) **303**: 810–11.

141 Law WM, Jr, Heath H, Familial benign hypercalcemia (hypocalciuric hypercalcemia). Clinical and pathogenetic studies in 21 families, *Ann Intern Med* (1985) **102**: 511–19.

142 Paterson CR, Gunn A, Familial benign hypercalcaemia, *Lancet* (1981) **ii**: 61–3.

143 Marx SJ, Attie MF, Levine MA et al, The hypocalciuric or benign variant of familial hypercalcemia: clinical and biochemical features in fifteen kindreds, *Medicine* (1981) **60**: 397–412.

144 Davies M, Klimiuk PS, Adams PH et al, Familial hypocalciuric hypercalcaemia and acute pancreatitis, *Br Med J* (1981) **282**: 1023–5.

145 Marx SJ, Attie MF, Spiegel AM et al, An association between neonatal severe primary hyperparathyroidism and familial hypocalciuric hypercalcemia in three kindreds, *N Engl J Med* (1982) **306**: 257–64.

146 Marx SJ, Brandi ML, In: Peck WA, ed., *Familial primary hyperparathyroidism* (Elsevier: Amsterdam 1987) 375–407.

147 Heath H, Purnell DC, Urinary cyclic 3′,5′-adenosine monophosphate responses to exogenous and endogenous parathyroid hormone in familial benign hypercalcemia and primary hyperparathyroidism, *J Lab Clin Med* (1980) **96**: 974–83.

148 Marx SJ, Familial hypocalciuric hypercalcemia, *N Engl J Med* (1980) **303**: 810–11.

149 Fu-Hsiung S, Sherrard DJ, Lithium-induced hyperparathyroidism: an alteration of the 'set-point', *Ann Intern Med* (1982) **96**: 63–5.

150 Jackson CE, Kleerekoper M, Hereditary hypocalciuric hypercalcemia is benign in 15 year followup, *Clin Res* (1981) **29**: abstr 409A.

151   Loeb RF, Chemical changes in the blood in Addison's disease, *Science* (1932) **76**: 420–5.

152   Leeksma CHW, De Graeff J, De Cock J, Hypercalcemia in adrenal insufficiency, *Acta Med Scand* (1957) **156**: 455–8.

153   Walser M, Robinson BHB, Duckett JW, The hypercalcemia of adrenal insufficiency, *J Clin Invest* (1963) **42**: 456–65.

154   Mundy GR, Rick ME, Turcotte R et al, Pathogenesis of hypercalcemia in lymphosarcoma cell leukemia—role of an osteoclast activating factor-like substance and mechanism of action for glucocorticoid therapy, *Am J Med* (1978) **65**: 600–6.

155   Bannister P, Roberts M, Sheridan P, Recurrent hypercalcaemia in a young man with mono-ostotic Paget's disease, *Postgrad Med* (1986) **62**: 481–3.

156   Benabe JE, Martinez-Maldonado M, Hypercalcemic nephropathy, *Arch Intern Med* (1978) **138**: 777–9.

157   Miach PJ, Dawborn JK, Douglas MC et al, Prolonged hypercalcemia following acute renal failure, *Clin Nephrol* (1975) **4**: 32–6.

158   de Torrente A, Berl T, Cohn PD et al, Hypercalcemia of acute renal failure, *Am J Med* (1976) **61**: 119–23.

159   Llach F, Felsenfeld AJ, Haussler MR, The pathophysiology of altered calcium metabolism in rhabdomyolysis-induced acute renal failure, *N Engl J Med* (1981) **305**: 117–23.

160   Akmal M, Goldstein DA, Telfer N et al, Resolution of muscle calcification in rhabdomyolysis and acute renal failure, *Ann Intern Med* (1986) **89**: 928–30.

161   Case Records of Massachusetts General Hospital, Weekly clinicopathological exercises, *N Engl J Med* (1963) **225**: 789–91.

162   Alfrey AC, LeGendre GR, Kaehny WD, The dialysis encephalopathy syndrome: possible aluminum intoxication, *N Engl J Med* (1976) **294**: 184–8.

163   Coburn JW, Brickman AS, Sherrard DJ et al, Use of 1,25(OH)$_2$D$_3$ to separate types of renal osteodystrophy. In: Robinson, Hawkins, Vereerstrachfen, Dialysis, transplantation, nephrology. *Proc Em Dialysis Transplant Assoc* (Pitman Medical: Kent 1977) 442.

164   Ott SM, Maloney NA, Coburn JW et al, The prevalence of bone aluminum deposition in renal osteodystrophy and its relation to the response to calcitriol therapy, *N Engl J Med* (1982) **307**: 709–13.

165   Hodsman AB, Sherrard DJ, Alfrey AC et al, Bone aluminum and histomorphometric features of renal osteodystrophy, *J Clin Endocrinol Metab* (1982) **54**: 539–46.

166   Malluche HH, Smith AJ, Abreo K et al, The use of deferoxamine in the management of bone aluminum accumulation in patients with renal failure, *N Engl J Med* (1984) **311**: 140–4.

167   Malluche HH, Abreo K, Faugere MC, Aluminium decreases turnover and mineralization of bone—an effect dependent on its localization in bone, *Calcif Tissue Int* (1984) **36**: 494.

168   Malluche HH, Faugere MC, Smith AJ et al, Aluminum intoxication of bone in renal failure—fact or fiction? *Kidney Int* (1986) **29**: S70–3.

169   Shike M, Sturtridge WC, Harrison JE et al, Bone disease in patients receiving long-term total parenteral nutrition, *Am J Clin Nutr* (1979) **32**: 16.

170   Shike M, Harrison JE, Strutridge WC et al, Metabolic bone disease in patients receiving long-term total parenteral nutrition, *Ann Intern Med* (1980) **92**: 343–50.

171   Klein GL, Ament ME, Bluestone R et al, Bone disease associated with total parenteral nutrition, *Lancet* (1980) **ii**: 1041–4.

172   Klein GL, Alfrey AC, Miller NI et al, Aluminium loading during total parenteral nutrition, *Am J Clin Nutr* (1982) **35**: 1425.

Chapter 10

1   Nagant de Deuxchaines C, Krane SM, Hypoparathyroidism. In: Avioli LV, Krane SM, eds, *Metabolic bone disease*, Vol. II (Academic Press: New York 1978) 217–445.

2   Parfitt MA, Surgical, idiopathic, and other varieties of parathyroid hormone-deficient hypoparathyroidism. In: DeGroot LJ et al, eds, *Endocrinology*, Vol. 2 (Grune and Stratton: New York 1979) 755–68.

3 Breslau NA, Pak CYC, Hypoparathyroidism, *Metabolism* (1979) **28**: 1261–76.

4 Frame B, Parfitt AM, Osteomalacia: current concepts, *Ann Intern Med* (1978) **89**: 966–82.

5 Hahn TJ, Hendin BA, Scharp CR et al, Effect of chronic anticonvulsant therapy on serum 25-hydroxycalciferol levels in adults, *N Engl J Med* (1972) **287**: 900–4.

6 Hahn TJ, Birge SJ, Scharp CR et al, Phenobarbital-induced alterations in vitamin D metabolism, *J Clin Invest* (1972) **51**: 741–8.

7 Dietrich JW, Mundy GR, Raisz LG, Inhibition of bone resorption in tissue culture by membrane-stabilizing drugs, *Endocrinology* (1979) **104**: 2644–8.

8 Fraser D, Kooh SW, Kind HP et al, Vitamin D-dependent rickets: an inborn error of vitamin D metabolism, *N Engl J Med* (1973) **289**: 817–22.

9 Liberman UA, Halase A, Samuel R et al, End-organ resistance to 1,25-dihydroxycholecalciferol, *Lancet* (1980) **i**: 504–6.

10 Marx SJ, Liberman UA, Eil C et al, Hereditary resistance to 1,25-dihydroxyvitamin D, *Recent Prog Horm Res* (1984) **40**: 589–620.

11 Neufeld M, Maclaren NK, Blizzard RM, Two types of autoimmune Addison's disease associated with different polyglandular autoimmune (PGA) syndromes, *Medicine* (1981) **60**: 355–62.

12 Boyle IT, Fogelman I, Boyce BF et al, 1-alpha-hydroxy-vitamin D in primary hyperparathyroidism, *Clin Endocrinol (Oxf)* (1977) **7**: 2155–225.

13 Drezner MK, Neelon FA, Pseudohypoparathyroidism. In: Stanbury JB, Wyngaarden JB, Fredrickson DS, Goldstein JL, Brown MS, eds, *The metabolic basis of inherited disease,* 5th edn (McGraw-Hill: New York 1983) 1508–28.

14 Albright F, Forbes AP, Hennemen PH, Pseudopseudohypoparathyroidism, *Trans Assoc Am Phys* (1952) **65**: 337–50.

15 Albright R, Burnett CH, Smith PH et al, Pseudohypoparathyroidism: an example of Seabright–Bantam syndrome: report of three cases, *Endocrinology* (1942) **30**: 922–32.

16 Van Dop C, Bourne HR, Neer RM, Father to son transmission of decreased Ns activity in pseudohypoparathyroidism type 1a, *J Clin Endocrinol Metab* (1984) **59**: 825–8.

17 Levine MA, Jap T, Mauseth RS et al, Activity of the stimulatory guanine nucleotide-binding protein is reduced in erythrocytes from patients with pseudohypoparathyroidism and pseudopseudohypoparathyroidism: biochemical, endocrine, and genetic analysis of Albright's hereditary osteodystrophy in six kindreds, *J Clin Endocrinol Metab* (1986) **62**: 497–502.

18 George FW, Noble JF, Wilson JD, Female feathering in Sebright cocks is due to conversion of testosterone to estradiol in skin, *Science* (1981) **213**: 557–9.

19 Chase LR, Melson GL, Aurbach GD, Pseudohypoparathyroidism: defective excretion of 3′,5′-AMP in response to parathyroid hormone, *J Clin Invest* (1969) **48**: 1832–44.

20 Drezner M, Neelon FA, Lebovitz HE, Pseudohypoparathyroidism type II: a possible defect in the reception of the cyclic AMP signal, *N Engl J Med* (1973) **289**: 1056–60.

21 Drezner MK, Burch WJ Jr, Altered activity of the nucleotide regulatory site in the parathyroid hormone-sensitive adenylate cyclase from the renal cortex of a patient with pseudohypoparathyroidism, *J Clin Invest* (1978) **62**: 1222–7.

22 Farfel Z, Brickman AS, Kaslow HR et al, Defect of receptor-cyclase coupling protein in pseudohypoparathyroidism, *N Engl J Med* (1980) **303**: 237–42.

23 Levine MA, Downs RW, Jr, Singer M et al, Deficient activity of guanine nucleotide regulatory protein in erythrocytes from patients with pseudohypoparathyroidism, *Biochem Biophys Res Commun* (1980) **94**: 1319–24.

24 Farfel Z, Abood ME, Brickman AS et al, Deficient activity of receptor-cyclase coupling protein in transformed lymphoblasts of patients with pseudohypoparathyroidism, Type I, *J Clin Endocrinol Metab* (1982) **55**: 113–17.

25 Levine MA, Downs RW, Moses AM et al, Resistance to multiple hormones in patients with pseudohypoparathyroidism: association with deficient activity of guanine nucleotide regulatory protein, *Am J Med* (1983) **74**: 545–56.

26 Bourne HR, Kaslow HR, Brickman AS et al, Fibro-

blast defect in pseudohypoparathyroidism, type I: reduced activity of receptor-cyclase coupling protein, *J Clin Endocrinol Metab* (1981) **53**: 636–40.

27 Downs RW Jr, Levine MA, Drezner MK et al, Deficient adenylate cyclase regulatory protein in renal membranes from a patient with pseudohypoparathyroidism, *J Clin Invest* (1983) **71**: 231–5.

28 Farfel ZVI, Friedman E, Mental deficiency in pseudohypoparathyroidism Type I is associated with Ns-protein deficiency, *Ann Intern Med* (1986) **105**: 197–9.

29 Carter A, Bardin C, Collins R et al, Reduced expression of multiple forms of the alpha subunit of the stimulatory GTP-binding protein in pseudohypoparathyroidism type Ia, *Proc Natl Acad Sci* (1987) **84**: 7266–9.

30 Levine MA, Ahn TG, Klupt SF et al, Genetic deficiency of the alpha subunit of the guanine nucleotide-binding protein G_s as the molecular basis for Albright hereditary osteodystrophy, *Proc Natl Acad Sci* (1988) **85**: 617–21.

31 Loveridge N, Fischer JA, Nagant de Deuxchaisnes C et al, Inhibition of cytochemical bioactivity of parathyroid hormone by plasma in pseudohypoparathyroidism type I, *J Clin Endocrinol Metab* (1982) **54**: 1274–5.

32 Loveridge N, Tschopp F, Born W et al, Separation of inhibitory activity from biologically active parathyroid hormone in patients with pseudohypoparathyroidism type I, *Biochim Biophys Acta* (1986) **889**: 117–22.

33 Goltzman D, Mitchell JA, Hendy GN, Pseudohypoparathyroidism: Studies of the pathogenesis of parathyroid hormone resistance. In: Kleerekoper M, Krane SM, eds, *Clinical Disorders of Bone and Mineral Metabolism* (Mary Ann Liebert, New York 1989) 341–51.

34 Mann JB, Alterman S, Hills AG, Albright's hereditary osteodystrophy comprising pseudohypoparathyroidism and pseudopseudohypoparathyroidism, *Ann Intern Med* (1962) **56**: 315–42.

35 Moses AM, Breslau N, Coulson R, Renal responses to PTH in patients with hormone-resistant (pseudo) hypoparathyroidism, *Am J Med* (1976) **61**: 184–9.

36 Drezner MK, Haussler MR, Normocalcemic pseudohypoparathyroidism—association with normal vitamin D_3 metabolism, *Am J Med* (1979) **66**: 503–8.

37 Zamkoff KW, Kirshner JJ, Marked hypophosphatemia associated with acute myelomonocytic leukemia: indirect evidence of phosphorus uptake by leukemic cells, *Arch Intern Med* (1980) **140**: 1523–4.

38 Schenkein DP, O'Neill C, Shapiro J et al, Accelerated bone formation causing profound hypocalcemia in acute leukemia, *Ann Intern Med* (1986) **105**: 375–8.

39 Zusman J, Brown DM, Nesbit ME, Hyperphosphatemia, hyperphosphaturia and hypocalcemia in acute lymphoblastic leukemia, *N Engl J Med* (1973) **289**: 1335–40.

40 Heaton FW, Fourman P, Magnesium deficiency and hypocalcaemia in intestinal malabsorption, *Lancet* (1965) **ii**: 50–2.

41 Chesney RW, McCarron DM, Haddad JG et al, Pathogenic mechanisms of the hypocalcemia of the staphylococcal toxic-shock syndrome, *J Lab Clin Med* (1983) **101**: 576–85.

42 Aderka D, Schwartz D, Dan M et al, Bacteremic hypocalcemia, a comparison between the calcium levels of bacteremic and nonbacteremic patients with infection, *Arch Intern Med* (1987) **147**: 232–6.

Chapter 11

1 Avioli LV, The therapeutic approach to hypoparathyroidism, *Am J Med* (1974) **57**: 34–42.

2 Parfitt MA, Surgical, idiopathic, and other varieties of parathyroid hormone-deficient hypoparathyroidism. In: DeGroot LJ et al., eds, *Endocrinology,* Vol. 2 (Grune and Stratton: New York 1979) 755–68.

3 Breslau NA, Pak CYC, Hypoparathyroidism, *Metabolism* (1979) **28**: 1261–76.

4 Porter RH, Cox BG, Heaney D et al, Treatment of hypoparathyroid patients with chlorthalidone, *N Engl J Med* (1978) **298**: 577–81.

5 Eisenberg E, Effects of serum calcium level and parathyroid extracts on phosphate and calcium excretion in hypoparathyroid patients, *J Clin Invest* (1965) **44**: 942–6.

6 Wells SA Jr, Gunnels JC, Shelburne JD et al, Transplantation of the parathyroid glands in man: clinical indications and results, *Surgery* (1975) **78**: 34–44.

7   Hickey RC, Samaan NA, Human parathyroid auto-
    transplantation: proved function by radioimmuno-
    assay of plasma parathyroid hormone, *Arch Surg*
    (1975) **110**: 892–5.

8   Niederle B, Roka R, Brennan MF, The transplantation
    of parathyroid tissue in man: development, indi-
    cations, techniques, and results, *Endocr Rev* (1982)
    3: 245–79.

INDEX

Page numbers in *italic* refer to the illustrations